Modeling with Data

Modeling with Data

Tools and Techniques for Scientific Computing

Ben Klemens

PRINCETON UNIVERSITY PRESS

PRINCETON AND OXFORD

Published by Princeton University Press
41 William Street, Princeton, New Jersey 08540

In the United Kingdom: Princeton University Press
6 Oxford Street, Woodstock, Oxfordshire, OX20 1TW

Klemens, Ben.
  Modeling with data : tools and techniques for scientific computing / Ben Klemens.
    p. cm.
  Includes bibliographical references and index.
  ISBN 978-0-691-13314-0 (hardcover : alk. paper)
  1. Mathematical statistics. 2. Mathematical models. I. Title.
  QA276.K546 2009
  519.5–dc22   2008028341

British Library Cataloging-in-Publication Data is available

This book has been composed in LATEX

The publisher would like to acknowledge the author of this volume for providing the camera-ready
copy from which this book was printed.

Printed on acid-free paper. ∞

press.princeton.edu

Printed in the United States of America

10 9 8 7 6 5 4 3 2 1

We believe that no one should be deprived of books for any reason.
—Russell Wattenberg, founder of the Book Thing

The author pledges to donate 25% of his royalties to the Book Thing of Baltimore, a non-profit that gives books to schools, students, libraries, and other readers of all kinds.

CONTENTS

PREFACE

Mathematics provides a framework for dealing precisely with notions of "what is."
Computation provides a framework for dealing precisely with notions of "how to."

—Alan J Perlis, in Abelson *et al.* (1985, p xvi)

SHOULD YOU USE THIS BOOK? This book is intended to be a complement to the standard stats textbook, in three ways.

First, descriptive and inferential statistics are kept separate beginning with the first sentence of the first chapter. I believe that the fusing of the two is the number one cause of confusion among statistics students.

Once descriptive modeling is given its own space, and models do not necessarily have to be just preparation for a test, the options blossom. There are myriad ways to convert a subjective understanding of the world into a mathematical model, including simulations, models like the Bernoulli/Poisson distributions from traditional probability theory, ordinary least squares, and who knows what else.

If those options aren't enough, simple models can be combined to form multilevel models to describe situations of arbitrary complexity. That is, the basic linear model or the Bernoulli/Poisson models may seem too simple for many situations, but they are building blocks that let us produce more descriptive models. The overall approach concludes with multilevel models as in, e.g., Eliason (1993), Pawitan (2001) or Gelman & Hill (2007).

Second, many stats texts aim to be as complete as possible, because completeness and a thick spine give the impression of value-for-money: you get a textbook *and* a reference book, so everything you need is guaranteed to be in there somewhere.

But it's hard to learn from a reference book. So I have made a solid effort to provide a narrative to the important points about statistics, even though that directly implies that this book is incomplete relative to the more encyclopædic texts. For example, moment generating functions are an interesting narrative on their own, but they are tangential to the story here, so I do not mention them.

Computation   The third manner in which this book complements the traditional stats textbook is that it acknowledges that if you are working with data full time, then you are working on a computer full time. The better you understand computing, the more you will be able to do with your data, and the faster you will be able to do it.

The politics of software

All of the software in this book is *free software*, meaning that it may be freely downloaded and distributed. This is because the book focuses on portability and replicability, and if you need to purchase a license every time you switch computers, then the code is not portable.

If you redistribute a functioning program that you wrote based on the GSL or Apophenia, then you need to redistribute both the compiled final program and the source code you used to write the program. If you are publishing an academic work, you should be doing this anyway. If you are in a situation where you will distribute only the output of an analysis, there are no obligations at all.

This book is also reliant on *POSIX*-compliant systems, because such systems were built from the ground up for writing and running replicable and portable projects. This does not exclude any current operating system (OS): current members of the Microsoft Windows family of OSes claim POSIX compliance, as do all OSes ending in X (Mac OS X, Linux, UNIX, ...).

People like to characterize computing as fast-paced and ever-changing, but much of that is just churn on the syntactic surface. The fundamental concepts, conceived by mathematicians with an eye toward the simplicity and elegance of pencil-and-paper math, have been around for as long as anybody can remember. Time spent learning those fundamentals will pay off no matter what exciting new language everybody happens to be using this month.

I spent much of my life ignoring the fundamentals of computing and just hacking together projects using the package or language of the month: C++, Mathematica, Octave, Perl, Python, Java, Scheme, S-PLUS, Stata, R, and probably a few others that I've forgotten. Albee (1960, p 30) explains that "sometimes it's necessary to go a long distance out of the way in order to come back a short distance correctly;" this is the distance I've gone to arrive at writing a book on data-oriented computing using a general and basic computing language. For the purpose of modeling with data, I have found C to be an easier and more pleasant language than the purpose-built alternatives—especially after I worked out that I could ignore much of the advice from books written in the 1980s and apply the techniques I learned from the scripting languages.

WHAT IS THE LEVEL OF THIS BOOK?   The short answer is that this is intended for the graduate student or independent researcher, either as a supplement to a standard first-year stats text or for later study. Here are a few more ways to answer that question:

- *Ease of use versus ease of initial use*: The majority of statistics students are just trying to slog through their department's stats requirement so they can never look at another data set again. If that is you, then your sole concern is ease of initial use, and you want a stats package and a textbook that focus less on full proficiency and more on immediate intuition.[1]
  Conversely, this book is not really about solving today's problem as quickly as physically possible, but about getting a better understanding of data handling, computing, and statistics. Ease of long-term use will follow therefrom.

- *Level of computing abstraction*: This book takes the fundamentals of computing seriously, but it is not about reinventing the wheels of statistical computing. For example, *Numerical Recipes in C* (Press *et al.*, 1988) is a classic text describing the algorithms for seeking optima, efficiently calculating determinants, and making random draws from a Normal distribution. Being such a classic, there are many packages that implement algorithms on its level, and this book will build upon those packages rather than replicate their effort.

- *Computing experience*: You may have never taken a computer science course, but do have some experience in both the basics of dealing with a computer and in writing scripts in either a stats package or a scripting language like Perl or Python.

- *Computational detail*: This book includes about 80 working sample programs. Code clarifies everything: English text may have a few ambiguities, but all the details have to be in place for a program to execute correctly. Also, code rewards the curious, because readers can explore the data, find out to what changes a procedure is robust, and otherwise productively break the code.
  That means that this book is not computing-system-agnostic. So if you are a devotee of a stats package not used here, then why look at this book? Although I do not shy away from C-specific details of syntax, most of the book focuses on the conceptual issues common to all computing environments. If you never look at C code again after you finish this book, you will still have a better grounding for effectively working in your preferred programming language.

- *Linear algebra*: You are reasonably familiar with linear algebra, such that an expression like $\mathbf{X}^{-1}$ is not foreign to you. There are a countably infinite number of linear algebra tutorials in books, stats text appendices, and online, so this book does not include yet another.

- *Statistical topics*: The book's statistical topics are not particularly advanced or trendy: OLS, maximum likelihood, or bootstrapping are all staples of first-year grad-level stats. But by creatively combining building blocks such as these, you will be able to model data and situations of arbitrary complexity.

---

[1] I myself learned a few things from the excellently written narrative in Gonick & Smith (1994).

Modeling with Data

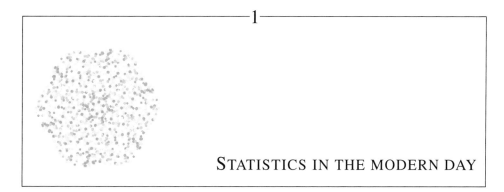

1

STATISTICS IN THE MODERN DAY

Retake the falling snow: each drifting flake
Shapeless and slow, unsteady and opaque,
A dull dark white against the day's pale white
And abstract larches in the neutral light.

—Nabokov (1962, lines 13–16)

Statistical analysis has two goals, which directly conflict. The first is to find patterns in static: given the infinite number of variables that one could observe, how can one discover the relations and patterns that make human sense? The second goal is a fight against *apophenia*, the human tendency to invent patterns in random static. Given that someone has found a pattern regarding a handful of variables, how can one verify that it is not just the product of a lucky draw or an overactive imagination?

Or, consider the complementary dichotomy of objective versus subjective. The objective side is often called *probability*; e.g., given the assumptions of the Central Limit Theorem, its conclusion is true with mathematical certainty. The subjective side is often called *statistics*; e.g., our claim that observed quantity A is a linear function of observed quantity B may be very useful, but Nature has no interest in it.

This book is about writing down subjective models based on our human understanding of how the world works, but which are heavily advised by objective information, including both mathematical theorems and observed data.[1]

---

[1]Of course, human-gathered data is never perfectly objective, but we all try our best to make it so.

The typical scheme begins by proposing a model of the world, then estimating the parameters of the model using the observed data, and then evaluating the fit of the model. This scheme includes both a descriptive step (describing a pattern) and an inferential step (testing whether there are indications that the pattern is valid). It begins with a subjective model, but is heavily advised by objective data.

Figure 1.1 shows a model in flowchart form. First, the descriptive step: data and parameters are fed into a function—which may be as simple as *a is correlated to b*, or may be a complex set of interrelations—and the function spits out some output. Then comes the testing step: evaluating the output based on some criterion, typically regarding how well it matches some portion of the data. Our goal is to find those parameters that produce output that best meets our evaluation criterion.

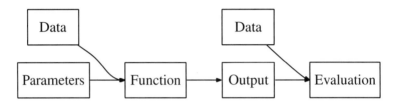

Figure 1.1 A flowchart for distribution fitting, linear regression, maximum likelihood methods, multilevel modeling, simulation (including agent-based modeling), data mining, non-parametric modeling, and various other methods. [Online source for the diagram: `models.dot`.]

The Ordinary Least Squares (OLS) model is a popular and familiar example, pictured in Figure 1.2. [If it is not familiar to you, we will cover it in Chapter 8.] Let $\mathbf{X}$ indicate the independent data, β the parameters, and $\mathbf{y}$ the dependent data. Then the function box consists of the simple equation $\mathbf{y}_{\text{out}} = \mathbf{X}\beta$, and the evaluation step seeks to minimize squared error, $(\mathbf{y} - \mathbf{y}_{\text{out}})^2$.

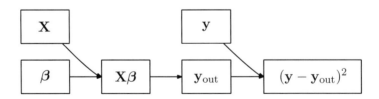

Figure 1.2 The OLS model: a special case of Figure 1.1.

For a simulation, the function box may be a complex flowchart in which variables are combined non-linearly with parameters, then feed back upon each other in unpredictable ways. The final step would evaluate how well the simulation output corresponds to the real-world phenomenon to be explained.

The key computational problem of statistical modeling is to find the parameters at the beginning of the flowchart that will output the best evaluation at the end. That is, for a given function and evaluation in Figure 1.1, we seek a routine to take in data and produce the optimal parameters, as in Figure 1.3. In the OLS model above, there is a simple, one-equation solution to the problem: $\beta_{\text{best}} = (\mathbf{X'X})^{-1}\mathbf{X'y}$. But for more complex models, such as simulations or many multilevel models, we must strategically try different sets of parameters to hunt for the best ones.

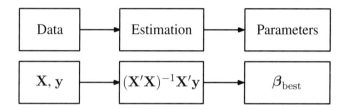

Figure 1.3 Top: the parameters which are the input for the model in Figure 1.1 are the output for the estimation routine.
Bottom: the estimation of the OLS model is a simple equation.

And that's the whole book: develop models whose parameters and tests may discover and verify interesting patterns in the data. But the setup is incredibly versatile, and with different function specifications, the setup takes many forms. Among a few minor asides, this book will cover the following topics, all of which are variants of Figure 1.1:

- Probability: how well-known distributions can be used to model data
- Projections: summarizing many-dimensional data in two or three dimensions
- Estimating linear models such as OLS
- Classical hypothesis testing: using the Central Limit Theorem (CLT) to ferret out apophenia
- Designing multilevel models, where one model's output is the input to a parent model
- Maximum likelihood estimation
- Hypothesis testing using likelihood ratio tests
- Monte Carlo methods for describing parameters
- "Nonparametric" modeling (which comfortably fits into the parametric form here), such as smoothing data distributions
- Bootstrapping to describe parameters and test hypotheses

THE SNOWFLAKE PROBLEM, OR A BRIEF   The simplest models in the above list
HISTORY OF STATISTICAL COMPUTING   have only one or two parameters, like
                                            a Binomial(n, p) distribution which is
built from n identical draws, each of which is a success with probability p [see
Chapter 7]. But draws in the real world are rarely identical—no two snowflakes
are exactly alike. It would be nice if an outcome variable, like annual income, were
determined entirely by one variable (like education), but we know that a few dozen
more enter into the picture (like age, race, marital status, geographical location, et
cetera).

The problem is to design a model that accommodates that sort of complexity, in
a manner that allows us to actually compute results. Before computers were common, the best we could do was analysis of variance methods (ANOVA), which
ascribed variation to a few potential causes [see Sections 7.1.3 and 9.4].

The first computational milestone, circa the early 1970s, arrived when civilian
computers had the power to easily invert matrices, a process that is necessary for
most linear models. The linear models such as ordinary least squares then became
dominant [see Chapter 8].

The second milestone, circa the mid 1990s, arrived when desktop computing power
was sufficient to easily gather enough local information to pin down the global optimum of a complex function—perhaps thousands or millions of evaluations of the
function. The functions that these methods can handle are much more general than
the linear models: you can now write and optimize models with millions of interacting agents or functions consisting of the sum of a thousand sub-distributions
[see Chapter 10].

The ironic result of such computational power is that it allows us to return to the
simple models like the Binomial distribution. But instead of specifying a fixed n
and p for the entire population, every observation could take on a value of n that is
a function of the individual's age, race, et cetera, and a value of p that is a different
function of age, race, et cetera [see Section 8.4].

The models in Part II are listed more-or-less in order of complexity. The infinitely
quotable Albert Einstein advised, "make everything as simple as possible, but not
simpler." The Central Limit Theorem tells us that errors often are Normally distributed, and it is often the case that the dependent variable is basically a linear or
log-linear function of several variables. If such descriptions do no violence to the
reality from which the data were culled, then OLS is the method to use, and using
more general techniques will not be any more persuasive. But if these assumptions
do not apply, we no longer need to assume linearity to overcome the snowflake
problem.

THE PIPELINE   A statistical analysis is a guided series of transformations of the data from its raw form as originally written down to a simple summary regarding a question of interest.

The flow above, in the statistics textbook tradition, picked up halfway through the analysis: it assumes a data set that is in the correct form. But the full pipeline goes from the original messy data set to a final estimation of a statistical model. It is built from functions that each incrementally transform the data in some manner, like removing missing data, selecting a subset of the data, or summarizing it into a single statistic like a mean or variance.

Thus, you can think of this book as a catalog of pipe sections and filters, plus a discussion of how to fit elements together to form a stream from raw data to final publishable output. As well as the pipe sections listed above, such as the ordinary least squares or maximum likelihood procedures, the book also covers several techniques for directly transforming data, computing statistics, and welding all these sections into a full program:

- Structuring programs using modular functions and the *stack* of *frames*
- Programming tools like the debugger and profiler
- Methods for reliability testing functions and making them more robust
- Databases, and how to get them to produce data in the format you need
- Talking to external programs, like graphics packages that will generate visualizations of your data
- Finding and using pre-existing functions to quickly estimate the parameters of a model from data.
- Optimization routines: how they work and how to use them
- Monte Carlo methods: getting a picture of a model via millions of random draws

To make things still more concrete, almost all of the sample code in this book is available from the book's Web site, linked from `http://press.princeton.edu/titles/8706.html`. This means that you can learn by running and modifying the examples, or you can cut, paste, and modify the sample code to get your own analyses running more quickly. The programs are listed and given a complete discussion on the pages of this book, so you can read it on the bus or at the beach, but you are very much encouraged to read through this book while sitting at your computer, where you can run the sample code, see what happens given different settings, and otherwise explore.

Figure 1.4 gives a typical pipeline from raw data to final paper. It works at a number of different *layers of abstraction*: some segments involve manipulating individual numbers, some segments take low-level numerical manipulation as given and op-

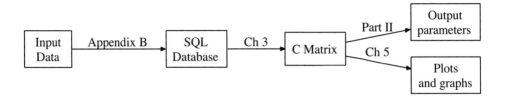

Figure 1.4 Filtering from input data to outputs. [Online source: `datafiltering.dot`]

erate on database tables or matrices, and some segments take matrix operations as given and run higher-level hypothesis tests.

The lowest level   Chapter 2 presents a tutorial on the C programming language itself. The work here is at the lowest level of abstraction, covering nothing more difficult than adding columns of numbers. The chapter also discusses how C facilitates the development and use of *libraries*: sets of functions written by past programmers that provide the tools to do work at higher and higher levels of abstraction (and thus ignore details at lower levels).[2]

For a number of reasons to be discussed below, the book relies on the C programming language for most of the pipe-fitting, but if there is a certain section that you find useful (the appendices and the chapter on databases comes to mind) then there is nothing keeping you from welding that pipe section to others using another programming language or system.

Dealing with large data sets   Computers today are able to crunch numbers a hundred times faster they did a decade ago—but the data sets they have to crunch are a thousand times larger. Geneticists routinely pull 550,000 genetic markers each from a hundred or a thousand patients. The US Census Bureau's 1% sample covers almost 3 million people. Thus, the next layer of abstraction provides specialized tools for dealing with data sets: databases and a query language for organizing data. Chapter 3 presents a new syntax for talking to a database, Structured Query Language (SQL). You will find that many types of data manipulation and filtering that are difficult in traditional languages or stats packages are trivial—even pleasant—via SQL.

---

[2]Why does the book omit a linear algebra tutorial but include an extensive C tutorial? Primarily because the use of linear algebra has not changed much this century, while the use of C has evolved as more libraries have become available. If you were writing C code in the early 1980s, you were using only the standard library and thus writing at a very low level. In the present day, the process of writing code is more about joining together libraries than writing from scratch. I felt that existing C tutorials and books focused too heavily on the process of writing from scratch, perpetuating the myth that C is appropriate only for low-level bit shifting. The discussion of C here introduces tools like package managers, the debugger, and the `make` utility as early as possible, so you can start calling existing libraries as quickly and easily as possible.

As Huber (2000, p 619) explains: "Large real-life problems always require a combination of database management and data analysis.... Neither database management systems nor traditional statistical packages are up to the task." The solution is to build a pipeline, as per Figure 1.4, that includes both database management and statistical analysis sections. Much of graceful data handling is in knowing where along the pipeline to place a filtering operation. The database is the appropriate place to filter out bad data, join together data from multiple sources, and aggregate data into group means and sums. C matrices are appropriate for filtering operations like those from earlier that took in data, applied a function like $(\mathbf{X'X})^{-1}\mathbf{X'y}$, and then measured $(\mathbf{y}_{\text{out}} - \mathbf{y})^2$.

Because your data probably did not come pre-loaded into a database, Appendix B discusses text manipulation techniques, so when the database expects your data set to use commas but your data is separated by erratic tabs, you will be able to quickly surmount the problem and move on to analysis.

Computation The GNU Scientific Library works at the numerical computation layer of abstraction. It includes tools for all of the procedures commonly used in statistics, such as linear algebra operations, looking up the value of F, t, χ^2 distributions, and finding maxima of likelihood functions. Chapter 4 presents some basics for data-oriented use of the GSL.

The Apophenia library, primarily covered in Chapter 4, builds upon these other layers of abstraction to provide functions at the level of data analysis, model fitting, and hypothesis testing.

Pretty pictures Good pictures can be essential to good research. They often reveal patterns in data that look like mere static when that data is presented as a table of numbers, and are an effective means of communicating with peers and persuading grantmakers. Consistent with the rest of this book, Chapter 5 will cover the use of Gnuplot and Graphviz, two packages that are freely available for the computer you are using right now. Both are entirely automatable, so once you have a graph or plot you like, you can have your C programs autogenerate it or manipulate it in amusing ways, or can send your program to your colleague in Madras and he will have no problem reproducing and modifying your plots.[3] Once you have the basics down, animation and real-time graphics for simulations are easy.

---

[3]Following a suggestion by Thomson (2001), I have chosen the gender of representative agents in this book by flipping a coin.

WHY C?   You may be surprised to see a book about modern statistical computing based on a language composed in 1972. Why use C instead of a specialized language or package like SAS, Stata, SPSS, S-Plus, SAGE, SIENA, SUDAAN, SYSTAT, SST, SHAZAM, J, K, GAUSS, GAMS, GLIM, GENSTAT, GRETL, EViews, Egret, EQS, PcGive, MatLab, Minitab, Mupad, Maple, Mplus, Maxima, MLn, Mathematica, WinBUGS, TSP, HLM, R, RATS, LISREL, LispStat, LIMDEP, BMDP, Octave, Orange, OxMetrics, Weka, or Yorick? This may be the only book to advocate statistical computing with a general computing language, so I will take some time to give you a better idea of why modern numerical analysis is best done in an old language.

One of the side effects of a programming language being stable for so long is that a mythology builds around it. Sometimes the mythology is outdated or false: I have seen professional computer programmers and writers claim that simple structures like linked lists always need to be written from scratch in C (see Section 6.2 for proof otherwise), that it takes ten to a hundred times as long to write a program in C than in a more recently-written language like R, or that because people have used C to write device drivers or other low-level work, it can not be used for high-level work.[4] This section is partly intended to dispel such myths.

Is C a hard language?   C *was* a hard language. With nothing but a basic 80s-era compiler, you could easily make many hard-to-catch mistakes. But programmers have had a few decades to identify those pitfalls and build tools to catch them. Modern compilers warn you of these issues, and debuggers let you interact with your program as it runs to catch more quirks. C's reputation as a hard language means the tools around it have evolved to make it an easy language.

Computational speed—really   Using a stats package sure beats inverting matrices by hand, but as computation goes, many stats packages are still relatively slow, and that slowness can make otherwise useful statistical methods infeasible.

R and Apophenia use the same C code for doing the Fisher exact test, so it makes a good basis for a timing test.[5] Listings 1.5 and 1.6 show programs in C and R (respectively) that will run a Fisher exact test five million times on the same data set. You can see that the C program is a bit more verbose: the steps taken in lines 3–8 of the C code and lines 1–6 of the R code are identical, but those lines are

---

[4]Out of courtesy, citations are omitted. This section makes frequent comparisons to R partly because it is a salient and common stats package, and partly because I know it well, having used it on a daily basis for several years.

[5]That is, if you download the source code for R's `fisher.test` function, you will find a set of procedures written in C. Save for a few minor modifications, the code underlying the `apop_test_fisher_exact` function is line-for-line identical.

```
1   #include <apop.h>
2   int main(){
3       int i, test_ct = 5e6;
4       double data[] = { 30, 86,
5                          24, 38 };
6       apop_data *testdata = apop_line_to_data(data,0,2,2);
7       for (i = 0; i< test_ct; i++)
8           apop_test_fisher_exact(testdata);
9   }
```

Listing 1.5 C code to time a Fisher exact test. It runs the same test five million times. Online source: `timefisher.c`.

```
1   test_ct <- 5e6
2   data <- c( 30, 86,
3              24, 38 )
4   testdata<- matrix(data, nrow=2)
5   for (i in 1:test_ct){
6       fisher.test(testdata)
7   }
```

Listing 1.6 R code to do the same test as Listing 1.5. Online source: `Rtimefisher`.

longer in C, and the C program has some preliminary code that the R script does not have.

On my laptop, Listing 1.5 runs in under three minutes, while Listing 1.6 does the same work in 89 minutes—about thirty times as long. So the investment of a little more verbosity and a few extra stars and semicolons returns a thirty-fold speed gain.[6] Nor is this an isolated test case: I can't count how many times people have told me stories about an analysis or simulation that took days or weeks in a stats package but ran in minutes after they rewrote it in C.

Even for moderately-sized data sets, real computing speed opens up new possibilities, because we can drop the (typically false) assumptions needed for closed-form solutions in favor of maximum likelihood or Monte Carlo methods. The Monte Carlo examples in Section 11.2 were produced using over a billion draws from t distributions; if your stats package can't produce a few hundred thousand draws per second (some can't), such work will be unfeasibly slow.[7]

---

[6]These timings are actually based on a modified version of `fisher.test` that omits some additional R-side calculations. If you had to put a Fisher test in a `for` loop without first editing R's code, the R-to-C speed ratio would be between fifty and a hundred.

[7]If you can produce random draws from t distributions as a batch (`draws <- rt(5e6, df)`), then R takes a mere 3.5 times as long as comparable C code. But if you need to produce them individually (`for (i in 1:5e6) {draw <- rt(1, df)}`), then R takes about fifteen times as long as comparable C code. On my laptop, R in

Simplicity   C is a super-simple language. Its syntax has no special tricks for poly-
           morphic operators, abstract classes, virtual inheritance, lexical scoping,
lambda expressions, or other such arcana, meaning that you have less to learn.
Those features are certainly helpful in their place, but without them C has already
proven to be sufficient for writing some impressive programs, like the Mac and
Linux operating systems and most of the stats packages listed above.

Simplicity affords stability—C is among the oldest programming languages in
common use today[8]—and stability brings its own benefits. First, you are reasonably assured that you will be able to verify and modify your work five or even
ten years from now. Since C was written in 1972, countless stats packages have
come and gone, while others are still around but have made so many changes in
syntax that they are effectively new languages. Either way, those who try to follow
the trends have on their hard drives dozens of scripts that they can't run anymore.
Meanwhile, correctly written C programs from the 1970s will compile and run on
new PCs.

Second, people have had a few decades to write good libraries, and libraries that
build upon those libraries. It is not the syntax of a language that allows you to easily
handle complex structures and tasks, but the vocabulary, which in the case of C is
continually being expanded by new function libraries. With a statistics library on
hand, the C code in Listing 1.5 and the R code in Listing 1.6 work at the same high
level of abstraction.

Alternatively, if you need more precision, you can use C's low-level bit-twiddling
to shunt individual elements of data. There is nothing more embarrassing than a
presenter who answers a question about an anomaly in the data or analysis with
'Stata didn't have a function to correct that.' [Yes, I have heard this in a real live
presentation by a real live researcher.] But since C's higher-level and lower-level
libraries are equally accessible, you can work at the level of laziness or precision
called for in any given situation.

Interacting with C scripts   Many of the stats packages listed above provide a pleas-
                        ing interface that let you run regressions with just a few
mouse-clicks. Such systems are certainly useful for certain settings, such as asking a few quick questions of a new data set. But an un-replicable analysis based
on clicking an arbitrary sequence of on-screen buttons is as useful as no analysis
at all. In the context of building a repeatable script that takes the data as far as
possible along the pipeline from raw format to final published output, developing

---

batch mode produced draws at a rate $\approx 424,000$/sec, while C produced draws at a rate $\approx 1,470,000$/sec.
    [8]However, it is not the oldest, an honor that goes to FORTRAN. This is noteworthy because some claim that
C is in common use today merely because of inertia, path dependency, et cetera. But C displaced a number of
other languages such as ALGOL and PL/I which had more inertia behind them, by making clear improvements
over the incumbents.

`script.do` for an interpreter and developing `program.c` for a compiler become about equivalent—especially since compilation on a modern computer takes on the order of 0.0 seconds.

With a debugger, the distance is even smaller, because you can jump around your C code, change intermediate values, and otherwise interact with your program the way you would with a stats package. Graphical interfaces for stats packages and for C debuggers tend to have a similar design.

But C is ugly!   C is by no means the best language for all possible purposes. Different systems have specialized syntaxes for communicating with other programs, handling text, building Web pages, or producing certain graphics. But for data analysis, C is very effective. It has its syntactic flaws: you will forget to append semicolons to every line, and will be frustrated that 3/2==1 while 3/2.==1.5. But then, Perl also requires semicolons after every line, and 3/2 is one in Perl, Python, and Ruby too. Type declarations are one more detail to remember, but the alternatives have their own warts: Perl basically requires that you declare the type of your variable (@, $, or #) with every use, and R will guess the type you meant to use, but will often guess wrong, such as thinking that a one-element list like {14} is really just an integer. C's `printf` statements look terribly confusing at first, but the authors of Ruby and Python, striving for the most programmer-friendly syntax possible, chose to use C's `printf` syntax over many alternatives that are easier on the eyes but harder to use.

In short, C does not do very well when measured by initial ease-of-use. But there is a logic to its mess of stars and braces, and over the course of decades, C has proven to be very well suited for designing pipelines for data analysis, linking together libraries from disparate sources, and describing detailed or computation-intensive models.

TYPOGRAPHY   Here are some notes on the typographic conventions used by this book.

✳ *Seeing the forest for the trees*   On the one hand, a good textbook should be a narrative that plots a definite course through a field. On the other hand, most fields have countless interesting and useful digressions and side-paths. Sections marked with a ✳ cover details that may be skipped on a first reading. They are not necessarily advanced in the sense of being somehow more difficult than unmarked text, but they may be distractions to the main narrative.

$\mathbb{Q}_{1.1}$ Questions and exercises are marked like this paragraph. The exercises are not thought experiments. It happens to all of us that we think we understand something until we sit down to actually do it, when a host of hairy details turn up. Especially at the outset, the exercises are relatively simple tasks that let you face the hairy details before your own real-world complications enter the situation. Exercises in the later chapters are more involved and require writing or modifying longer segments of code.

Notation

$\mathbf{X}$: boldface, capital letters are matrices. With few exceptions, data matrices in this book are organized so that the rows are each a single observation, and each column is a variable.

$\mathbf{x}$: lowercase boldface indicates a vector. Vectors are generally a column of numbers, and their transpose, $\mathbf{x}'$, is a row. $\mathbf{y}$ is typically a vector of dependent variables (the exception being when we just need two generic data vectors, in which case one will be x and one y).

x: A lowercase variable, not bold, is a scalar, i.e., a single real number.

$\mathbf{X}'$ is the transpose of the matrix $\mathbf{X}$. Some authors notate this as $\mathbf{X^T}$.

$\underline{\mathbf{X}}$ is the data matrix $\mathbf{X}$ with the mean of each column subtracted, meaning that each column of $\underline{\mathbf{X}}$ has mean zero. If $\mathbf{X}$ has a column of ones (as per most regression techniques), then the constant column is left unmodified in $\underline{\mathbf{X}}$.

n: the number of observations in the data set under discussion, which is typically the number of rows in $\mathbf{X}$. When there is ambiguity, n will be subscripted.

$\mathbf{I}$: The *identity matrix*. A square matrix with ones along its diagonal and zeros everywhere else.

β: Greek letters indicate parameters to be estimated; if boldface, they are a vector of parameters. The most common letter is β, but others may slip in, such as...

σ, μ: the standard deviation and the mean. The variance is σ^2.

$\hat{\sigma}, \hat{\beta}$: a carat over a parameter indicates an empirical estimate of the parameter derived from data. Typically read as, e.g., *sigma hat, beta hat*.

$\epsilon \sim \mathcal{N}(0, 1)$: Read this as *epsilon is distributed as a Normal distribution with parameters 0 and 1*.

$P(\cdot)$: A probability density function.

$LL(\cdot)$: The log likelihood function, $\ln(P(\cdot))$.

$S(\cdot)$: The Score, which is the vector of derivatives of $LL(\cdot)$.

$\mathbb{I}(\cdot)$: The information matrix, which is the matrix of second derivatives of $LL(\cdot)$.

$E(\cdot)$: The expected value, aka the mean, of the input.

$P(x|\beta)$: The probability of x given that β is true.

$P(x,\beta)|_x$: The probability density function, holding x fixed. Mathematically, this is simply $P(x,\beta)$, but in the given situation it should be thought of as a function only of β.[9]

$E_x(f(x,\beta))$: Read as *the expectation over x* of the given function, which will take a form like $\int_{\forall x} f(x,\beta)P(x)dx$. Because the integral is over all x, $E_x(f(x,\beta))$ is not itself a function of x.

`teletype typeface` indicates text that can be typed directly into a text file and understood as a valid shell script, C commands, SQL queries, et cetera.

`cat` *`sample_file`*: Slanted teletype text indicates a placeholder for text you will insert—a variable name rather than text to be read literally. You could read the code here as, 'let *`sample_file`* be the name of a file on your hard drive. Then type `cat` *`sample_file`* at the command prompt'.

$a \equiv b$: Read as 'a is equivalent to b' or 'a is defined as b'.

$a \propto b$: Read as 'a is proportional to b'.

2.3e6: Engineers often write scientific notation using so-called *exponential* or *E notation*, such as $2.3 \times 10^6 \equiv 2.3$e6. Many computing languages (including C, SQL, and Gnuplot) recognize E-notated numbers.

---

[9]Others use a different notation. For example, Efron & Hinkley (1978, p 458): "The log likelihood function $l_\theta(x)$...is the log of the density function, thought of as a function of θ." See page 329 for more on the philosophical considerations underlying the choice of notation.

Every section ends with a summary of the main points, set like this paragraph. There is much to be said for the strategy of flipping ahead to the summary at the end of the section before reading the section itself.
The summary for the introduction:

➤ This book will discuss methods of estimating and testing the parameters of a model with data.

➤ It will also cover the means of writing for a computer, including techniques to manage data, plot data sets, manipulate matrices, estimate statistical models, and test claims about their parameters.

Credits   Thanks to the following people, who added higher quality and richness to the book:

- Anjeanette Agro for graphic design suggestions.
- Amber Baum for extensive testing and critique.
- The Brookings Institution's Center on Social and Economic Dynamics, including Rob Axtell, Josh Epstein, Carol Graham, Emily Groves, Ross Hammond, Jon Parker, Matthew Raifman, and Peyton Young.
- Dorothy Gambrel, author of *Cat and Girl*, for the Lonely Planet data.
- Rob Goodspeed and the National Center for Smart Growth Research and Education at the University of Maryland, for the Washington Metro data.
- Derrick Higgins for comments, critique, and the Perl commands on page 414.
- Lucy Day Hobor and Vickie Kearn for editorial assistance and making working with Princeton University Press a pleasant experience.
- Guy Klemens, for a wide range of support on all fronts.
- Anne Laumann for the tattoo data set (Laumann & Derick, 2006).
- Abigail Rudman for her deft librarianship.

I

COMPUTING

2

C

This chapter introduces C and some of the general concepts behind good programming that script writers often overlook. The function-based approach, stacks of frames, debugging, test functions, and overall good style are immediately applicable to virtually every programming language in use today. Thus, this chapter on C may help you to become a better programmer with any programming language.

As for the syntax of C, this chapter will cover only a subset. C has 32 keywords and this book will only use 18 of them.[1] Some of the other keywords are basically archaic, designed for the days when compilers needed help from the user to optimize code. Other elements, like bit-shifting operators, are useful only if you are writing an operating system or a hardware device driver. With all the parts of C that directly manipulate hexadecimal memory addresses omitted, you will find that C is a rather simple language that is well suited for simulations and handling large data sets.

An outline This chapter divides into three main parts. Sections 2.1 and 2.2 start small, covering the syntax of individual lines of code to make assignments, do arithmetic, and declare variables. Sections 2.3 through 2.5 introduce functions, describing how C is built on the idea of modular functions that are each independently built and evaluated. Sections 2.6 through 2.8 cover *pointers*, a somewhat C-specific means of handling computer memory that complements C's means of handling functions and large data structures. The remainder of the chapter offers some tips on writing bug-free code.

---

[1]For comparison, C^{++} has 62 keywords as of this writing, and Java has an even 50.

Tools   You will need a number of tools before you can work, including a C *compiler*,
a *debugger*, the *make* facility, and a few libraries of functions. Some systems
have them all pre-installed, especially if you have a benevolent system adminis-
trator taking care of things. If you are not so fortunate, you will need to gather
the tools yourself. The online appendix to this book, at the site linked from `http:`
`//press.princeton.edu/titles/8706.html`, will guide you through the process of putting together a complete C development environment and using the tools
for gathering the requisite libraries.[2]

Check your C environment by compiling and running "Hello, world," a classic first program adapted from Kernighan & Ritchie (1988).

- Download the sample code for this book from the link at `http://`
  `press.princeton.edu/titles/8706.html`.

- Decompress the `.zip` file, go into the directory thus created, and compile the program with the command `gcc hello_world.c`. If you are
  using an *IDE*, see your manual for compilation instructions.

- If all went well, you will now have a program in the directory named
  either `a.out` or `hello_world`. From the command line, you can execute it using `./a.out` or `./hello_world`.

- You may also want to try the `makefile`, which you will also find in
  the code directory. See the instructions at the head of that file.

If you need troubleshooting help, see the online appendix, ask your local
computing guru, or copy and paste your error messages into your favorite
search engine.

2.1 LINES   The story begins at the smallest level: a single line of code. Most
of the work on this level will be familiar to anyone who has written
programs in any language, including instructions like assignments, basic arithmetic, if-then conditions, loops, and comments. For such common programming
elements, learning C is simply a question of the details of syntax. Also, C is a
typed language, meaning that you will need to specify whether every variable and
function is an integer, a real, a vector, or whatever. Thus, many of the lines will be
simple type declarations, whose syntax will be covered in the next section.

---

[2]A pedantic note on standards: this book makes an effort to comply with the ISO C99 standard and the IEEE
POSIX standard. The C99 standard includes some features that do not appear in the great majority of C textbooks
(like designated initializers), but if your compiler does not support the features of C99 used here, then get a new
compiler—it's been a long while since 1999. The POSIX standard defines features that are common to almost
every modern operating system, the most notable of which is the *pipe*; see Appendix B for details.

   The focus is on `gcc`, because that is what I expect most readers will be using. The command-line switches
for the `gcc` command are obviously specific to that compiler, and users of other compilers will need to check
the compiler manual for corresponding switches. However, all C code should compile for any C99- and POSIX-
compliant compiler. Finally, the `gcc` switch most relevant to this footnote is `-std=gnu99`, which basically puts
the compiler in C99 + POSIX mode.

ASSIGNMENT Most of the work you will be doing will be simple assignments. For example,

$$\left[\text{ratio = a / b;} \right.$$

will find the value of a divided by b and put the value in `ratio`. The = indicates an assignment, not an assertion about equality; on paper, computer scientists often write this as `ratio` ← `a/b`, which nicely gets across an image of `ratio` taking on the value of a/b. There is a semicolon at the end of the line; you will need a semicolon at the end of everything but the few exceptions below.[3] You can use all of the usual operations: +, -, /, and *. As per basic algebraic custom, * and / are evaluated before + and -, so `4 + 6 / 2` is seven, and `(4 + 6)/2` is five.

※ **TWO TYPES OF DIVISION** There are two ways to answer the question, "What is 11 divided by 3?" The common answer is that $11/3 = 3.\overline{66}$, but some say that it is three with a remainder of two. Many programming languages, including C, take the second approach. Dividing an integer by an integer gives the answer with the fractional part thrown out, while the modulo operator, %, finds the remainder. So `11/3` is 3 and `11%3` is 2.

Is k an even number? If it is, then `k % 2` is zero.[4]

Splitting the process into two parts provides a touch of additional precision, because the machine can write down integers precisely, but can only approximate real numbers like $3.\overline{66}$. Thus, the machine's evaluation of `(11.0/3.0)*3.0` may be ever-so-slightly different from `11.0`. But with the special handling of division for integers, you are guaranteed that for any integers a and b (where b is not zero), `(a/b)*b + a%b` is exactly a.

But in most cases, you just want $11/3 = 3.\overline{66}$. The solution is to say when you mean an integer and when you mean a real number that happens to take on an integer value, by adding a decimal point. `11/3` is 3, as above, but `11./3` is `3.66...` as desired. Get into the habit of adding decimal points now, because integer division is a famous source of hard-to-debug errors. Page 33 covers the situation in slightly more detail, and in the meantime we can move on to the more convenient parts of the language.

---

[3]The number one cause of compiler complaints like "line 41: syntax error" is a missing semicolon on line 40.
[4]In practice, you can check evenness with `GSL_IS_EVEN` or `GSL_IS_ODD`:
```
#include <gsl/gsl_math.h>
...
if (GSL_IS_EVEN(k))
    do_something();
```

INCREMENTING   It is incredibly common to have an operation of the form
a = a + b;—so common that C has a special syntax for it:

> a += b;

This is slightly less readable, but involves less redundancy. All of the above arithmetic operators can take this form, so each of the following lines show two equivalent expressions:

> a −= b; /*is equivalent to*/   a = a − b;
> a *= b; /*is equivalent to*/   a = a * b;
> a /= b; /*is equivalent to*/   a = a / b;
> a %= b; /*is equivalent to*/   a = a % b;

The most common operation among these is incrementing or decrementing by one, and so C offers the following syntax for still less typing:[5]

> a++;   /*is equivalent to*/   a = a + 1;
> a−−;   /*is equivalent to*/   a = a − 1;

CONDITIONS   C has no need for FALSE and TRUE keywords for Boolean operations: if an expression is zero, then it is false, and otherwise it is true. The standard operations for comparison and Boolean algebra all appear in somewhat familiar form:

> (a > b)   // a is greater than b
> (a < b)   // a is less than b
> (a >= b)   // a is greater than or equal to b
> (a <= b)   // a is less than or equal to b
> (a == b)   // a equals b
> (a != b)   // a is not equal to b
> (a && b)   // a and b
> (a || b)   // a or b
> (! a)   // not a

- All of these evaluate to either a one or a zero, depending on whether the expression in parens is true or false.

---

[5]There is also the pre-increment form, ++a and −−a. Pre- and post-incrementing differ only when they are being used in situations that are bad style and should be avoided. Leave these operations on a separate line and stick to whichever form looks nicer to you.

- The comparison for equality involves *two* equals signs in a row. One equals sign (a = b) will assign the value of b to the variable a, which is not what you had intended. Your compiler will warn you in most of the cases where you are probably using the wrong one, and you should heed its warnings.

- The && and || operators have a convenient feature: if a is sufficient to determine whether the entire expression is true, then it won't bother with b at all. For example, this code fragment—

$$((a < 0) \;||\; (\mathrm{sqrt}(a) < 3))$$

—will never take the square root of a negative number. If a is less than zero, then the evaluation of the expression is done after the first half (it is true), and evaluation stops. If a>=0, then the first part of this expression is not sufficient to evaluate the whole expression, so the second part is evaluated to determine whether $\sqrt{a} < 3$.

Why all the parentheses? First, parentheses indicate the order of operations, as they do in pencil-and-paper math. Since all comparisons evaluate to a zero or a one, both ((a>b)||(c>d)) and (a>(b||c)>d) make sense to C. You probably meant the first, but unless you have the order-of-operations table memorized, you won't be sure which of the two C thinks you mean by (a>b||c>d).[6]

Second, the primary use of these conditionals is in flow control: causing the program to repeat some lines while a condition is true, or execute some lines only if a condition is false. In all of the cases below, you will need parentheses around the conditions, and if you forget, you will get a confusing compiler error.

IF-ELSE STATEMENTS   Here is a fragment of code (which will not compile by itself) showing the syntax for conditional evaluations:

```
1   if (a > 0)
2          { b = sqrt(a); }
3   else
4          { b = 0; }
```

If a is positive, then b will be given the value of a's square root; if a is zero or negative, then b is given the value zero.

- The condition to be evaluated is always in parentheses following the if statement, and there should be curly braces around the part that will be evaluated when the

---

[6]The order-of-operations table is available online, but you are encouraged to not look it up. [If you must, try man operator from the command prompt]. Most people remember only the basics like how multiplication and division come before addition and subtraction; if you rely on the order-of-operations table for any other ordering, then you will merely be sending future readers (perhaps yourself) to check the table.

condition is true, and around the part that will be evaluated when the condition is false.

- You can exclude the curly braces on lines two and four if they surround exactly one line, but this will at some point confuse you and cause you to regret leaving them out.

- You can exclude the `else` part on lines three and four if you don't need it (which is common, and much less likely to cause trouble).

- The `if` statement and the line following it are smaller parts of one larger expression, so there is no semicolon between the `if(...)` clause and what happens should it be true; similarly with the `else` clause. If you do put a semicolon after an `if` statement—if (a > 0);—then your `if` statement will execute the null statement—/*do nothing*/;—when a > 0. Your compiler will warn you of this.

$Q_{2.2}$ Modify `hello_world.c` to print its greeting if the expression (1 || 0 && 0) is true, and print a different message of your choosing if it is false. Did C think you meant ((1 || 0) && 0) (which evaluates to 0) or (1 || (0 && 0)) (which evaluates to 1)?

LOOPS   Listing 2.1 shows three types of loop, which are slightly redundant.

```
1   #include <stdio.h>
2   int main(){
3       int i = 0;
4       while (i < 5){
5           printf("Hello.\n");
6           i ++;
7       }
8
9       for (i=0; i < 5; i++){
10          printf("Hi.\n");
11      }
12
13      i = 0;
14      do {
15          printf("Hello.\n");
16          i ++;
17      } while (i < 5);
18
19      return 0;
20  }
```

Listing 2.1 C provides three types of loop: the `while` loop, the `for` loop, and the `do-while` loop. Online source: `flow.c`.

The simplest is a `while` loop. The interpretation is rather straightforward: while the expression in parentheses on line four is true (mustn't forget the parentheses), execute the instructions in brackets, lines five and six.

Loops based on a counter ($i = 0$, $i = 1$, $i = 2$, ...) are so common that they get their own syntax, the `for` loop. The `for` loop in lines 9–11 is exactly equivalent to the `while` loop in lines 3–7, but gathers all the instructions about incrementing the counter onto a single line.

You can compare the `for` and `while` loop to see when the three subelements in the parentheses are evaluated: the first part (`i=0`) is evaluated before the loop runs; the second part (`i<5`) is tested at the beginning of each iteration of the loop; the third part (`i++`) is evaluated at the end of each loop. After the section on arrays, you will be very used to the `for (i=0; i<limit; i++)` form, and will recognize it to mean *step through the array*. There may even be a way to get your text editor to produce this form with one or two keystrokes.

Finally, if you want to guarantee that the loop will run at least once, you can use a `do-while` loop (with a semicolon at the end of the `while` line to conclude the thought). The `do-while` loop in Listing 2.1 is equivalent to the `while` and `for` loops. But say that you want to iteratively evaluate a function until it converges to within 1×10^{-3}. Naturally, you would want to run the function at least once. The form would be something like:

```
do {
    error = evaluate_function();
} while (error > 1e−3);
```

Example: the birthday paradox   The birthday paradox is a staple of undergraduate statistics classes.[7] The professor writes down the birth date of every student in the class, and finds that even though there is a 1 in 365 chance that any given pair of students have the same birthday, the odds are good that there is a match in the class overall.

Listing 2.2 shows code to find the likelihood that another student shares the first person's birthday, and the likelihood that any two students share a birthday.

• Most of the world's programs never need to take a square root, so functions like `pow` and `sqrt` are not included in the standard C library. They are in the separate math library, which you must refer to on the command line. Thus, compile the program with

---

[7]It even mystifies TV talk show hosts, according to Paulos (1988, p 36).

```
1    #include <math.h>
2    #include <stdio.h>
3
4    int main(){
5       double no_match = 1;
6       double matches_me;
7       int ct;
8          printf("People\t Matches me\t Any match\n");
9          for (ct=2; ct<=40; ct ++){
10            matches_me = 1− pow(364/365., ct−1);
11            no_match *= (1 − (ct−1)/365.);
12            printf("%i\t %.3f\t\t %.3f\n", ct, matches_me, (1−no_match));
13         }
14         return 0;
15   }
```

Listing 2.2 Print the odds that other students share my birthday, and that any two students in the room share a birthday. Online source: `birthday.c`.

```
gcc birthday.c −lm −o birthday
```

where `-lm` indicates the math library and `-o` indicates that the output program will be named `birthday` (rather than the default `a.out`). More on linking and libraries will follow below.

- Lines 1–7 are introductory material, to be discussed below, including a preface #include-ing a few external files, and a list of the dramatis personæ: variables named `no_match`, `matches_me`, and `ct`.

- Line 8 prints a header line labeling the columns of numbers the `for` loop will be producing; it is easy to read once you know that `\t` means *print a tab* and `\n` means *newline*.

- Line 9 tells us that the counter `ct` will start at two, and count up until it reaches 40.

- As for the math itself, it is easier to calculate the complement—the odds that nobody shares a birthday. The odds that one person does not share the first person's birthday is $364/365$; the odds that two people both do not share the first person's birthday is $(364/365)^2$, et cetera.[8] Thus, the odds that among `ct-1` additional people, none have the same birthday as the first person is $1 − (364/365)^{ct−1}$. You can see this calculation on line ten.

- As above, the odds that the second person does not share the first person's birthday is $\left(\frac{364}{365}\right)$. The odds that an additional person shares no birthday with the first two given that the first two do not share a birthday is $\left(\frac{363}{365}\right)$, so the odds that the first

---

[8]We assume away leap years, and the fact that the odds of being born on any given day are not exactly $1/365$—more children are born in the summer.

three do not share a birthday is

$$\left(\frac{364}{365}\right)\left(\frac{363}{365}\right). \tag{2.1.1}$$

This expression is best produced incrementally. In the introductory material, `no_-match` was initialized at 1, and on line 11, another element of the sequence headed by Expression 2.1.1 gets multiplied in to `no_match` at each step of the `for` loop.

• Line 12 prints the results. The first input to the `printf` function will be discussed in detail below, but the next inputs indicate what is to be printed: the counter, `matches_me`, and `1-(no_match)`.

Q₂.₃ Modify the `for` loop to verify that the program prints the correct values for a class of one student.

COMMENTS Put a long block of comments at the head of a file and at the head of each function to describe what the file or function does, using complete sentences. Describe what the function expects to come in, and what the function will put out. The common wisdom indicates that these comments should focus on why your code is doing what it is doing, rather than how, which will be self-explanatory in clearly-written code.[9]

The primary audience of your comment should be you, six months from now. When you are shopping for black boxes to plug in to your next project, or re-auditing your data after the referee finally got the paper back to you, a note to self at the head of each function will pay immense dividends.

> /* Long comments begin with a slash−star,
>   continue as long as you want, and end
>   at the first star−slash.
>   */

The stars and slashes are also useful for *commenting out* code. If you would like to temporarily remove a few lines from your program to see what would happen, but don't want to delete them entirely, simply put a /* and a */ around the code, and the compiler will think it is a comment and ignore it.

However, there is a slight problem with this approach: what if there is a comment in what you had just commented out? You would have a sequence like this in your code:

---

[9]The sample code for this book attempts to be an example of good code in most respects, but it has much less documentation than real-world code should have, because this book is the documentation.

```
/* Line A;
   /* Line B */
   Line C;
*/
```

We had hoped that all three lines would be commented out now, but the compiler will ignore everything from the first /* until it sees the first */. That means Line A and Line B will be ignored, but

```
   Line C;
*/
```

will be read as code—and malformed code at that.[10]

You will always need to watch out for this when commenting out large blocks of code. But for small blocks, there is another syntax for commenting individual lines of code that deserve a note.

```
this_is_code; //Everything on a line
              //after two slashes
              //will be ignored.
```

Later, we will meet the preprocessor, which modifies the program's text before compilation. It provides another solution for commenting out large blocks that may have comments embedded. The compiler will see none of the following code, because the preprocessor will skip everything between the #if statement which evaluates to zero and the #endif:

```
#if 0
/*This function does nothing. */
void do_nothing(){ }
#endif
```

PRINTING   C prints to anything—the screen, a string of text in memory, a file—using the same syntax. The formatting works much like the Mad Libs party game (Price & Stern, 1988). First, there is a format specifier, showing what the output will be, but with blanks to be filled in:

---

[10] Q: If C didn't have this quirk, and allowed comments inside comments, what different quirk would you have to watch out for instead?

My _____ is very _____.
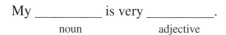
      noun   adjective

Then, the user provides a specific instance of the noun and adjective to be filled in (which in this case is left as an exercise for the reader). Since C is a programming language and not a party game, the syntax is a little more terse. Instead of

_____ is number _____ in line.
   string   int

`printf` uses:

%s is number %i in line.

Here is a complete example:

```
1  #include <stdio.h>
2
3  int main(){
4      int position = 3;
5      char name[] = "Steven";
6      printf("%s is number %i in line\n", name, position);
7      return 0;
8  }
```

The `printf` function is not actually defined by default—its definition in the standard input/output header must be #included, which is what line one does. Lines four and five are variable declarations, defining the *types* of the variables; these lines foreshadow the next section.

Finally, line six is the actual print statement, which will insert `Steven` into the first placeholder (`%s`), and insert 3 into the second placeholder (`%i`). It will thus print `Steven is number 3 in line` (plus an invisible newline).

Here are the odd characters you will need for almost all of your work.

%i	insert an integer here
%g	insert a real number in general format here
%s	insert a string of text here
%%	a plain percent sign
\n	begin a new line
\t	tab
\"	a quote that won't end the text string
\(newline)	continue the text string on the next line

There are many more format specifiers, which will give you a great deal of control; you may want them when printing tables, for example, and can refer to any of a number of detailed references when you need these, such as man 3 printf from your command line.

At this point, you may want to flip through this book to find a few examples of printf and verify that they will indeed print what they promise to.

➤ Assignment uses a single equals sign: assignee = value;.

➤ The usual arithmetic works: ten = 2*3+8/2;.

➤ Conditions such as (a > b), (a <= b), and (a == b) (two equals signs) can be used to control flow.

➤ Conditional flow uses the form: if (condition) {do_if_true;} else {do_if_false;}.

➤ The basic loop is a while loop: while (this_is_true) {do_-this;}.

➤ When iterating through an array, a for loop makes the iteration clearer: for (j=0; j< limit; j++) {printf("processing item %i\n", j);}.

➤ Write comments for your future self.

2.2   VARIABLES AND THEIR DECLARATIONS

Having covered the verbs that a line of code will execute, we move on to the nouns—variables.

You would never use x or z in a paper without first declaring, say, 'let $x \in \mathbb{R}^2$ and $z \in \mathbb{C}$'. You could leave the reader to guess at what you mean by x by its first use, but some readers would misunderstand, and your referee would wonder why you did not just come out and declare x. C is a strict referee, and requires that you declare the type of every variable before using it. The declaration consists of listing the type of the variable and then the variable name, e.g.

```
int a_variable, counter=0;
double stuff;
```

• This snippet declared three variables in two lines.
• We could initialize counter to zero as it is declared.

- The other variables (such as a_variable) have unknown values right now. Assume nothing about what is contained in a declared but uninitialized value.[11]

- Since the first step in using a variable is typically an assignment, and you can declare and initialize on the same line, the burden of declaring types basically means putting a type at the head of the first line where the variable is used, as on lines four and five of the sample code on page 27.

Here is a comprehensive list of the useful basic types for C.

int	an integer: $-1, 0, 3$
double	a real number: $2.4, -1.3e8, 27$
char	a character: 'a', 'b', 'C'

An int can only count to about $2^{32} \approx 4.3$ billion; you may have a simulation that involves five billion agents or other such uses for counting into the trillions, in which case you can use the long int type.[12]

There are ways to extend or shrink the size of the numbers, which are basically not worth caring about. A double counts up to about $\pm 1e308$, which is already significantly more than common estimates of the number of atoms in the universe (circa 1e80), but there is a long double type in case you need more precision or size.[13] Section 4.5 offers detailed notes about how numbers are represented.

Finally, notice that the variable names used throughout are words, not letters.[14] Using English variable names is the number one best thing you could do to make your code readable. Imagine how much of your life you have spent flipping back through journal articles trying to remember what μ, M, and m stood for. Why impose that on yourself?

---

[11]I am reluctant to mention this, but later you will see the distinction between global, static, and local variables. Global and static variables are automatically initialized to zero (or NULL), while local variables are not. But you will suffer fewer painful debugging sessions if you ignore this fact and get into the habit of explicitly initializing everything that needs initialization.

[12]The int type on most 64-bit systems is still 32 bits. Though this norm will no doubt change in the future, the safe bet is to write code under the assumption that an int counts to 2^{32}.

[13]The double name is short for "double-precision floating-point number," and the float thus has half the precision and range of a double. Why *floating point*? The computer represents a real using a form comparable to scientific notation: $n \times 10^k$, where n represents the number with the decimal point in a fixed location, and k represents the location of the decimal point. Multiplying by ten doesn't change the number n, it just causes the decimal point to float to a different position.

The float type is especially not worth bothering with because the GSL's matrices and vectors default to holding doubles, and C's floating-point functions internally operate on doubles. For example, the atof function (ASCII text to floating-point number) actually returns a double.

[14]The exception are indices for counters and for loops, which are almost always i, j, or k.

Arrays   Much of the art of describing the real world consists of building aggregates of these few basic types into larger structures. The simplest such aggregate is an *array*, which is simply a numbered list of items of the same type. To declare a list of a hundred integers, you would use:

> **int** a_list[100];

Then, to refer to the items of the list, you would use the same square brackets. For example, to assign the value seven to the last element of the array, you would use: `a_list[99]= 7;`. Why is 99 the last element of the list? Because the index is an *offset* from the first element. The first element is zero items away from itself, so it is `a_list[0]`, not `a_list[1]` (which is the second element). The reasoning behind this system will become evident in the section on pointers.

2-D arrays simply require more indices—`int a_2d_list[100][100]`. But there are details in implementation that make 2-D arrays difficult to use in practice; Chapter 4 will introduce the `gsl_matrix`, which provides many advantages over the raw 2-D array.

Just as you can initialize a scalar with a value at declaration, you can do the same with an array, e.g.:

> **double** data[ ] = {2,4,8,16,32,64};

You do not have to bother counting how many elements the array has; C is smart enough to do this for you.

Q2.4   Write a program to create an array of 100 integers, and then fill the array with the squares (so `the_array[7]` will hold 49). Then, print a message like "7 squared is 49." for each element of the array. Use the Hello World program as a template from which to start.

Q2.5   The first element of the Fibonacci sequence is defined to be 0, the second is defined to be 1, and then element n is defined to be the sum of elements $n-1$ and $n-2$. Thus, the third element is 0+1=1, the fourth is 1+1=2, the fifth is 1+2=3, then 2+3=5, then 3+5=8, et cetera.
The ratio of the nth element over the $(n-1)$st element converges to a value known as the golden ratio.
Demonstrate this convergence by producing a table of the first 20 elements of the sequence and the ratio of the nth over the $(n-1)$st element for each n.

Declaring types   You can define your own types. For example, these lines will first declare a new type, `triplet`, and then declare two such triplets, `tri1` and `tri2`:

```
typedef double triplet[3];
triplet tri1, tri2;
```

This is primarily useful for designing complex data types that are collections of many subelements, in conjunction with the `struct` keyword. For example:

```
typedef struct {
    double real;
    double imaginary;
} complex;

complex a, b;
```

You now have two variables of type `complex` and can now use `a.real` or `b.imaginary` to refer to the appropriate constituents of these complex numbers.

Listing 2.3 repeats the birthday example, but stores each class size's data in a `struct`.

- Lines 4–7 define the structure: it will hold one variable indicating the probability of somebody matching the first person's birthday, and one variable giving the probability that no two people share a birthday.
- Those lines only defined a type; line 11 declares a variable, `days`, which will be of this type. Since there is a number in brackets after the name, this is an array of `bday_structs`.
- In line 12, the `none_match` element of `days[1]` is given a value. Lines 14 and 15 assign values to the elements of `days[2]` through `days[40]`. Having calculated the values and stored them in an organized manner, it is easy for lines 18–20 to print the values.

Initializing   As with an array, you can initialize most or all of the elements of a struct to a value on the declaration line. The first option, comparable to the array syntax above, is to remember the order of the `struct`'s elements and make the assignments in that order. For example,

```
complex one = {1, 0};
complex pioverfour = {1, 1};
```

```
1    #include <math.h>
2    #include <stdio.h>
3
4    typedef struct {
5        double one_match;
6        double none_match;
7    } bday_struct;
8
9    int main(){
10       int ct, upto = 40;
11       bday_struct days[upto+1];
12       days[1].none_match = 1;
13       for (ct=2; ct<=upto; ct ++){
14           days[ct].one_match = 1− pow(364/365., ct−1);
15           days[ct].none_match = days[ct−1].none_match ∗ (1 − (ct−1)/365.);
16       }
17       printf("People\t Matches me\t Any match\n");
18       for (ct=2; ct<=upto; ct ++){
19           printf("%i\t %.3f\t\t %.3f\n", ct, days[ct].one_match, 1−days[ct].none_match);
20       }
21       return 0;
22   }
```

Listing 2.3 The birthday example (Listing 2.2) rewritten using a `struct` to hold each day's data. Online source: `bdaystruct.c`.

would initialize `one` to $1 + 0i$ and `pioverfour` to $1 + 1i$. This is probably the best way to initialize a `struct` where there are few elements and they have a well-known order.

The other option is to use *designated initializers*, which are best defined by example. The above two initializations are equivalent to:

```
complex one = {.real = 1};
complex pioverfour = {.imaginary = 1, .real = 1};
```

In the first case, the imaginary part is not given, so it is initialized to zero. In the second case, the elements are out of order, which is not a problem. Designated initializers will prove to be invaluable when dealing with structures like the `apop_-model`, which has a large number of elements in no particular order.

Two final notes on designated initializers. They can also be used for arrays, and they can be interspersed with unlabeled elements. The line

```
int isprime[] = {[1]=1, 1, 1, [5]=1, [7]=1, [11]=1};
```

initializes an array from zero to eleven (the length is determined by the last initialized element), setting the elements whose index is a prime number to one. The two ones with no label will go into the 2 and 3 slot, because their index will follow sequentially after the last index given.

Structs are syntactically simple, so there is little to say about them, but much of good programming goes in to designing structs that make sense and are a good reflection of reality. This book will be filled with them, including both purpose-built structures like the `bday_struct` and structures defined by libraries like the GSL, such as the `gsl_matrix` mentioned above.

✳ **TYPE CASTING**   There are a few minor complications when assigning a value of one type to a variable of a different type. When you assign a `double` value to an integer, such as `int i = 3.2`, everything after the decimal point would be dropped. Also, the range of `floats`, `doubles`, and `ints` do not necessarily match, so you may get unpredictable results with large numbers even when there is nothing after the decimal point.

If you are confident that you want to assign a variable of one type to a variable of another, then you can do so by putting the type to re-cast into in parentheses before the variable name. For example, if `ratio` is a `double`, `(int) ratio` will cast it to an integer. If you want to accept the truncation and assign a floating-point real to an integer, say `int n`, then explicitly tell the compiler that you meant to do this by making the cast yourself; e.g., `n = (int) ratio;`.

Type casting solves the division-by-integers problem from the head of this chapter. If `num` and `den` are both `ints`, then `ratio = (double) num / den` does the division of a real by an integer, which will produce a real number as expected.

There are two other ways of getting the same effect: `(num + 0.0)` is an `int` plus a `double`, which is a `double`. Then `(num + 0.0)/den` is division of a real by an integer, which again works as expected (but don't forget the parens). And as above, if one of the numbers is a constant, then just add a decimal point, because `2` is an `int`, while `2.` is a floating-point real number.

Finally, note that when casting from `double` to `int`, numbers are truncated, not rounded. As a lead-in to the discussion of functions, here is a function that uses type casting to correctly round off numbers:[15]

---

[15]In the real world, use `rint` (in `math.h`) to round to integer: `rounded_val = rint(unrounded_number)`.

```
int round(double unrounded){
/* Input a real number and output the number
   rounded to the nearest integer. */

    if (unrounded > 0)
            return (int) (unrounded + 0.5);
    else
            return (int) (unrounded − 0.5);
}
```

➤ All variables must be declared before the first use.

➤ Until a variable is given a value, you know nothing about its value.

➤ You can assign an initial value to the variable on the declaration line, such as int i = 3;.

➤ Arrays are simply declared by including a size in brackets after the variable name: int array[100];. Refer to the elements using an *offset*, so the first element is array[0] and the last is array[99].

➤ You can declare new types, including structures that amalgamate simpler types: typedef struct {double length, waist; int leg_ct;} pants;.

➤ After declaring a variable as a structure, say pants cutoffs;, refer to structure elements using a dot: cutoffs.leg_ct = 1;.

➤ An integer divided by an integer is an integer: 9/4 == 2. By putting a decimal after a whole number, it becomes a real number, and division works as expected: 9./4 == 2.25.

2.3 FUNCTIONS   The instruction *take the inverse of the matrix* is six words long, but refers to a sequence of steps that typically require several pages to fully describe.

Like many fields, mathematics progresses through the development of new vocabulary like the phrase *take the inverse*. We can comprehensibly express a complex statement like *the variance is* $\sigma^2(\mathbf{X'X})^{-1}$ because we didn't need to write out exactly how to do a squaring, a transposition $(\mathbf{X'})$ and an inverse.

Similarly, most of the process of writing code is not about describing the procedures involved, but building a specialized vocabulary to make describing the procedures trivial. Adding new nouns to the vocabulary is a simple task, discussed above using both basic nouns and `structs` that aggregate them to larger concepts. This section covers functions, which are single verbs that encapsulate a larger procedure.

```
1    #include <math.h>
2    #include <stdio.h>
3
4    typedef struct {
5        double one_match;
6        double none_match;
7    } bday_struct;
8
9    int upto = 40;
10   void calculate_days(bday_struct days[]);
11   void print_days(bday_struct days[]);
12
13   int main(){
14       bday_struct days[upto+1];
15       calculate_days(days);
16       print_days(days);
17       return 0;
18   }
19
20   void calculate_days(bday_struct days[]){
21     int ct;
22       days[1].none_match = 1;
23       for (ct=2; ct<=upto; ct ++){
24           days[ct].one_match = 1 − pow(364/365., ct−1);
25           days[ct].none_match = days[ct−1].none_match ∗ (1 − (ct−1)/365.);
26       }
27   }
28
29   void print_days(bday_struct days[]){
30     int ct;
31       printf("People\t Matches me\t Any match\n");
32       for (ct=2; ct<=upto; ct ++){
33           printf("%i\t %.3f\t\t %.3f\n", ct, days[ct].one_match, 1−days[ct].none_match);
34       }
35   }
```

Listing 2.4 The birthday example broken into logical functions. Online source: `bdayfns.c`.

The second birthday example, Listing 2.3 can be hard to read, with its mess of multiple `for` loops. Listing 2.4 re-presents the program using one function to do the math and one to print the output. The `main` function (lines 13–18) now describes the procedure with great clarity: declare an array of `bday_struct`s, calculate values for the days, print the values, and exit. The functions to which `main` refers—on lines 20–27 and 29–35—are short, and so are easier to read than the long string of code in Listing 2.3. Simply put, the functions provide structure to what had been a relatively unstructured mess.

Structure takes up space—you can see that this listing is more lines of code than the unstructured version. But consider the format of the book you are reading right now: it uses such stylistic features as paragraphs, chapter headings, and indentation, even though they take up space. Brevity is a good thing, which means that it is typically worth the effort to minimize redundancy and search for simple and brief algorithms. But brevity should never come at the cost of clarity. By eliminating intermediate variables and not using subfunctions, you can sometimes reduce an entire program into a single line of code, but that one-liner may be virtually impossible to debug, modify, or simply understand. No trees have to be killed to add a few lines of white space or a few function headers to your on-screen code, and the additional structure will save you time when dealing with your code later on.

Functional form   Have a look at the function headers—the first line of each function, on lines 13, 20 and 29. In parens are the inputs to the function (aka the *arguments*), and they look like the familiar declarations from before. The `main` function takes no arguments, while you will see that many functions take several arguments, in which case the argument declarations are a comma-separated list.

Or consider the function declaration for the `round` function above:

> **int** round (**double** unrounded)

If we ignore the argument list in parens, `int round` looks like a declaration as well—and it is. It indicates that this function will return an integer value, that can be used anywhere we need an integer. For example, you could assign the function output to a variable, via `int eight = round(8.3)`.

Declaring a function   You can declare the existence of a function separately from the function itself, as per lines 10 and 11 of Listing 2.3. The `main` function thus has an idea of what to expect when it comes across these functions on lines 15 and 16, even though the functions themselves appear later. You will see below that the compiler gets immense mileage out of the declaration of functions, because it can compile `main` knowing only what the other functions take in and return, leaving the inner workings as a black box.

✳ *The* `void` *type*   If a function returns nothing, declare it as type `void`. Such functions will be useful for side effects such as changing the values of the inputs (like `calculate_days`) or printing data to the screen or an external file (like `print_days`). You can also have functions which take no inputs, so any of the following are valid declarations for functions:

```
void do_something(double a);
double do_something_else(void);
double do_something_else();
```

The last two are equivalent, but you can't forget the parentheses entirely—then the compiler would think you are declaring a variable instead of a function.

Q2.6 Write a function with header `void print_array(int in_array[ ], int array_size)` that takes in an integer array and the size of the array, and prints the array to the screen. Modify your square-printing program from earlier to use this function for output.

How to write a program   Given a blank screen and a program to write, how should you begin? Write an outline, based on function headers.

For example, in the birthday example, you could begin by writing the `main` function, which describes the broad outline of calculating probabilities and then printing to the screen. In writing the outline, you will need to write down the inputs, outputs, and intent of a number of functions. Then you can begin filling in each function. When writing a function's body, you can put the rest of the program out of your mind and focus on making sure that the black box you are working on does exactly what it should to produce the right output. When all of the functions correctly do their job, and the main outline is fully fleshed out, you will have a working program.

You want your black boxes to be entirely predictable and error-free, and the best way to do this is to keep them small and autonomous. Flip through this book and have a look at the structure of the longer sample programs. You will notice that few functions run for more than about fifteen lines, especially after discounting the introductory material about declaring variables and checking inputs.

FRAMES   The manner in which the computer evaluates functions also abides by the principle of encapsulating functions, focusing on the context of one function at a time. When a function is called, the computer creates a *frame* for that function. Into that frame are placed any variables that are declared at the top of the file in which the function is defined, including those in files that were `#included` (see below); and copies of variables that are passed as arguments.

The function is then run, using the variables it has in that frame, blithely ignorant of the rest of the program. It does its math, making a note of the return value it calculates (if any), and then destroys itself entirely, erasing all of the variables created in the frame and copies of variables that had been put in to the frame. Variables that were not passed as an argument but were put in the frame anyway (*global variables*; see below) come out unscathed, as does the return value, which is sent back to the function which had called the frame into existence.

```
1    #include <stdio.h> //printf
2    double globe=1; //a global variable.
3
4    double factorial (int a_c){
5        while (a_c){
6            globe *= a_c;
7            a_c −−;
8        }
9         return globe;
10   }
11
12   int main(void){
13       int a = 10;
14       printf("%i factorial is %f.\n", a, factorial(a));
15       printf("a= %i\n", a);
16       printf("globe= %f\n", globe);
17       return 0;
18   }
```

Listing 2.5 A program to calculate factorials. Online source: `callbyval.c`.

One way to think about this is in terms of a *stack* of frames. The base of the stack is always the function named `main`. For example, in the program in Listing 2.5, the computer at first ignores the function `factorial`, instead starting its work at line twelve, where it finds the `main` function. It creates a `main` frame and then starts working, reading the declaration of a, and creating that variable in the `main` frame. The global variable declared on line two is also put into the `main` frame.

Then, on line 14, it is told to print the value of `factorial(a)`, which means that it will have to evaluate that expression. This is a *function call*, which commands the program to halt whatever it is doing and start working on evaluating the function `factorial`. So the system freezes the `main` frame, generates a frame for the `factorial` function, and jumps to line four. Think of the new frame as being put on top of the first, leaving only the topmost frame visible and active. The value of a = 10 will be *copied* into a_c, the global variable globe will be put into the frame, and the function does its math and returns the calculated value of `globe`.[16]

---

[16]Why is `globe` a double, when the factorial is always an integer? Because the `double` uses exponential notation when necessary, so its range is much larger than that of the `int`, which does not. You could also try using a `long int` (and replacing printf's int placeholder `%i` with the `long int` placeholder `%li`), but even that fails after about a=31.

Having returned a value, the `factorial` frame and its contents are discarded. Since a copy of the value a = 10 was sent into the frame, a still has the value 10 when the function returns, even though the copy a_c was decremented to zero.

The `main` frame is now at the top of the stack, so it can pick up where it had left off, printing the value of a and 10! to the screen, using the calls to the `printf` function—which each create their own frames in turn. Finally, the `main` function finishes its work, and its frame is destroyed, leaving an empty stack and a finished program.

Call-by-value   A common error is to forget that global variables are put in all function frames, but only *copies* of the variables in the function's argument list are put in the frame.

When the `factorial` function is called, the system puts a copy of a into a_c, and then the function modifies the copy, a_c. Meanwhile, `globe` is not a copy of itself, but the real thing, so when it is changed inside the function, it changes globally. This is why the output you got when you ran the program showed a=10, not a=0.

On the one hand, the `factorial` function could mangle a_c without affecting the original; on the other hand, we sometimes want functions to change their inputs. This may make global variables tempting to you, but resist. Section 2.6 will give a better alternative (and explain why the `bday_struct` examples worked).

※ *Static variables*   There is one exception to the rule that all local variables are destroyed with their frame: you can define a `static` variable. When the function's frame is destroyed, the program makes a note of the value of the static variable, and when the function is called again, the static variable will start off with the same value as before. This provides continuity within a function, but only that function knows about the variable.

Static variable declarations look just like other declarations but with the word `static` before them. You can also put an initialization on the declaration line, which will be taken into consideration only the first time the function is called. Here is a sample function to enter data points into an array. It assumes that the calling function knows the length of `survey_data` and does bounds-checking accordingly.

```
void add_a_point(double number, double survey_data[]){
    static int count_so_far = 0;
    survey_data[count_so_far] = number;
    count_so_far++;
}
```

The first time this function is called, `count_so_far` will be initialized at zero, the number passed in will be put in `survey_data[0]`, and `count_so_far` will be incremented to one. The second time the function is called, the program will remember that `count_so_far` is one, and will thus put the second value in `survey_-data[1]`, where we would want it to be.

※ *The* `main` *function*   All programs must have one and only one function named `main`, which is where the program will begin executing—the base of the stack of frames. The consistency checks are now with the operating system that called the program, which will expect `main` to be declared in one of two forms:

> **int** main(**void**);
> **int** main(**int** argc, **char** ∗∗argv);

```
#include <stdlib.h>
#include <stdio.h>

int main(int argc, char **argv){
    if (argc==1){
        printf("Give me a command to run.\n");
        return 1;
    }
    int return_value = system(argv[1]);
    printf("The program returned %i.\n", return_value);
    return return_value;
}
```

Listing 2.6 A *shell* is a program that is primarily intended for the running of other programs; this is a very rudimentary one. Online source: `simpleshell.c`.

The second form will be discussed on page 206, but for now, Listing 2.6 provides a quick example of the use of inputs to `main`. It uses C's `system` function to call a program. The usage of this program would be something like

> ./simpleshell "ls /a_directory"

Conceptually, there is little difference between calling a function that you wrote and calling the `main` function of a foreign program. In this case, the `system` function would call `ls`, effectively putting the `ls` program's `main` function on top of the current stack. More generally, you can think of your computer's entire functioning, from boot to shutdown, as the evaluation of a set of stacks of frames. At boot, the system starts a C program named `init`, and every other program is a child of `init` (or a child of a child of init, or a child of a child of a child, et cetera).

There is also a `fork` function that generates a second stack that runs concurrently with the parent, which is how a system runs several programs at once.

The `system` function will pass back the return value of the subprogram's `main`. The general custom is that if `main` returns 0 then all went well, while a positive integer indicates a type of error.[17] Because so many people are not concerned with the return value of `main`, the current C standard assumes that `main` returns zero if no indication is given otherwise, which is how most of the programs in this book get away with not having a `return` statement in their `main` function.

SCOPE   When one function is running, only the variables in that frame are visible: all of the variables in the rest of the program are dormant and inaccessible. This is a good thing, since you don't want to have to always bear in mind the current state of all the variables in your program, the GNU Scientific Library, the standard library, and who knows what else.

A variable's *scope* is the set of functions that can see the variable. A variable declared inside a function is visible only inside that function. If a variable is declared at the top of a file, then that variable is *global* to the file, and any function in that file can see that variable. If declared in a *header file* (see below, including an important caveat on page 50), then any function in a file that `#includes` that header can see the variable.

The strategy behind deciding on the scope of a variable is to keep it as small as possible.

Block scope

You can declare variables inside a loop.
- `while(...){int i;... }` works as you expect, declaring i only once.
- `while(...){int i=1;... }` will re-set i to one for every iteration of the loop.
- `for (int i=0; i<max; i++){...}` works, but gcc may complain about it unless you specify the `-std=c99` or `-std=gnu99` flags for the compiler.
- A variable declared inside a bracketed block (or at the header for a pair of curly braces, like a `for` loop) is destroyed at the end of the bracketed code. This is known as *block scope*. Function-level scope could be thought of a special case of block scope.
Block scope is occasionally convenient—especially the `for (int i;...)` form—but bear in mind that you won't be able to refer to a block-internal variable after the loop ends.

- If only one function uses a variable, then by all means declare the variable inside the function (possibly as a `static` variable).
- If a variable is used by only a few functions, then declare the variable in the `main` function and pass it as an argument to the functions that use it.
- If a variable is used throughout a single file and is infrequently changed, then let it be globally available throughout the file, by putting it at the top of the file, outside all the function bodies.

[17]In the bash shell (the default on many POSIX systems), the return value from the last program run is stored in $?, so `echo $?` will print its value.

- Finally, if a variable is used throughout all parts of a program consisting of multiple files, then declare it in a header file, so that it will be globally available in every file which #includes that header file (see page 50).[18]

There is often the temptation to declare every variable as global, and just not worry about scope issues. This makes maintaining and writing the code difficult: are you sure a tweak you made to the black box named function_a won't change the workings inside the black box named function_b? Next month, when you want to use function_a in a new program you have just written, you will have to verify that nothing in the rest of the program affects it, so what could have been a question of just cutting and pasting a black box from one file to another has now become an involved analysis of the original program.

> ➤ Good coding form involves breaking problems down into functions and writing each function as an independent entity.

> ➤ The header of a function is of the form *function_type function_name(p1_type p1_name, p2_type p2_name, ...)*.

> ➤ The computer evaluates each function as an independent entity. It maintains a stack of frames, and all activity is only in the current top frame.

> ➤ When a program starts, it will first build a frame for the function named main; therefore a complete program must have one and only one such function.

> ➤ Global variables are passed into a new frame, but only copies of parameters are passed in. If a variable is not in the frame, it is out of scope and can not be accessed.

---

[18]This is the appropriate time to answer a common intro-to-C question: What is the difference between C and C++? There is much confusion due to the almost-compatible syntax and similar name—when explaining the name C-double-plus, the language's author references the Newspeak language used in George Orwell's *1984* (Orwell, 1949; Stroustrup, 1986, p 4).

The key difference is that C++ adds a second scope paradigm on top of C's file- and function-based scope: object-oriented scope. In this system, functions are bound to objects, where an object is effectively a struct holding several variables and functions. Variables that are *private* to the object are in scope only for functions bound to the object, while those that are *public* are in scope whenever the object itself is in scope.

In C, think of one file as an object: all variables declared inside the file are private, and all those declared in a header file are public. Only those functions that have a declaration in the header file can be called outside of the file.

But the real difference between C and C++ is in philosophy: C++ is intended to allow for the mixing of various styles of programming, of which object-oriented coding is one. C++ therefore includes a number of other features, such as yet another type of scope called *namespaces*, templates and other tools for representing more abstract structures, and a large standard library of templates. Thus, C represents a philosophy of keeping the language as simple and unchanging as possible, even if it means passing up on useful additions; C++ represents an all-inclusive philosophy, choosing additional features and conveniences over parsimony.

2.4 THE DEBUGGER   The *debugger* is somewhat mis-named. A better name would perhaps be the *interrogator*, because it lets you interact with and ask questions of your program: you can look at every line as it is being executed, pause to check the value of variables, back up or jump ahead in the program, or insert an extra command or two. The main use of these powers is to find and fix bugs, but even when you are not actively debugging, you may still want to run your program from inside the debugger.

The joy of segfaults   There are a few ways in which your program can break. For example, if you attempt to calculate 1/0, there is not much for the computer to do but halt.

Or, say that you have declared an array, `int data[100]` and you attempt to read to `data[1000]`. This is a location somewhere in memory, 901 `ints`' distance past the end of the space allocated for the array. One possibility is that `data[1000]` happens to fall on a space that has something that can be interpreted as an integer, and the computer processes whatever junk is at that location as if nothing were wrong. Or, `data[1000]` could point to an area of protected memory, such as the memory that is being used for the operating system or your dissertation. In this case, referring to `data[1000]` will halt the program with the greatest of haste, before it destroys something valuable. This is a *segmentation fault* (*segfault* for short), since you attempted to refer to memory outside of the segment that had been allocated for your program. Below, in the section on pointers, you will encounter the *null pointer*, which by definition points to nothing. Mistakenly trying to read the data a null pointer is pointing to halts the program with the complaint *attempting to dereference a null pointer*.

A segfault is by far the clearest way for the computer to tell you that you mis-coded something and need to fire up the debugger.[19] It is much like refusing to compile when you refer to an undeclared variable. If you declared `receipts` and `data[100]`, then setting `reciepts` or `data[999]` to a value is probably an error. A language that saves you the trouble of making declarations and refuses to segfault will just produce a new variable, expand the array, and thus insert errors into the output that you may or may catch.

---

[19]In fact, you will find that the worst thing that can happen with an error like the above read of `data[1000]` would be for the program to *not* segfault, but to continue with bad data and then break a hundred lines later. This is rare, but when such an event becomes evident, you will need to use a memory debugger to find the error; see page 214.

The debugging process   To debug the program *run_me* under the debugger, type
                             gdb *run_me* at the command line. You will be given the
gdb prompt.[20]

You need to tell the compiler to include the names of the variables and functions
in the compiled file, by adding the -g flag on the compiler command line. For
example, instead of gcc hello.c, use gcc -g hello.c. If the debugger complains that it can't find any debugging symbols, then that means that you forgot
the -g switch. Because -g does not slow down the program but makes debugging
possible, you should use it every time you compile.

If you know the program will segfault or otherwise halt, then just start gdb as
above, run the program by typing run at gdb's prompt, and wait for it to break.
When it does, you will be returned to gdb's prompt, so you can interrogate the
program.

The first thing you will want to know is where you are in the program. You can
do this with the backtrace command, which you can abbreviate to either bt or
where. It will show you the stack of function calls that were pending when the
program stopped. The first frame is always main, where the program started. If
main called another function, then that will be the next frame, et cetera. Often,
your program will break somewhere in the internals of a piece of code you did
not write, such as in a call to mallopt. Ignore those. You did not find a bug in
mallopt. Find the topmost frame that is in the code that you wrote.

At this point, the best thing to do is look at a listing of your code in another window
and look at the line the debugger pointed out. Often, simply knowing which line
failed is enough to make the error painfully obvious.

If the error is still not evident, then go back to the debugger and look at the variables. You need to be aware of which frame you are working in, so you know
which set of variables you have at your disposal. You will default to the last frame
in the stack; to change to frame number three, give the command frame 3 (or f 3
for short). You can also traverse the stack via the up and down commands, where
up goes to a parent function and down goes to the child function.

Once you are in the frame you want, get information about the variables. You
can get a list of the local variables using info locals, or information about the
arguments to the function using info args (though the argument information is
already in the frame description). Or, you can print any variable that you think
may be in the frame using print var_name, or more briefly, p var_name. You

---

[20]Asking your favorite search engine for *gdb gui* will turn up a number of graphical shells built around gdb.
Some are stand-alone programs like ddd and others are integrated into IDEs. They will not be discussed here
because they work exactly like gdb, except that they involve using the mouse more.
    Also, GDB itself offers many conveniences not described here. See Stallman *et al.* (2002) for the full story.

can print the value of any expression in scope: p sqrt(var) or p apop_show_-matrix(m) will display the square root of var and the matrix m, provided the variables and functions are available to the scope in which you are working. Or, p stddev = sqrt(var) will set the variable stddev to the given value and print the result. Generally, you can execute any line of C code that makes sense in the given context via the print command.

GDB has a special syntax for viewing several elements of an array at once. If you would like to see the first five elements of the array items, then use: p *items@5.

Breaking and stepping   If your program is doing things wrong but is not kind enough to segfault, then you will need to find places to halt the program yourself. Do this with the break command. For a program with only one file of code, simply give a line number: break 35 will stop the program just before line 35 is evaluated.

- For programs based on many files, you may need to specify a file name: break file2.c:35.
- Or, you can specify a function name, and the debugger will stop at the first line after the function's header. E.g, break calculate_days.
- You may also want the program to break only under certain conditions, such as when an iterator reaches 10,000. I.e., break 35 if counter > 10000.[21]
- All breakpoints are given a number, which you can list with info break. You can delete break point number three with the command del 3.

Once you have set the breakpoints, run (or just r) will run the program until it reaches a break point, and then you can apply the interrogation techniques above. You may want to carefully step through from there:

- s will step to the next line to be evaluated, which could mean backing up in the current function or going to a subfunction.
- next or n will step through the function (which may involve backtracking) but will run without stopping in any subframes which may be created (i.e., if subfunctions are called).
- until or u will keep going until you get to the next line in the function, so the debugger will run through subfunctions and loops until forward progress is made in the current function.
- c will continue along until the next break point or the end of the program.[22]

---

[21] You can also set *watchpoints*, which tell gdb to watch a variable and halt if that variable changes, e.g., watch myvar. Watchpoints are not as commonly used as breakpoints, and sometimes suffer from scope issues.

[22] A mnemonic device for remembering which is which: s is the slowest means of stepping, n slightly faster, u still faster, and c the fastest. In this order, they spell snuc, which is almost an English word with implications of stepping slowly.

- `jump lineno` will jump to the given line number, so you can repeat a line with some variables tweaked, or skip over a few lines. Odd things will happen if you jump out of the frame in which you are working, so use this only to jump around a single function.
- `return` will exit the given frame and resume in the parent frame. You can give a return value, like `return` *var.*
- Just hitting <enter> will repeat the last command, so you won't have to keep hitting n to step through many lines.

Q2.7

Break `bdayfns.c` (from Listing 2.4) and debug it.

- Modify line 13 from `days[1]` to `days[-1]`.
- Recompile. Be sure to include the `-g` flag.
- Run the program and observe the output (if any).
- Start the debugger. If the program segfaulted, just type `run` and wait for failure; otherwise, insert a breakpoint, `break calculate_days`, and then `run`.
- Check the backtrace to see where you are on the stack. What evidence can you find that things are not right?

A note on debugging strategy   Especially for numeric programs, the strategy in debugging is to find the first point in the chain of logic where things look askew. Below, you will see that your code can include *assertions* that check that things have not gone astray, and the debugger's break-and-inspect system provides another means of searching for the earliest misstep. But if there are no intermediate steps to be inspected, debugging becomes very difficult.

Say you are writing out the roots of the quadratic equation, $x = \frac{-b \pm \sqrt{b^2 - 4ac}}{2a}$, and erroneously code the first root as:

firstroot = −b + sqrt(b*b − 4 *a *c)/2*a; *//this is wrong.*

There are basically no intermediate steps: you put in a, b, and c, and the system spits out a bad value for `firstroot`. Now say that you instead wrote:

```
1   firstroot = −b;
2   firstroot += sqrt(b*b − 4*a*c);
3   firstroot = firstroot/2*a; //still wrong.
```

If you know that the output is wrong, you can interrogate this sequence for clues about the error. Say that a=2. As you step through, you find that the value of `firstroot` does not change after line three runs. From there, the error is obvious: the line should have been either `firstroot = firstroot/(2*a)` or `firstroot /= 2*a`. Such a chain of logic would be impossible with the one-line version of the routine.[23]

However, for the typical reader, the second version is unattractively over-verbose. A first draft of code should err on the side of inelegant verbosity and easy debugability. You can then incrementally tighten the code as it earns your trust by repeatedly producing correct results.

Q2.8

A triangular number is a number like 1 ($\cdot$), 1+2=3 ($\because$), 1+2+3=6 ($\therefore\!\cdot$), 1+2+3+4=10 ($\vdots$), et cetera. Fermat's polygonal number theorem states that any natural number can be expressed as the sum of at most three triangular numbers. For example, 13 = 10+3 and 19=15+3+1. Demonstrate this via a program that finds up to three triangular numbers for every number from 1 to 100.

- Write a function `int triangular(int i)` that takes in an index and returns the ith triangular number. E.g., `triangular(5)` would return 1+2+3+4+5=15. Write a `main` to test it.

- Use that function to write a function `int find_next_-triangular(int in)` that returns the index of the smallest triangular number larger than its input. Modify `main` to test it.

- Write a function `void find_triplet(int in, int out[])` that takes in a number and puts three triangular numbers that sum to it in out. You can use `find_next_triangular` to find the largest triangular number to try, and then write three nested `for` loops to search the range from zero to the maximum you found. If the loop-in-a-loop-in-a-loop finds three numbers that sum to `in`, then the function can `return`, thus cutting out of the loops.

- Finally, write a `main` function that first declares an array of three `int`s that will be filled by `find_triplet`, and then runs a `for` loop that calls `find_triplet` for each integer from 1 to 100 and prints the result.

---

[23]The one-line version also has a second error; spotting it is left as a quick exercise for the reader.

➤ The debugger will allow you to view intermediate results at any point along a program's execution.

➤ You can either wait for the program to segfault by itself, or use `break` to insert breakpoints.

➤ You can execute and print any expression or variable using p `variable`.

➤ Once the program has stopped, use s, n, and u to step through the program at various speeds.

2.5 COMPILING AND RUNNING   The process of compiling program text into machine-executable instructions relies heavily on the system of frames. If function A calls function B, the compiler can write down the instructions for creating and executing function A without knowing anything about function B beyond its declaration. It will simply create a frame with a series of instructions, one of which is a call to function B. Since the two frames are always separate, the compiler can focus on creating one at a time, and then link them later on.

What to type   To this point, you have been using a minimal command line to compile programs, but you can specify much more. Say that we want the compiler to

- include symbols for debugging (`-g`),
- warn us of all potential coding errors (`-Wall`),
- use the C99 and POSIX standards (`-std=gnu99`),
- compile using two source files, `file1.c` and `file2.c`, plus
- the sqlite3 and standard math library (`-lsqlite3 -lm`), and finally
- output the resulting program to a file named `run_me` (`-o run_me`).

You could specify all of this on one command line:

```
gcc -g -Wall -std=gnu99 file1.c file2.c -lsqlite3 -lm -o run_me
```

This is a lot to type, so there is a separate program, `make`, which is designed to facilitate compiling. After setting up a makefile to describe your project, you will be able to simply type `make` instead of the mess above. You may benefit from reading Appendix A at this point. Or, if you decide against using `make`, you could write yourself an alias in your shell, write a batch file, or use an IDE's compilation features.

Multiple windows also come in handy here: put your code in one window and compile in another, so you can see the inevitable compilation errors and the source code at the same time. Some text editors and IDEs even have features to compile from within the program and then step you through the errors returned.

The components   Even though we refer to the process above as *compilation*, it actually embodies three separate programs: a preprocessor, a compiler, and a linker.[24]

The three sub-programs embody the steps in developing a set of frames: the preprocessor inserts header files declaring functions to be used, the compilation step uses the declarations to convert C code into machine instructions about how to build and execute a standalone frame, and the linking step locates all the disparate frames, so the system knows where to look when a function call is made.

THE PREPROCESSING STEP   The *preprocessor* does nothing but take text you wrote and convert it into more text. There are a dozen types of text substitutions the preprocessor can do, but its number one use is dumping the contents of header files into your source files. When the preprocessor is processing the file main.c and sees the lines

```
#include <gsl/gsl_matrix.h>
#include "a_file.h"
```

it finds the gsl_matrix.h and the a_file.h *header files*, and puts their entire contents verbatim at that point in the file. You will never see the expansion (unless you run gcc with the -E flag); the preprocessor just passes the expanded code to the compiler. For example, the gsl_matrix.h header file declares the gsl_matrix type and a few dozen functions that act on it, and the preprocessor inserts those declarations into your program, so you can use the structure and its functions as if you'd written them yourself.

The angle-bracket form, #include <gsl/gsl_matrix.h> indicates that the preprocessor should look at a pre-specified include path for the header; use this for the headers of library files, and see Appendix A for details. The #include "a_file.h" form searches the current directory for the header; use this for header files you wrote yourself.[25]

---

[24] As a technical detail which you can generally ignore in practice, the preprocessor and compiler are typically one program, and the linker is typically a separate program.

[25] The #include "a_file.h" form searches the include path as well, so you could actually use it for both home-grown and system #includes. In practice, the two forms serve as an indication of where one can find the given header file, so most authors use the <> form even though it is redundant.

Header aggregation

The Apophenia library provides a convenience header that aggregates almost every header you will likely be using. By placing
`#include <apop.h>`
at the top of your file, you should not need to include any of the other standard headers that one would normally include in a program for numerical analysis (`stdio.h`, `stdlib.h`, `math.h`, `gsl_anything.h`). This means that you could ignore the headers at the top of all of the code snippets in this chapter.

Of course, you will still need to include any headers you have written, and if the compiler complains about an undeclared function, then its header is evidently not included in `apop.h`.

Many programming languages have a way to declare variables as having *global scope*, meaning that every function everywhere can make use of the variable. Technically, C has no such mechanism. Instead, the best you can do is what I will call *file-global scope*, meaning that every function in a single file can see any variable declared above it in that file.

Header files allow you to simulate truly global scope, but with finer control if you want it. If some variables should be global to your entire program, then create a file named `globals.h`, and put all declarations in that file (but see below for details). By putting `#include "globals.h"` at the top of every file in your project, all variables declared therein are now project-global. If the variables of `process.c` are used in only one or two other code files, then project-global scope is overkill: `#include "process.h"` only in those few code files that need it.

In prior exercises, you wrote a program with one function to create an array of numbers and their squares (page 30), and another function, `print_array`, to print those values (page 37).

- Move `print_array` to a new text file, `utility_fns.c`.

- Write the corresponding one-line file `utility_fns.h` with `print_array`'s header.

- `#include "utility_fns.h"` in the main square-printing program.

- Modify the square-calculating code to call `print_array`.

- Compile both files at once, e.g., `gcc your_main.c utility_fns.c`.

- Run the compiled program and verify that it does what it should.

❋ *Variables in headers*   The system of putting file-scope variables in the base `.c` file and global-scope variables in the `.h` file has one ugly detail. A function declaration is merely advice to the compiler that your function has certain inputs and outputs, so when multiple files reread the declaration, the program suffers only harmless redundancy. But a variable declaration is a command to the compiler to allocate space as listed. If `head.h` includes a declaration `int x;`, and `file1.c` includes `head.h`, it will set aside an `int`'s worth of memory and name it `x`; if `file2.c` includes the same header, it will also set aside some memory named `x`. So which bit of memory are you referring to when you use `x` later in your code?

C's solution is the `extern` keyword, which tells the compiler that the declaration that follows is not for memory allocation, but is purely informative. Simply put it in front of the normal declaration: `extern int x; extern long double y;` will both work. Then, in one and only one `.c` file, declare the variables themselves, e.g., `int x; long double y = 7.0`. Thus, all files that `#include` the header know what to make of the variable `x`, so `x`'s scope is all files with the given header, but the variable is allocated only once.

To summarize: function declarations and `typedefs` can go into a header file that will be included in multiple `.c` files. Variables need to be declared as usual in one and only one `.c` file, and if you want other `.c` files to see them, re-declare them in a header file with the `extern` keyword.

THE COMPILATION STEP   The *compilation* stage consists of taking each `.c` file in turn and writing a machine-readable *object file*, so `file1.c` will result in `file1.o`, and `file2.c` will compile to `file2.o`. These object files are self-encapsulated files that include a table of all of the symbols declared in that file (functions, variables, and types), and the actual machine code that tells the computer how to allocate memory when it sees a variable and how to set up and run a frame when it sees a function call. The preprocessor inserted declarations for all external functions and variables, so the compiler can run consistency checks as it goes.

The instructions for a function may include an instruction like *at this point, create a frame for* `gsl_matrix_add` *with these input variables*, but executing that instruction does not require any knowledge of what `gsl_matrix_add` looks like—that is a separate frame in which the current frame has no business meddling.

THE LINKING STEP   After the compilation step, you will have on hand a number of standalone frames. Some are in `.o` files that the compiler just created, and some are in libraries elsewhere on the system. The *linker* collects all of these elements into a single executable, so when one function's instructions tell the computer to go evaluate `gsl_matrix_add`, the computer will have no problem locating and loading that function. Your primary interaction with the linker will be in telling it where to find libraries, via -l commands on the compiler command line (-lgsl -lgslcblas -lm, et cetera).

Note well that a library's header file and its object file are separate entities—meaning that there are two distinct ways in which a call to a library function can go wrong. To include a function from the standard math library like sqrt, you will need to (1) tell the preprocessor to include the header file via #include <math.h> in the code, and (2) tell the linker to link to the math library via a -l flag on the command line, in this case -lm. Appendix A has more detail on how to debug your #include statements and -l flags.

※ *Finding libraries*   An important part of the art of C programming is knowing how to find libraries that will do your work for you, both online and on your hard drive.

- The first library to know is the standard library. Being standard, this was installed on your computer along with the compiler. If the documentation is not on your hard drive (try info glibc), you can easily find it online. It is worth giving the documentation a skim so you know which wheels to not reinvent.
- The GNU/UNESCO website (gnu.org) holds a hefty array of libraries, all of which are free for download.
- sourceforge.net hosts on the order of 100,000 projects (of varying quality). To be hosted on Sourceforge, a project must agree to make its code public, so you may fold anything you find there into your own work.
- Finally, you can start writing your own library, since next month's project will probably have some overlap with the one you are working on now. Simply put all of your functions relating to *topic* into a file named *topic.c*, and put the useful declarations into a separate header file named *topic.h*. You already have a start on creating a utility library from the exercise on page 50.

$\sum$

➤ Compilation is a three-step process. The first step consists of text expansions like replacing #include <*header.h*> with the entire contents of *header.h*.

➤ Therefore, put public variable declarations (with the extern keyword) and function declarations in header files.

➤ The next step consists of compilation, in which each source (.c) file is converted to an object (.o) file.   ≫

≫

➤ Therefore, each source file should consist of a set of standalone functions that depend only on the file's contents and any declarations included via the headers.

➤ The final step is linking, in which references to functions in other libraries or object files are reconciled.

➤ Therefore, you can find and use libraries of functions for any set of tasks you can imagine.

2.6 POINTERS   Pointers will change your life. If you have never dealt with them before, you will spend some quantity of time puzzling over them, wondering why anybody would need to bother with them. And then, when you are finally comfortable with the difference between data and the location of data, you will wonder how you ever wrote code without them.

Pointers embody the concept of the *location of data*—a concept with which we deal all the time. I know the location `http://nytimes.com`, and expect that if I go to that location, I will get information about today's events. I gave my colleagues an email address years ago, and when they have new information, they send it to that location. When so inclined, I can then check that same location for new information. Some libraries are very regimented about where books are located, so if you need a book on probability (Library of Congress classification QA273) the librarian will tell you to go upstairs to the third bookshelf on the left. The librarian did not have to know any information about probability, just the location of such information.

Returning to the computer for a moment, when you declare `int k`, then the computer is going to put `k` somewhere in memory. Perhaps with a microscope, you could even find it: there on the third chip, two hundred transistors from the bottom. You could point to it.

Lacking a finger with which to point, the computer will use an illegible hexadecimal location, but you will never have to deal with the hexadecimal directly, and lose nothing by ignoring the implementation and thinking of pointers as just a very precise finger, or a book's call number.

The confusing part is that the location of data is itself data. After all, you could write "QA273" on as slip of paper as easily as "$P(A \cap B) = P(A|B)P(B)$."

Further, the location of information may itself have a location. Before computers took over, there was a card catalog somewhere in the library, so you would have to go to the card catalog—the place where location data is stored—and then look up the location of your book. It sometimes happens that you arrive at the QA273 shelf and find a wood block with a message taped to it saying "oversized books are at the end of the aisle."

In these situations, we have no problem distinguishing between information that is just the location of data and data itself. But in the computing context, there is less to guide us. Is 8,049,588 just a large integer (data), or a memory address (the location of data)?[26] C's syntax will do little to clear up the confusion, since a variable like k could be integer data or the location of integer data. But C uses the location of data to solve a number of problems, key among them being function calls that allow inputs to be modified and the implementation of arrays.

Call-by-address v call-by-value   First, a quick review of how functions are called: when you call a function, the computer sets up a separate frame for the function, and puts into that frame *copies* of all of the variables that have been passed to the function. The function then does its thing and produces a return value. Then, the entire frame is destroyed, including all of the copies of variables therein. A copy of the return value gets sent back to the main program, and that is all that remains of the defunct frame.

This setup, known as call-by-value since only values are passed to the function, allows for a more stable implementation of the paradigm of standalone frames. But if k is an array of a million `doubles`, then making a copy every time you call a common function could take a noticeable amount of time. Also, you will often want your function to change the variables that get sent to it.

Pointers fix these problems. The trick is that instead of sending the function a copy of the variable, we send in a copy of the location of the variable: we copy the book's call number onto a slip of paper and hand that to the function. In Figure 2.7, the *before* picture shows the situation before the function call, in the main program: there is a pointer to a location holding the number six. Then, in the *during* picture, a function is called with the pointer as an argument, via a form like `fn_-call(pointer)`. There are now two fingers, original and copy, pointing to the same spot, but the function knows only about the copy. Given its copy of a finger, it is easy for the function to change the value pointed to to seven. When the function returns, in the *after* picture, the copy of a finger is destroyed but the changes are not undone. The original finger (which hasn't changed and is pointing to the same place it was always pointing to) will now be pointing to a modified value.

---

[26]This number is actually an address lifted from my debugger, where it is listed as 0x08049588. The 0x prefix indicates that the number is represented in hexadecimal.

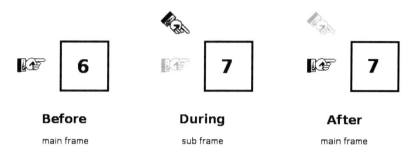

Before
main frame

During
sub frame

After
main frame

Figure 2.7 Before, during, and after a function call that modifies a pointed-to value

Returning to C syntax, here are the rules for using pointers:

- To declare a pointer to an integer, use int *k.
- Outside the declarations, to refer to the integer being pointed to, use *k.
- Outside the declarations, to refer to the pointer itself, use k.

The declaration int *p, i means that p will be a pointer to an integer and i is an integer, but in a non-declaration line like i = *p, *p refers to the integer value that p points to. There is actually a logical justification for the syntax, which I will not present here because it tends to confuse more often than it clarifies. Instead, just bear in mind that the star effectively means something different in declarations than in non-declaration use.

To give another example, let us say that we are declaring a new pointer p2 that will be initialized to point to the same address as p. Then the declaration would be int *p2 = p, because p2 is being *declared* as a pointer, and p is being *used* as a pointer.

The spaces around our stars do not matter, so use whichever of int *k, int* k, or int * k you like best. General custom prefers the first form, because it minimizes the chance that you will write int * k, b (allocate a pointer named k and an int named b) when you meant int *k, *b (allocate two pointers). The star also still means multiply. There is never ambiguity, but if this bothers you, use parentheses.

Listing 2.8 shows a sample program that uses C's pointer syntax to implement the call-by-address trick from Figure 2.7.

```
1   #include <stdio.h> //printf
2   #include <malloc.h> //malloc
3
4   int globe=1; //a global variable.
5
6   int factorial (int *a_c){
7       while (*a_c){
8           globe *= *a_c;
9           (*a_c) --;
10      }
11        return globe;
12  }
13
14  int main(void){
15    int *a = malloc(sizeof(int));
16        *a = 10;
17        printf("%i factorial ...", *a);
18        printf(" is %i.\n", factorial(a));
19        printf("*a= %i\n", *a);
20        printf("globe= %i\n", globe);
21        free(a);
22        return 0;
23  }
```

Listing 2.8 A version of the factorial program using call-by-address. Online source: `callbyadd.c`.

- In the `main` function, a is a pointer—the address of an integer—as indicated by the star in its declaration on line 15; the `malloc` part is discussed below.
- To print the integer being pointed to, as on line 19, we use `*a`.
- The header for a function is a list of declarations, so on line 6, `factorial(int *a_c)` tells us that the function takes a pointer to an integer, which will be named `a_c`.
- Thus, in non-declaration use like lines eight and nine, `*a_c` is an integer.

Now for the call-by-address trick, as per Figure 2.7. When the call to `factorial` is made on Line 18, the pointer a gets passed in. The computer builds itself a frame, using a copy of a—that is, a copy of the location of an integer. Both a (in the `main` frame) and `a_c` (in the `factorial` frame) now point to the same piece of data. Line 9, `(*a_c)--`, tells the computer to go to the address `a_c` and decrement the value it finds there. When the frame is destroyed (and `a_c` goes with it), this will not be undone: that slot of memory will still hold the decremented value. Because `*a`—the integer a points to—has changed as a side effect to calling the `factorial` function, you saw that line 19 printed `*a=0` when you ran the program.

Dealing with memory   Finally, we must contend with the pointer initialization on line 15:

```
int *a = malloc(sizeof(int));
```

Malloc, a function declared in `stdlib.h`, is short for *memory allocate*. By the library metaphor, `malloc` builds bookshelves that will later hold data. Just as we have no idea what is in an `int` variable before it is given a value, we have no idea what address a points to until we initialize it. The function `malloc()` will do the low-level work of finding a free slot of memory, claiming it so nothing else on the computer uses it, and returning that address. The input to `malloc` is the quantity of memory we need, which in this case is the size of one integer: `sizeof(int)`.

There are actually three characteristics to a given pointer: the location (where the finger is pointing), the type (here, `int`), and the amount of memory which has been reserved for the pointer (`sizeof(int)` bytes—enough for one integer). The location is up to the computer—you should never have to look at hexadecimal addresses. But you need to bear in mind the type and size of your pointer. If you treat the data pointed to by an `int` pointer as if it is pointing to a `double`, then the computer will read good data as garbage, and if you read twenty variables from a space allocated for fifteen, then the program will either read garbage or segfault.

The ampersand

Every variable has an address, whether you declared it as a pointer or not. The ampersand finds that address: if `count` is an integer, then `&count` is a pointer to an integer. The ampersand and star are inverses: `*(&count) == count`, which may imply that they are symmetric, but the star will appear much more often in your code than the ampersand, and an ampersand will never appear in a declaration or a function header. As a mnemonic, *ampersand*, *and sign*, and *address of* all begin with the letter A.

By the way, `int *k = 7` will fail—the initialization on the declaration line is for the pointer, not the value the pointer holds. Given that k is a pointer to an integer, all of these lines are correct:

```
int *k = malloc(sizeof(int));
*k = 7;
k = malloc(sizeof(int));
```

One convenience that will help with allocating pointers is `calloc`, which you can read as *clear and allocate*: it will run `malloc` and return the appropriate address, and will also set everything in that space to zero, running `*k = 0` for you. Sample usage:

```
int *k = calloc(1, sizeof(int));
```

The syntax requires that we explicitly state that we want one space, the size of an integer. You need to give more information than `malloc` because the process of putting a zero in a `double` pointer may be different from putting a zero in a `int` pointer. Thus `calloc` requires two arguments: `calloc(element_count, sizeof(element_type))`.

Finally, both allocation and de-allocation are now your responsibility. The de-allocation comes simply by calling `free(k)` when you are done with the pointer k. When the program ends, the operating system will free all memory; some people free all pointers at the end of `main` as a point of good form, and some leave the computer to do the freeing.

Q2.10
Write a function named `swap` that takes two pointers to `int` variables and exchanges their values.

- First, write a `main` function that simply declares two `int`s (not pointers) `first` and `second`, gives them values, prints the values and returns. Check that it compiles.

- Then, write a `swap` function that accepts two pointers, but does nothing. That is, write out the header but let the body be `{ }`.

- Call your empty function from `main`. Do you need to use `&first` (as per the box on page 57), `*first`, or just `first`? Check that the program still compiles.

- Finally, write the swap function itself. (*Hint*: include a local variable `int temp`.) Add a `printf` to the end of `main` to make sure that your function worked.

Q2.11
Modify your swap program so that the two variables in `main` are now pointers to `int`s.

- Add allocations via `malloc`, either in the declaration itself or on another line.

- Which of `&first`, `*first`, or `first` will you need to send to `swap` now? Do you need to modify the `swap` function itself?

Q2.12
Modify `callbyadd.c` so that the declaration on line 15 is for an integer, not a pointer. That is, replace the current `int *a = malloc(...);` with `int a;`. Make the necessary modifications to `main` to get the program running. Do not modify the `factorial` function.

➤ A variable can hold the address of a location in memory.

➤ By passing that address to a function, the function can modify that location in memory, even though only copies of variables are passed into functions.

➤ A star in a declaration means the variable is a pointer, e.g., int *k. A star in a non-declaration line indicates the data at the location to which the pointer points, e.g., two_times = *k + *k.

➤ The space to which a pointer is pointing needs to be prepared using malloc, e.g., int *integer_address = malloc(sizeof(int));.

➤ When you are certain a pointer will not be used again, free it, e.g., free(integer_address).

2.7 ARRAYS AND OTHER POINTER TRICKS

You can use a pointer as an array: instead of pointing to a single integer, for example, you can point to the first of a sequence of integers. Listing 2.9 shows some sample code to declare an array and fill it with square numbers:

```c
#include <stdlib.h>
#include <stdio.h>

int main(){
    int array_length=1000;
    int *squares = malloc (array_length * sizeof(int));
        for (int i=0; i < array_length; i++)
            squares[i] = i * i;
}
```

Listing 2.9 Allocate an array and fill it with squares. Online source: squares.c.

The syntax for declaring the array exactly matches that of allocating a single pointer, except we needed to allocate a block of size 1000 * sizeof(int) instead of just a single sizeof(int). Referring to an element of squares uses identical syntax to the automatically-declared arrays at the beginning of this chapter. Internally, both types of array are just a sequence of blocks of memory holding a certain data type.

Q$_{2.13}$ The listing in `squares.c` is not very exciting, since it has no output. Add a second `for` loop to print a table of squares to the screen, by printing the index `i` and the value in the `squares` array at position `i`.

Q$_{2.14}$ After the `for` loop, `squares[7]` holds a plain integer (49). Thus, you can refer to that integer's address by putting a `&` before it. Extend your version of `squares.c` to use your swap function to swap the values of `squares[7]` and `squares[8]`.

But despite their many similarities, arrays and pointers are not identical: one is automatically allocated memory and the other is manually allocated. Given the declarations

```
double a_thousand_doubles[1000];
// and
double *a_thousand_more_doubles = malloc(1000 * sizeof(double));
```

the first declares an automatically allocated array, just as `int i` is automatically allocated, and therefore the allocation and de-allocation of the variable is the responsibility of the compiler. The second allocates memory that will be at the location `a_thousand_doubles` until you decide to free it.

In function arguments, you can interchange their syntaxes. These are equivalent:

```
int a_function(double *our_array);
//and
int a_function(double our_array[]);
```

But be careful: if your function `frees` an automatically allocated array passed in from the parent, or assigns it a new value with `malloc`, then you are stepping on C's turf, and will get a segfault.

ARRAYS OF STRUCTS   Before, when we used the `struct` for complex numbers, we referred to its elements using a dot, such as `a.real` or `b.imaginary`. For a pointer to a structure, use `->` instead of a dot. Here are some examples using the definition of the `complex` structure from page 31.

```
complex *ptr_to_cplx = malloc (sizeof(complex));
ptr_to_cplx->real = 2;
ptr_to_cplx->imaginary = -2;
complex *array_of_cplxes = malloc (30 * sizeof(complex));
array_of_cplexes[15]->real = 3;
```

If you get an error like *request for member 'real' in something not a structure or union* then you are using a dot where you should be using -> or vice versa. Use that feedback to understand what you misunderstood, then switch to the other and try again.

REALLOCATING   If you know how many items you will have in your array, then you probably won't bother with pointers, and will instead use the `int` `fixed_list[300]` declaration, so you can leave the memory allocation issues to the computer. But if you try to put 301 elements in the list (which, you will recall, means putting something in `fixed_list[300]`), then you will be using memory that the machine hadn't allocated for you—a segfault.

If you are not sure about the size of your array, then you will need to expand the array as you go. Listing 2.10 is a program to find *prime numbers*, with a few amusing tricks thrown in. Since we don't know how many primes we will find, we need to use `realloc`. The program runs until you hit <ctrl-c>, and then dumps out the complete list to that point.

- Line 13: `SIGINT` is the signal that hitting <ctrl-c> sends to your program. By default, it halts your program immediately, but line 13 tells the system to call the one-line function on line 9 when it receives this signal. Thus, the `while` loop beginning at line 14 will keep running until you hit <ctrl-c>; then the program will continue to line 26.
- Lines 16–17: Check whether `testme` is evenly divisible by `primes[i]`. The second element of the `for` loop, the run-while condition, includes several conditions at once.
- Line 19: Computers in TV and movies always have fast-moving counters on the screen giving the impression that something important is happening, and line 19 shows how it is done. The newline `\n` is actually two steps: a carriage return (go to beginning of line) plus a line feed (go to next line). The `\r` character is a carriage return with no line feed, so the current line will be rewritten at the next `printf`. The `fflush` function tells the system to make sure that everything has been written to the screen.
- Lines 20–24: Having found a prime number, we add it to the list, which is a three-step process: reallocate the list to be one item larger, put the element in the last space, and add one to the counter holding the size of the array. The first argument to `realloc` is the pointer whose space needs resizing, and the second argument is the new size. The first part of this new block of memory will be the `primes` array so far, and the end will be an allocated but garbage-filled space ready for us to fill with data.

```c
1    #include <math.h>
2    #include <stdio.h>
3    #include <signal.h>
4    #include <malloc.h>
5
6    int ct =0, keepgoing = 1;
7    int *primes = NULL;
8
9    void breakhere(){ keepgoing = 0; }
10
11   int main(){
12      int i, testme = 2, isprime;
13       signal(SIGINT, breakhere);
14       while(keepgoing){
15           isprime = 1;
16           for (i=0; isprime && i< sqrt(testme) && i<ct; i++)
17               isprime = testme % primes[i];
18           if (isprime){
19               printf("%i \r", testme); fflush(NULL);
20               primes = realloc(primes, sizeof(int)*(ct+1));
21               primes[ct] = testme;
22               ct ++;
23           }
24           testme ++;
25       }
26      printf("\n");
27      for (i=0; i< ct; i++)
28         printf("%i\t", primes[i]);
29      printf("\n");
30   }
```

Listing 2.10 Find prime numbers and put them in an array. Online source: `primes.c`.

SOME COMMON FAUX PAS   Figure 2.11 shows two errors in pointer handling. The first step picks up from the second step of Figure 2.7: a function has been called with a pointer as an argument. Then, the function frees what the copy of a pointer is pointing to—and thus frees what the original finger was pointing to. Next, it allocates new space, moving the copy of a finger to point to a new location. But when the function finishes, depicted in the final step, the copy of a pointer is destroyed, so there is no way to refer to the `malloc`ed space in the main program. We now have a pointer in the main frame with no space and a space with no pointer.

Returning to the library metaphor for a moment, the calling function wrote down a call number on a slip of paper, handed it to the function, and the function then went into the shelves and moved things around. But since the function has no mechanism of telling the caller where it put things, the shelves are now a mess.

Listing 2.12 is a repeat of the prime-finding code, with the cute tricks removed and

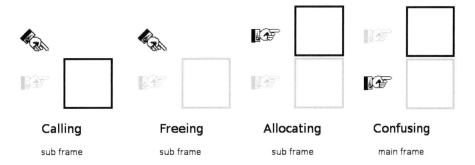

| Calling | Freeing | Allocating | Confusing |
| sub frame | sub frame | sub frame | main frame |

Figure 2.11 How to mess up your pointers

```
1   #include <math.h>
2   #include <stdio.h>
3   #include <malloc.h>
4
5   void add_a_prime(int addme, int *ct, int **primes){
6       *primes = realloc(*primes, sizeof(int)*(*ct+1));
7       (*primes)[*ct] = addme;
8       (*ct)++;
9   }
10
11  int main(){
12      int ct =0, i, j, testme = 2, isprime, max = 1000;
13      int *primes = NULL;
14      for (j=0; j< max; j++){
15          isprime = 1;
16          for (i=0; isprime && i< sqrt(testme) && i<ct; i++)
17              isprime = testme % primes[i];
18          if (isprime)
19              add_a_prime(testme, &ct, &primes);
20          testme ++;
21      }
22      for (i=0;i< ct; i++)
23          printf("%i\t", primes[i]);
24      printf("\n");
25  }
```

Listing 2.12 Find prime numbers and put them in an array. Online source: `primes2.c`.

the process of adding a prime relegated to a separate function (as it should be). The wrong way to implement the function in lines 5–9 would be

```
void add_a_prime_incorrectly(int addme, int *ct, int *primes){
    primes = realloc(primes, sizeof(int)*(*ct+1));
    primes[*ct] = addme;
    (*ct)++;
}
```

This commits the above faux pas of changing the value of the *copy* of `primes` and allocating new data at the new location—but whatever called that function has no idea about these changes in `primes`, and will keep on pointing to what may now be an invalid location.

The correct method is shown in Listing 2.12. It passes a pointer to the `primes` pointer—the location of the location of data. The calling function sends in the location of the card catalog, and the called function can then revise the locations listed in the card catalog when it makes changes on the shelves.

The syntax may seem confusing, but compare it to how `ct` is treated. Because the function will modify it, line 19 sends in its location, `&ct`, and the function header has an extra star: instead of `int ct`, it refers to the address `int *ct`. Similarly, the function call sends in the array's address, `&primes`, and the function adds a star to the header, `int **primes`.

Q2.15

Kernighan & Pike (1999) point out that `realloc` can be slow, so you are better off not calling it every time an array is extended. Instead, they suggest that every time an array of length n is `realloc`ed, its size be doubled to $2n$. Thus, if an array will eventually have ten elements, it will be `realloc`ed when adding the first, second, fourth, and eighth data points, for a total of four reallocations instead of ten.

Rewrite the code in figure `primes2` to implement this method of selective reallocation.

Σ

➤ Arrays are internally represented as pointers. The `int sarray[100]` form creates an automatically-allocated array, where the computer creates and destroys the array; the `int *harray= malloc(sizeof(int) * 100);` form creates a manually-allocated array that is yours to allocate and deallocate.

➤ Refer to array elements in both cases using the same square-brackets notation, such as `harray[14]`.

➤ Arrays of structs can be declared just as with arrays of basic variables. Refer to the elements in a pointer-to-struct using `->`.

➤ You can expand a manually-allocated array using `realloc`.

➤ Just as copies of normal variables are passed to functions, copies of pointers are sent in. Therefore, be careful when modifying a pointer in a subfunction.

2.8 **STRINGS**   C's handling of text *strings* is simple, elegant, unpleasant and
                       awkward. Although this book is oriented toward programs about
manipulating numbers rather than building Web pages or other such text, words
and phrases are inevitable in even the mathiest of programs. So this section will
cover the basics of how a system deals with variable-length text.

There are three approaches to dealing with C's awkwardness regarding text. You
can (and should) skim the standard library documentation, to get to know what
functions are always available for the most common string operations. You can
use a higher-level library-provided data type for text, such as Glib's GString type
(see Chapter 6 on Glib). Or, you can leave C entirely and do text manipulation
via a number of command-line tools or a text-focused language like Ruby or Perl
(see Appendix B). Nonetheless, all of these methods are based on C's raw string-
handling at their core, and the raw C methods are always at hand, so it is worth
getting to know C's string-handling even if you prefer higher-level methods.

C implements lines of text such as "hello" as arrays of individual characters,
followed by an invisible null character, written as \0. That is, you may think of the
above single word as a shorthand for the array {'h', 'e', 'l', 'l', 'o',
'\0'}. This means that you have to think in terms of arrays when dealing with
strings of characters.

The first implication of the elegant use of arrays to represent text is that your
expectations about assignment won't work. Here are some examples:

```
1   char hello[30];
2   char hello2[] = "Hi.";
3   hello = hello2; //This is probably not what you meant.
4   hello = "Hi there"; //Nor is this.
```

- Line one shows that strings are declared with array-style declarations, either of the
  static form here or via malloc.
- Line two shows that, as with arrays of integers, we can specify a list to put in to
  the array when we initialize the array, but not later.
- But it is a common error to think that line three will copy the text "Hi." into
  hello, but as with pointers to integers, the actual function of third line is a pointer
  operation: instead of copying the data pointed to by hello2 to the location pointed
  to by hello, it simply copies the location of hello2. When you change the text
  at one pointer later on, the other (now pointing to the same location) will change
  as well. Along a similar vein, line four also does not behave as you may expect.[27]

---

[27]Line four will compile, because the string "Hi there." is held in memory somewhere, so the pointer
hello can point to it. However, a pointer to literal text is a const pointer, meaning that the first time you try to
change hello, the program will crash.

Instead, there are a series of functions that take in strings to copy and otherwise handle strings.

strlen   There are two lengths associated with a string pointer: the first is the space malloced for the pointer, and the second is the number of characters until the string-terminating '\0' character appears. Size of free memory is your responsibility, but strlen(*your_string*) will return the number of text characters in *your_string*.

strncpy   Continuing the above example, this is the right way to copy data into hello:

```
#include <string.h>
strncpy(hello, "Hi there.", 30);
strncpy(hello, hello2, 30);
```

The third argument is the total space malloced to *hello*, not strlen(*hello*).

strncat   Rather than overwriting one string with another, you can also append (i.e., concatenate) one string to another, using

```
strncat(base_string, addme, freespace);
```

For example, strncat(hello, hello2, 30) will leave "Hi there. Hello." in hello.

Q2.16   The key problem with pointers-as-strings is that editing a string often becomes a three-step problem: measure the length the string will have after being changed, then realloc the string to the appropriate size, then finally make the change to the string.
Write a function astrncpy(char **base, char *copyme), that will copy copyme to *base. Internally, it will use strlen, realloc, and strncpy to execute the three steps of string extension. All of the above discussion regarding re-pointing pointers inside a function applies here, which is why the function needs to take in a pointer-to-pointer as its first argument. Once this function is working, write a function astrncat that executes the same procedure for string concatenation. After you've tested both functions, add them to your library of utilities.

snprintf   The snprintf function works just like the printf statements above, but
            prints its output to a string instead of the screen.[28] The fill-in-the-blanks
syntax for printf works in exactly the same manner with strings.

```
#include <string.h>
...
int string_length = 1000;
char write_to_me[string_length];
char name[] = "Steven";
int position = 3;
snprintf(write_to_me, string_length, "person %s is number %i in line\n", name, position);
```

However, the three-step process of measuring the new string, calling realloc,
and then finally modifying the string is especially painful for snprintf, because
it is hard to know how long the printf-style format specifier will be after all its
blanks are filled in. One way to make sure there is enough room for adding more
text would be to simply allocate an absurd amount of space for each string, like
just under a megabyte of memory: char hello[1000000]. You won't notice the
wasted memory on a modern computer, but this method is also error-prone: what
if you have a brilliant idea about a for loop that will add a little text for each of a
million data points into the string?

asprintf   If you are using the GNU C library, BSD's standard C library, or Apo-
            phenia, you could use asprintf, which allocates a string that is just big
enough to handle the inputs, and then runs snprintf. Here is a simple example,
with no memory allocation in sight, to print to line.

```
char *line;
asprintf(&line, "%s is number %i in line.", "Steven", 7);
```

Once again, because asprintf will probably move line in memory, we need to
send the location of the pointer, not the pointer itself.

You can comfortably put asprintf into a for loop without worrying about overflow. Here is a snippet to write a string that counts to a hundred:[29]

---

[28]The other nice feature of snprintf is that it is more secure: other common functions like strcpy and
sprintf do not check the length of the input, and so make it easy for you to inadvertently overwrite important
bits of memory with the input string, including the location of the next instruction to be executed. Unsafe string-
handling functions are thus a common security risk, allowing the execution of malicious code.

[29]Because asprintf does not free the current space taken up by string before reallocating the new version,
this for loop is a memory leak. In many situations it isn't enough of a leak to matter, but if it is, you will need to
use:
```
for (int i =0; i< 100; i++){
    char *tmp = string;
    asprintf(&string, "%s %i", string, i);
    free(tmp);
}
```

```
char *string = NULL;
asprintf(&string, ""); //initialize to empty, non−NULL string.
for (int i =0; i< 100; i++)
    asprintf(&string, "%s %i", string, i);
```

If you don't have `asprintf` on hand, you can hack your way through by guessing the final length of the filled-in string and running the measure/`realloc`/write
procedure directly, e.g.:[30]

```
char *string = NULL;
char int_as_string[10000];
for (int i =0; i< 100; i++){
    int newlen = strlen(string) + 10000;
    string = realloc(string, newlen);
    snprintf(int_as_string, 10000, " %i", i);
    strncat(string, int_as_string, newlen);
}
```

• The C standard defines `sizeof(char)==1`, so it is safe to write `newlen` in the
place of `sizeof(char)*newlen`.[31]

See Listing 3.7, page 112, for another example of extending a string inside a loop.

$Q_{2.17}$ Modify `primes2.c` to write to a string named `primes_so_far` rather than
printing to the screen. As the last step of the program, `printf(primes_-
so_far)`.

$Q_{2.18}$ Modify the program in the last exercise to print only primes whose last digit
is seven. (*Hint*: if you have written a number to `pstring`, you can compare
the last element in `pstring`'s array of characters to the single character
`'7'`.)

strcmp   Because strings are arrays, the form `if("Joe"=="Jane")` does not make
sense. Instead, the `strcmp` function goes through the two arrays of characters you input, and determines character-by-character whether they are equal.

You are encouraged to *not* use `strcmp`. If `s1` and `s2` are identical strings, then
`strcmp(s1, s2)==0`; if they are different, then `strcmp(s1, s2)!=0`. There is

---

[30]It would be nice if we could use `snprintf(string, newlen, "%s %i", string, i)`, but using
`snprintf` to insert `string` into `string` behaves erratically. This is one more reason to find a system with
`asprintf`.

[31]ISO C standard, Committee Draft, ISO/IEC 9899:TC2, §6.5.3.4, par 3.

a rationale for this: the `strcmp` function effectively subtracts one string from the other, so when they are equal, then the difference is zero; when they are not equal, the difference is nonzero. But the great majority of humans read if (strcmp(s1, s2)) to mean 'if s1 and s2 are the same, do the following.' To translate that English sentiment into C, you would need to use if (!strcmp(s1, s2)). Experienced programmers the world over regularly get this wrong.

$Q_{2.19}$ Add a function to your library of convenience functions to compare two strings and return a nonzero value if the strings are identical, and zero if the strings are not identical. Feel free to use `strcmp` internally.

Σ

➤ Strings are actually arrays of `chars`, so you must think in pointer terms when dealing with them.

➤ There are a number of functions that facilitate copying, adding to, or printing to strings, but before you can use them, you need to know how long the string will be after the edit.

2.9 ✳ ERRORS

The compiler will warn you of syntax errors, and you have seen how the debugger will help you find runtime errors, but the best approach to errors is to make sure they never happen. This section presents a few methods to make your code more robust and reliable. They dovetail with the notes above about writing one function at a time and making sure that function does its task well.

TESTING THE INPUTS   Here is a simple function to take the mean of an input array.[32]

```
double find_means(double *in, int length){
   double mean = in[0];
      for (int i=1; i < length, i++)
          mean += in[i];
      return mean/length;
}
```

What happens if the user calls the function with a NULL pointer? It crashes. What happens if `length==0`? It crashes. You would have an easy enough time pulling

---

[32]Normally, we'd just assign `double mean=0` at first and loop beginning with `i=0`; I used this slightly odd initialization for the sake of the example. How would the two approaches differ for a zero-length array?

out the debugger and drilling down to the point where you sent in bad values, but it would be easier if the program told you when there was an error.

Below is a version of `find_means` that will save you trips to the debugger. It introduces a new member of the `printf` family: `fprintf`, which prints to files and *streams*. Streams are discussed further in Appendix B; for now it suffices to note that writing to the `stderr` stream with `fprintf` is the appropriate means of displaying errors.

```
double find_means(double *in, int length){
    if (in==NULL){
        fprintf(stderr, "You sent a NULL pointer to find_means.\n");
        return NAN;
    }
    if (length<=0){
        fprintf(stderr, "You sent an invalid length to find_means.\n");
        return NAN;
    }
    double mean = in[0];
    for (int i=1; i < length, i++)
        mean += in[i];
    return mean/length;
}
```

This took more typing, and does not display the brevity that mathematicians admire, but it gains in clarity and usability. Listing conditions on the inputs provides a touch of additional documentation on what the function expects and thus what it will do. If you misuse the function, you will know the error in a heartbeat.

The `&&` and `||` are perfect for inserting quick tests, because the left-hand side can test for validity and the right-hand side will execute only if the validity test passes. For example, let us say that the user gives us a list of element indexes, and we will add them to a counter only if the chosen array elements are even. The quick way to do this is to simply use the `%` operator:

```
if (!(array[i] % 2))
    evens += array[i];
```

But if the array index is invalid, this will break. So, we can add tests before testing for evenness:

```
if (i > 0 && i < array_len && !(array[i]%2))
    evens += array[i];
```

If `i` is out of bounds, then the program just throws `i` out and moves on. When failing silently is OK, these types of tests are perfect; when the system should complain loudly when it encounters a failure, then move on to the next tool in our list: `assert`.

`assert`   The `assert` macro makes a claim, and if the claim is false, the program halts at that point. This can be used for both mathematical assertions and for housekeeping like checking for `NULL` pointers. Here is the above input-checking function rewritten using `assert`:

```
#include <assert.h>

double find_means(double *in, int length){
    assert (in!=NULL);
    assert (length>0);
    double mean = in[0];
    for (int i=1; i < length, i++)
        mean += in[i];
    return mean/length;
}
```

If your assertion fails, then the program will halt, and a notice of the failure will print to the screen. On a gcc-based system, the error message would look something like

```
assert: your_program.c:4: find_means: Assertion 'length > 0' failed.
Aborted
```

Some people comment out the assertions when they feel the program is adequately debugged, but this typically saves no time, and defeats the purpose of having the assertions to begin with—are you *sure* you'll never find another bug? If you'd like to compare timing with and without assertions, the -DNDEBUG flag to the compiler (just add it to the command line) will compile the program with all the `assert` statements skipped over.

Q2.20   The method above for taking a mean runs risks of overflow errors: for an array of a million elements, `mean` will grow to a million times the average value before being divided down to its natural scale.
Rewrite the function so that it calculates an incremental mean as a function of the mean to date and the next element. Given the sequence $x_1, x_2, x_3, \ldots$, the first mean would be $\mu_1 = x_1$, the second would be $\mu_2 = \frac{\mu_1}{2} + \frac{x_2}{2}$, the third would be $\mu_3 = \frac{2\mu_2}{3} + \frac{x_3}{3}$, et cetera. Be sure to make the appropriate assertions about the inputs. For a solution, see the GSL `gsl_vector_mean` function, or the code in `apop_db_sqlite.c`.

TEST FUNCTIONS   The best way to know whether a function is working correctly is
to test it, via a separate function whose sole purpose is to test the
main function.

A good test function tries to cover both the obvious and strange possibilities: what
if the vector is only one element, or has unexpected values? Do the *corner cases*,
such as when the input counter is zero or already at the maximum, cause the
function to fail? It may also be worth checking that the absolutely wrong inputs,
like find_means(array4, -3) will fail appropriately. Here is a function to run
find_means through its paces:

```
void test_find_means(){
    double array1[] = {1,2,3,4};
    int length = 4;
        assert(find_means(array1, length) == 2.5);
    double array2[] = {INFINITY,2,3,4};
        assert(find_means(array2, length) == INFINITY);
    double array3[] = {−9,2,3,4};
        assert(find_means(array3, length) == 0);
    double array4[] = {2.26};
        assert(find_means(array4, 1) == 2.26);
}
```

Writing test functions for numerical computing can be significantly harder than
writing them for general computing, but this is no excuse for skipping the testing
stage. Say you had to write a function to invert a ten-by-ten matrix. It would take
a tall heap of scrap paper to manually check the answer for the typical matrix. But
you do know the inverse of the identity matrix (itself), and the inverse of the zero
matrix (NaN). You know that $\mathbf{X} \cdot \mathbf{X}^{-1} = \mathbf{1}$ for any $\mathbf{X}$ where $\mathbf{X}^{-1}$ is defined. Errors
may still slip through tests that only look at broad properties and special cases, but
that may be the best you can do with an especially ornery computation, and such
simple diagnostics can still find a surprising number of errors.

Some programmers actually write the test functions first. This is one more manner
of writing an outline before filling in the details. Write a comment block explaining
what the function will do, then write a test program that gives examples of what
the comment block described in prose. Finally, write the actual function. When the
function passes the tests, you are done.

Once you have a few test functions, you can run them all at once, via a supple-
mentary test program. Right now, it would be a short program that just calls the
test_find_means function, but as you write more functions and their tests, they
can be added to the program as appropriate. Then, when you add another test, you
will re-run all your old tests at the same time. Peace of mind will ensue. For ulti-
mate peace of mind, you can call your test functions at the beginning of your main

analysis. They should take only a microsecond to run, and if one ever fails, it will be much easier to debug than if the function failed over the course of the main routine.

Q2.21

Write a test function for the incremental mean program you'd written above. Did your function pass on the first try?

Some programmers (Donald Knuth is the most famous example) keep a bug log listing errors they have committed. If your function didn't pass its test the first time, you now have your first entry for your bug log.

Σ

➤ Before you have even written a function, you will have expectations about how it will behave; express those in a set of tests that the function will have to pass.

➤ You also have expectations about your function's behavior at run time, so `assert` your expectations to ensure that they are met.

Q2.22

This chapter stuck to the standard library, which is installed by default with any C compiler. The remainder of the book will rely on a number of libraries that are commonly available but are not part of the POSIX standard, and must therefore be installed separately, including Apophenia, the GNU Scientific Library, and SQLite. If you are writing simulations, you will need the GLib library for the data structures presented in Chapter 6.

Now that you have compiled a number of programs and C source is not so foreign, this is a good time to install these auxiliary libraries. Most will be available via your package manager, and some may have to be installed from C source code. See the online appendix (linked from the book's web site, `http://press.princeton.edu/titles/8706.html`) for notes on finding and installing these packages, and Appendix A for notes on preparing your environment.

DATABASES

There is a way between voice and presence
Where information flows.

—Rumi (2004, p 32)

Structured Query Language (*SQL*[1]) is a specialized language that deals only with the flow of information. Some things, like joining together multiple data sets, are a pain using traditional techniques of matrix manipulation, but are an easy *query* in a database language. Meanwhile, operations like matrix multiplication or inversion just can not be done via SQL queries. With both database tables and C-side matrices, your data analysis technique will be unstoppable.

As a broad rule, try to do data manipulation, like pulling subsets from the data or merging together multiple data tables, using SQL. Then, as a last step, pull the perfectly formatted data into an in-memory matrix and do the statistical analysis.

Because SQL is a specialized language that deals only with information flows, it is not nearly as complex as C. Here is some valid SQL: `select age, gender, year from survey`. That's almost proper English. It goes downhill from there in terms of properness, but at its worst, it is still not difficult to look at an SQL query and have some idea of what the rows and columns of the output table will look like.

---

[1] Some people pronounce SQL as *sequel* and some as *ess queue ell*. The official ISO/IEC standard has no comment on which is correct.

Like C, SQL is merely a language, and it is left to the programmers of the world to write code that can parse SQL and return data from SQL queries. Just as this book leans toward `gcc` to interpret C code, it recommends the SQLite library, by D Richard Hipp, to interpret code written in SQL. SQLite provides a library of functions that parse SQL queries and uses those instructions to read and write a specific format of file [see binary trees in Chapter 6]. Any program that uses the SQLite function library is reading and writing the same file format, so SQLite files can be traded among dozens of programs.

As with C utilities, the only problem is selecting which SQLite database viewer to use among the many options. The SQLite library comes with a command-line program, `sqlite3`, but there are many other alternatives that are more reminiscent of the table view in the standard stats package or spreadsheet; ask your search engine for *sqlite browser* or *sqlite GUI*. These programs will give you immediate feedback about any queries you input, and will let you verify that the tables you are creating via C code are as you had expected.

Why is SQLite *lite*? Because most SQL-oriented databases are designed to be used by multiple users, such as a firm's customers and employees. With multiple users come issues of simultaneous access and security, that add complications on top of the basic process of querying data. SQLite is designed to be used by one user at a time, which is exactly right for the typical data analysis project. If you hope to use another database system, you will need to learn the (typically vendor-specific) commands for locking and permissions.

This chapter will primarily consist of an overview of SQL, with which you can follow along using any of the above tools. Section 3.5 will describe the Apophenia library functions that facilitate using an SQL database such as SQLite or mySQL from within a C program.

Check that both the SQLite executable and development libraries are correctly installed. In the online code supplement, you will find an SQLite-formatted database named `data-wb.db` listing the 2005 GDP and population for the countries of the world. Verify that you can open the database using one of the above tools (e.g., `sqlite3 data-wb.db` from the command prompt), and that you can execute and view the results of the query `select * from pop;`.

Once you have a working query interpreter, you can follow along with the discussion in this chapter. For your cutting and pasting convenience, most of the queries in this chapter are also in the `queries` file in the online code supplement.

Data format   A database holds one or more tables. Each column in a table repre-
                 sents a distinct variable. For example, a health survey would include
columns such as subject's age, weight, and height. Expect the units to be different
from column to column.

Each row in a table typically represents one observation. For example, in a survey,
each row would be data about a single person. There is no mechanism in SQL for
naming a row, although it is common enough to have a plain column named `row_-`
`name`, or another identifier such as `social_security_no` that serves this purpose.

The asymmetry between columns and rows will be very evident in the syntax for
SQL below. You will select columns using the column name, and there is no real
mechanism for selecting an arbitrary subset of columns; you will select rows by
their characteristics, and there is no real mechanism to select rows by name.[2]

Your C-side matrices will generally be expected to have a similar format; see page
147 for further notes.

Most of the world's data sets are already in this format. If your data set is not,
your best bet is to convert it rather than fighting SQL's design; see the notes on
crosstabs, page 101, for tips on converting from the most common alternative data
format.

3.1 BASIC QUERIES   SQL's greatest strength is selecting subsets of a data set.
                          If you need all of the data for those countries in the World
Bank data set (`data-wb.db`) with populations under 50 million, you can ask for it
thusly:

```
select *
from pop
where population <= 50;
```

You can read this like English (once you know that ∗ means 'all columns'): it will
find all of the rows in a table named pop where `population` in that row is less
than or equal to 50, and return all the columns for those rows.

---

[2]If there is a `row_name` variable, then you could select rows `where row_name = 'Joe'`, but that is simply
selecting rows with the characteristic of having a `row_name` variable whose value is `'Joe'`. That is, column
names are *bona fide* names; row names are just data.

Generally, the `select` statement gives a list of columns that the output table will have; the `from` clause declares where the source data comes from; and the `where` clause lists restrictions on the rows to be output. And that's it. Every query you run will have these three parts in this order: column specification, data source, row specification.[3] This simple means of specifying rows, columns, and source data allows for a huge range of possibilities.

> **Commas and semicolons**
>
> In SQL, semicolons are *terminators* for a given command. You can send two SQL commands at once, each ending with a semicolon. Many SQLite-based programs will forgive you for omitting the final semicolon.
>
> Commas are *separators*, meaning that the last element in a comma-separated list must not have a comma after it. For example, if you write a query like `select country, pop, from population` then you will get an error like "syntax error near `from`" which is referring to the comma just before `from` that is not separating two columns.

Select The `select` clause will specify the columns of the table that will be output. The easiest list is `*`, which means 'all the columns'. Other options:

- Explicitly list the columns:
  `select country, population`
- Explicitly mention the table(s) from which you are pulling data:
  `select pop.population, gdp.country`
  This is unnecessary now, but will become essential when dealing with multiple tables below.
- Rename the output columns:
  `select pop.country as country, gdp as gdp_in_millions_usd`
  If you do not alias `pop.country as country`, then you will need to use the name `pop\.country` in future queries, which is a bit annoying.
- Generate your own new columns. For example, to convert GDP in dollars to GDP in British pounds using the conversion rate as of this writing:
  `select country, gdp*0.506 as gdp_in_GBP`
  The `as gdp_in_GBP` subclause is again more-or-less essential if you hope to refer to this column in the future.

From The `from` clause specifies the tables from which you will be pulling data. The simplest case is a single table: `from data_tab`, but you can specify as many tables as necessary: `from data_tab1, data_tab2`.

You can alias the tables, for easier reference. The clause `from data_tab1 d1,`

---

[3]You may have no row restrictions, in which case your query will just have the first two parts and a null third part.

`data_tab2 d2` gives short names to both tables, which can be used for lines like `select d1.age, d2.height`.

Another option is to take data from subqueries; see below.

Borrowing C's annoyances

SQL accepts C-style block comments of the form `/* ... */`. It has the same trouble with nested block comments as C (see p 25). With one-line comments, everything after two dashes, `--`, is ignored, comparable to the two slashes, `//`, in C. [mySQL users will need two dashes and a space: `-- `.]

Also following C's lead, dividing two integers produces an integer, not the real number we humans expect. Thus, rather than calculating, say, `count1/count2`, cast one of the columns to a real number by adding `0.0`: `(count1+0.0)/count2` will return the real number it should. The add-zero trick also works to turn the string `"1990"` into the number 1990. [SQL has a `cast` keyword, but it is much easier to just use the trick of adding `0.0`.]

Aliasing is generally optional but convenient, but one case where it is necessary arises when you are joining a table to itself. For now, simply note the syntax: `from data t1, data t2` will let you refer to the `data` table as if it were two entirely independent tables.

Notice, by the way, that when we aliased something in the `select` section, the form was `select long_col_description as lcd`, while in the `from` section there is no `as`: `from long_file_name lfn`.[4]

Where   The `where` clause is your chance to pick out only those rows that interest you. With no `where` clause, the query will return one line for every line in your original table (and the columns returned will match those you specified in the `select` clause). For example, try `select 1 from gdp` using the `data-wb.db` database.

You can use the Boolean operators you know and love as usual: `where ((d1.age > 13) or (d2.height >= 175)) and (d1.weight = 70)`. SQL does not really do assignments to variables, so the clause `(d1.weight = 70)` is a test for equality, not an assignment. SQLite is easygoing, and will also accept the C-format `(d1.weight == 70)`; other SQL parsers (like mySQL) are less forgiving and consider the double-equals to be an error.

- You can select based on text the same way you select on a number, such as `where country = 'United States'`. Any string that is not an SQL keyword or the name of a table or column must be in 'single-tick' quotation marks.[5]

---

[4]The `as` is actually optional in the `select` clause, but it improves readability.

[5]Again, SQLite is forgiving, and will also accept C-style "double-tick" quotation marks. However, it is beneficial that SQL uses single-ticks while C uses double-ticks, because `snprintf(q, 100, "select * where country = 'Qatar'")` requires no unsightly backslashes, while double-tick quotation marks do: `snprintf(q, 100, "select * where country = \"Qatar\"")`.

- Case matters: 'United States' != 'united states'. However, there is an out should you need to be case-insensitive: the `like` keyword. The clause `where country like 'united states'` will match the fully-capitalized country name as well as the lower case version. The `like` keyword will even accept two wild cards: `_` will match any single character, and `%` will match any set of characters. Both `country like 'unit%ates'` and `country like 'united_states'` will match 'United States'.

- The `where` clause refers to the root data, not the output, meaning that you can readily refer to columns that you do not mention in the `select` clause.

$\mathbb{Q}_{3.2}$ | Use a `where` clause and the `population` table to find the current population of your home country. Once you know this amount, select all of the countries that are more populous than your country.

Generalizing from equality and inequalities, you may want a group of elements or a range. For this, there are the `in` and `between` keywords. Say that we want only the United States and China in our output. Then we would ask only for columns where the country name is in that short list:

```
select *
from gdp
where country in ("United States", "China")
```

The `in` keyword typically makes sense for text data; for numeric data you probably want a range. Here are the countries with GDP between $10 and $20 billion:

```
select *
from gdp
where gdp between 10000 and 20000
```

$\mathbb{Q}_{3.3}$ | Write a query using `<=` and `>=` to replicate the above query that used `between`.

> ➤ A query consists of three parts: the columns to be output, the data source, and the rows to be output.

> ➤ The columns are specified in the `select` statement. You can pull all the columns from the data using `select *`, or you can specify individual columns like `select a, b, (a+0.0)/b as ratio`.   ⋙

$\gg$

$\sum$

> ➤ The data source is in the `from` clause, which is typically a list of tables.

> ➤ The row specification, generally in the `where` clause, is a list of conditions that all rows must meet. It can be missing (and so all possible rows are returned) or it can include a series of conditions, like `where (a = b) and (b <= c)`.

3.2 ❋ **DOING MORE WITH QUERIES**   Beyond the basic `select` - `from` - `where` format, a `select` query can include several auxiliary clauses to refine the output further. Here is the complete format of a `select` query, which this section will explore clause by clause.

```
select [distinct] columns
from tables
where conditions
group by columns
having group_conditions
order by columns
limit n offset n
```

PRUNING ROWS WITH `distinct`   The `data-metro.db` file includes a listing of all stations and the color of the subway line(s) on which the station lies. The query `select line from lines` produces massive redundancy, because there are a few dozen stations on every line, so each color appears a few dozen times in the table.

The `distinct` keyword will tell the SQL engine that if several rows would be exact duplicates, to return only one copy of that row. In this case, try

```
select distinct line
from lines
```

The `distinct` word prunes the rows, but is placed in the `select` portion of the program. This reads more like English, but it breaks the story above that the `select` statement specifies the columns and the `where` statement specifies the rows.

AGGREGATION   Here is how to get the number of rows in the gdp table of `data-wb.db`:

```
select count(*) as row_ct
from gdp;
```

This produces a table with one column and one row, listing the total number of rows in the `data` table.

Q.3.4   How many rows does `select * from pop, gdp` produce? The explanation for the answer will appear in the section on *joins*, below.

You probably want more refinement than that; if you would like to know how much data you have in each region, then use the `group by` clause to say so:

```
select class, count(*) as countries_per_class
from classes
group by class;
```

After `count`, the two most common aggregation commands are `sum()` and `avg()`. These take an existing row as an argument. For example, the `data-tattoo.db` database has a single table representing a telephone survey regarding tattoos. To get the average number of tattoos per person broken down by race, you could use this query:

```
select race, avg(tattoos.'ct tattoos ever had')
from tattoos
group by race;
```

Feel free to specify multiple `group by` clauses. For example, you could modify the above query to sort by race and age by changing `group by race` to `group by race, tattoos.'year of birth'`. When you want to analyze the output, you will be very interested in the `apop_db_to_crosstab` function; see page 101.

Q.3.5   In the `precip` table of the `data-climate.db` database, the `yearmonth` column encodes dates in forms like 199608 to mean August, 1996. Fortunately, the SQL-standard `round()` function can be used to produce a plain year: `round(199608./100.) == 1996.0`. Use `round`, `group by`, and `avg` to find the average precipitation (`pcp`) in each year.

You can use `count` with the `distinct` keyword to find out how many of each row you have in a table. This is useful for producing weights for each observation type,

Function	Standard SQL	mySQL	SQLite via Apophenia
abs, avg, count, max, min, round,[a] sum	○	○	○
acos, asin, atan, cos, exp, ln, log10, pow, rand, sin, sqrt, $stddev_s$, tan, $variance_p$, std_p, $stddev\_pop_p$, stddev_-$samp_s$, var_samp_s, var_pop_p		○	○
ran, var_s, $skew_s$, $kurtosis_s$, $kurt_s$			○

[a]Round is not part of the SQL standard, which instead provides `floor` and `ceil`.

Table 3.1 Standard SQL offers very few mathematical functions, so different systems offer different extensions. The p and s subscripts indicate functions for populations or for samples (see box on page 222.

as in this query to produce a tabulation of respondents to the tattoo survey by race and birth year:

```
select distinct race, tattoos.'year of birth' as birthyear, count(*) as weight
from tattoos
group by race, birthyear
```

With a group by command, you have two levels of elements, items and groups, and you may want subsets of each. As above, you can get a subset of the items with a where clause. Similarly, you can exclude some groups from your query using the having keyword. For example, the above query produced a lot of low-weighted groups. What groups have a count(*) > 4? We can't answer this using where weight > 4, because there is no weight column in the data table, only in the post-aggregation table. This is where the having keyword comes in:

```
select distinct race, tattoos.'year of birth' as birthyear, count(*) as weight
from tattoos
group by race, birthyear
having weight > 4
```

✳ *SQL extensions*   That's all the aggregators you get in standard SQL. So implementers of the SQL standard typically add additional functions beyond the standard; see Table 3.1 for a list, including both aggregation functions like var and $\mathbb{R} \to \mathbb{R}$ functions like `log`. The table focuses on numeric functions, and the standard and mySQL both include several functions for manipulation of text, dates, and other sundry types of data; see the online references for details.

Bear portability in mind when using these functions, and be careful to stick to the SQL standard if you ever hope to use your queries in another context. If you want to stay standard, call your data into a C-side vector or matrix and use `apop_vector_log`, `apop_vector_exp`, `apop_vector_skew`, `apop_vector_var`, ..., to get the desired statistics on the matrix side.

SORTING   To order the output table, add an `order` by clause. For example, to view the list of country populations in alphabetical order, use

```
select *
from pop
order by country
```

- You may have multiple elements in the clause, such as `order by country, pop`. If there are ties in the first variable, they are broken by the second.
- The keyword `desc`, short for *descending*, will reverse the order of the variable's sorting. Sample usage: `order by country desc, pop`.

GETTING LESS   Especially when interactively interrogating a database, you may not want to see the whole of the table you have constructed with a `select` clause. The output may be a million lines long, but twenty should be enough to give you the gist of it, so use a `limit` clause. For example, the following query will return only the first twenty rows of the pop table:

```
select *
from pop
limit 20
```

You may want later rows, and so you can add the `offset` keyword. For example,

```
select *
from pop
limit 5 offset 3
```

will return the first five rows, after discarding the first three rows. Thus, you will see rows 4–8. Beyond making interactive querying easier, `limit - offset` clauses can also be used to break tables that are somehow giving you problems into more manageable pieces, probably via a C-side `for` loop.

- You get one `limit`/`offset` per query, which must be the last thing in the query.
- If you are using `union` and family to combine `select` statements (see below), your `limit` clause should be at the end of all of them, and applies only to the aggregate table.

※ *Random subsets*   The `limit` clause gives you a sequential subset of your data, which may not be representative. If this is a problem, you can take a random draw of some subset of your data. Ideally, you could provide a query like `select * from data where rand() < 0.14` to draw 14% of your data.

SQLite-via-Apophenia and mySQL provide a `rand` function that works exactly as above.[6] For every call to the function (and thus, for every row), it draws a uniform random number between zero and one.[7]

CREATING TABLES   There are two ways to create a table. One is via a `create` statement and then an `insert` statement for every single row of data. The `create` statement requires a list of column names;[8] the `insert` statement requires a list of one data element for each column.

```
begin;
create table newtab(name, age);
insert into newtab values("Joe", 12);
insert into newtab values("Jill", 14);
insert into newtab values("Bob", 14);
commit;
```

The `begin-commit` wrapper, by the way, means that everything will happen in memory until the final commit. The program may run faster, but if the program

---

[6]Standard SQL's `random` function is absolutely painful. SQLite's version currently produces a number between $\pm 9{,}223{,}372{,}036{,}854{,}775{,}807$, which the reader will recognize as $\pm(2^{63}-1)$. So we need to pull a random number, divide by $2^{63}-1$, shift it to the familiar $[0,1]$ range, and then compare it to a limit. Standard SQL does not even provide exponentiation, so doing this requires the bit-shifting operator which I had promised you would never need; read `1<<x` as 2^x. That said, `select * from data where (random()/(-(1<<63)-1.0)+1)/2 < 0.14` will pull approximately 14% of the data set.

[7]After you read Section 11.1, you will wonder about the stream of random numbers produced in the database. There is one stream for the database, which Apophenia maintains internally. To initialize it with a seed of seven, use `apop_db_rng_init(7)`. If you do not call this function, the database RNG auto-allocates at first use with seed zero.

[8]SQLite has the pleasant property that its columns are basically type-less. Other database engines insist on table declarations that look a little like C functions, e.g., `create table newtab(name varchar[30], age int)`; see your database engine documentation for details.

crashes in the middle, then you will have lost everything. The optimal speed/security trade-off is left as an exercise for the reader.

If you have hundreds or thousands of `inserts`, you are almost certainly better off putting the data in a text file and using either the C function `apop_text_to_db` or the command-line program with the same name. The form above is mostly useful in situations where you are creating the table in mid-program, as in the example on page 108.

The other method of creating a table is by saving the results of a query. Simply put `create table` *newtab_name* `as` at the head of the query you would like to save:

> **create table** tourist_traps **as**
>   **select** country
>   **from** lonely_planet
>   **where** (0.0+pp) > 600

 Q3.6   The `riders` table of the `data-metro.db` database includes the average boardings in each station of the Washington Metro system, every year since its opening. Create a `riders_per_year` table with one column for the year and one column for total average boardings across the system for the given year.

DROPPING A TABLE   The converse of table creation is table dropping:

> **drop table** newtab;

See also `apop_table_exists` on the C-side (p 108), which can also delete tables if desired.

ROWID   Sometimes, you need a unique identifier for each output row. This would be difficult to create from scratch, but SQLite always inserts such a row, named `rowid`. It is a simple integer counting from one up to the number of rows, and does not appear when you query `select * from table`. But if you query `select rowid, * from table`, then the hidden row numbers will appear in the output.[9]

---

[9]mySQL users will need to explicitly ask for such a column when creating the table. A statement like `create table` *newtab* (*id_column* `int auto_increment`, *info1* `char(30)`, *info2* `double`, ...) will create the table with the typical columns that you will fill, plus an *id_column* that the system will fill. After `insert into newtab values ("Joe", 23); insert into newtab values ("Jane" 21.8);`, the table will have one row for Joe where *id_column*==1 and one for Jane where *id_column*==2.

$\underset{3.7}{\text{Q}}$ | Using `order by` and `rowid`, find the rank of your home country's GDP among countries in the World Bank database.

METADATA   What tables are in the database? What are their column names? Standard SQL provides no easy way to answer these questions, so every database engine has its own specific means. SQLite gives each database a table named `sqlite_master` that provides such information. It includes the type of object (either index or table, in the `type` column), the name (in the `name` column), and the query that generated the object (in the `sql` column). MySQL users, see page 106.

In practical terms, this table is primarily good for getting the lay of an unfamiliar database—a quick `select * from sqlite_master;` when you first open the database never hurts. If you are using the SQLite command line, there is a `.table` command that does exactly what this program does. Thus, the command `sqlite3` *mydb.db* `.table` just lists available tables, and the `.schema` command gives all of the information from `sqlite_master`.

MODIFYING TABLES   SQL is primarily oriented toward the filtering style of program design: e.g., have one query to filter a data table to produce a new table with bad data removed, then have another query to filter the resulting table to produce an aggregate table, then select some elements from the aggregate table to produce a new table, et cetera.

But you will often want to modify a table in place, rather than sending it through a filter to produce a new table (especially if the table is several million entries long). SQL provides three operations that will modify a table in place.

`delete`   Unlike `drop`, which acts on an entire table, `delete` acts on individual rows of a database. For example, to remove the columns with missing GDP data, you could use this query [—but before you destroy data in the sample databases, make a copy, e.g., via `create table gdp2 as select * from gdp`]:

```
delete from gdp
where gdp='..'
```

`insert`   The obvious complement to deleting lines is inserting them. You already saw `insert` used above in the context of creating a table and then inserting elements item-by-item. You can also insert via a query, via the form `insert into` *existing_table* `select * from ....`

update   The update query will replace the data in a column with new data. For
           example, the World Bank refrained from estimating Iraq's 2006 population,
but the US Central Intelligence Agency's *World Factbook* for 2006 estimates it at
26,783,383. Here is how to change Iraq's population (in the pop table) from .. to
26783:

```
update pop
set population=26783
where country='Iraq'
```

$\sum$

➤ You can limit your queries to fewer rows using a `limit` clause, which
  gives you a sequential snippet, or via random draws.

➤ The SQL standard includes a few simple aggregation commands:
  `avg()`, `sum()`, and `count()`, and most SQL implementations provide a few more nonstandard aggregators for queries called using its
  functions.

➤ When aggregating, you can add a `group` by clause to indicate how
  the aggregation should be grouped.

➤ Sort your output using an `order` by clause.

➤ You can create tables using the `create` and `insert` commands, but
  you are probably better off just reading the table from a text file. Use
  `drop` to delete a table.

➤ SQLite gives every row a `rowid`, though it is hidden unless you ask
  for it explicitly.

3.3 JOINS AND SUBQUERIES   So far, we have been cutting one table down, ei-
                                  ther by selecting a subset of rows or by group-
ing rows. SQL's other great strength is in building up tables by joining together
data from disparate sources. The joining process is not based on a `join` keyword,
but simply specifying multiple data sources in the `from` section of your query and
describing how they mesh together in the `where` section.

If you specify two tables in your `from` line, then, lacking any restrictions, the
database will return one joined line for every pair of lines. Let table 1 have one
column with data $\begin{bmatrix} 1 \\ 2 \\ 3 \end{bmatrix}$ and table 2 have one column with data $\begin{bmatrix} a \\ b \\ c \end{bmatrix}$; then `select`

`* from table1, table2` will produce an output table with every combination, $3 \times 3 = 9$ rows:

```
1 a
1 b
1 c
2 a
2 b
2 c
3 a
3 b
3 c.
```

Such a product quickly gets overwhelming: in the exercise on page 81, you saw how joining the 208 countries in the World Bank data's pop table with the same 208 countries in the gdp table produces a few hundred pages of rows.

Thus, the `where` clause becomes essential. Its most typical use for a join arises when one column in each table represents identical information. Out of the 43,264 rows from the above join, including those that matched Qatar with Ghana and Cameroon with Zimbabwe, we are interested only in those that match Qatar with Qatar, Cameroon with Cameroon, and so on. That is, we want only those rows where `pop.country = gdp.country`, and so the query makes sense only when that restriction is added in:

> **select** pop.country, pop.population, gdp.GDP
>   **from** pop, gdp
>   **where** pop.country = gdp.country

You can see that using the table-dot-column format for the column names is now essential. In the `select` clause specifying the output columns, you can use either `pop.country` or `gdp.country`, since the two will be by definition identical, or if you are unconcerned with the country names and just want the numeric data you can omit names entirely.

Q3.8   Add a calculation to the `select` portion of the above query to find the GDP per capita of each country. Be sure to give the calculated column a name, like `gdp_per_cap` so you can `order by gdp_per_cap`.

Q_{3.9}
The World Bank data includes a classification for each country. Countries receiving World Bank assistance (what the WB calls *client countries*) are classed by region (e.g., Middle East and North Africa), while other countries are binned into a generic class like "Lower-middle-income economies." Find the total GDP per capita for each World Bank grouping. Here, you will join using the country columns in the gdp and classes table, and by the country columns in the pop and classes table. Add up total GDP in the region, and divide by total population in the region.

Example: a time lag   The form above, where two columns match, is by far the most common type of join, but there are other creative uses of joins. For example, it is common in time series analysis to include the value of a variable at time $t - 1$ as data that influenced the value at time t.

The data-climate.db database includes a table of the deviation from the century-long norm for aggregate worldwide temperatures (see Smith & Reynolds (2005) for methods, caveats, and discussion). A quick select * from temp will show that there is an upward trend in the data: the first few years are all below zero; the last few years hover around 0.5.[10]

What does the month-to-month change look like? The first step is dealing with the fact that there are separate year and month columns. One solution would be to deal only with year + month/12., which moves through time in smooth increments of $\frac{1}{12}$. This creates its own problem, because comparing floating-point values is not reliable: 1900 + 1./12. - 1./12. could wind up as something like 1900.00001, and a test whether this value exactly equals 1900 will fail. As a variant that solves this problem, instead of dividing months by 12, multiply years by 12, so that we are comparing only integers:

```
select R.year+R.month/12., R.temp − L.temp
from temp L, temp R
where R.year*12 +R.month = L.year*12 +L.month +1;
```

The salient feature of this data set is that not much happens. The long-term shift is the result of a large number of very small month-to-month changes.

---

[10]Chapter 5 will cover graphing, but for now, try apop_plot_query data-climate.db "select temp from temp" from your command line to get a visual indication of the trend.

Perhaps we would see a larger change via a larger time span. Calculate the year-to-year differences.

- Create an `annualized` table with two columns: the year and average `temp` over all months for the year.

- Join that table with itself lagged by one year. You won't have to worry about unreliable float comparisons, but recall that if SQLite thinks `year` is a string, then it will treat `year+0.0` as a number.

$Q_{3.10}$

Having looked at year-long differences, try decades.

- Create a decades table with the average for each decade. (*Hint*: group by round(year/10).)

- Join the table with itself lagged by ten years. Are the differences beginning to show a pattern?

Given that the data is sorted, we could also have done the matching of rows using the `rowid`:

> **select** L.temp − R.temp
> **from** temp L, temp R
> **where** R.rowid+0.0=L.rowid−1;

SPEEDING IT UP   Now that you have seen how to join tables, we now cover how to avoid joining tables. If two tables have a million elements each, then joining them using a clause like `where a=b` requires 1e6 × 1e6 = 1e12 (a trillion) comparisons. This is impossibly slow, so there are a number of tricks to avoid making all those 1e12 comparisons.[11]

Indices   You can ask the SQL engine to create an index for a table that you intend to use in a join later. The commands:

> **create index** pop_index **on** population(country)
> **create index** gdp_index **on** gdp(country)

---

[11] Say that you mean to join a million subjects via ID number, via `select t1.*, t2.* from t1, t2 where t1.id = t2.id`, but you forget to include the `where` clause. Then you just asked the system to create a trillion-entry table, which will take from several hours to weeks. Thus, the first step in speeding up an inordinately slow query is not to try the tricks in this section, but to make sure that you actually wrote the query you had intended to write.

would index the `pop` and `gdp` tables on the `country` column. The name of the index, such as `pop_index`, is basically irrelevant and can be any gibberish that sounds nice to you. Once you have created this index, a join using any of the indexed columns goes *much* faster, because the system no longer has to do 1e12 comparisons. Basically, it can look at the first value of *var* in the left table—say it is 17—and then check the right table's index for the list of elements whose value is 17. That is, instead of one million comparisons to join the first element, it only has to do one index lookup. The lookup and the process of building the tree took time as well, but these processes are on the order of millions of operations, not millions squared. The tree is internally structured as a binary tree; see Chapter 6 for discussion of b-trees.

There is standard SQL syntax for indexing multiple columns, e.g., `create index pop_index2 on pop(country, population)`, which goes by *lexicographic order*. This is just an index on the first item (`country`) with the second column (`population`) as a backup ordering; if you want to join by the second column, you should prepare by creating another index that puts that column in the first (or the only) position.

Subqueries   Among SQL's nicest tricks is that it allows for the input tables to be queries themselves. For example: how large is the average World Bank grouping? Answering this question is a two-step process: get a `count(*)` for each category, and then get an average of that. You could run a query to produce a table of counts, save the table, and then run a query on that table to find the averages.

```
create table temptab as
    select count(*) as ct
        from classes
        group by class;
select avg(ct)
    from temptab
```

But rather than generating a temporary table, SQL allows you to simply insert the `select` statement directly into the query where it is used:

```
select avg(ct)
    from (select count(*) as ct
        from classes
        group by class)
```

The query inside the `from` clause will return a table, and even though that table has no name, it can be used as a data source like any other table. If the query output needs a name, you can alias the result as usual: `from (select ...) t1` will allow you to refer to the query's output as `t1` elsewhere in the query.

Q3.11 | On page 79, you first found your home country's population, then the countries with populations greater than this. Use a subquery to do this in one query. (*Hint*: you can replace a number with a query that returns one element.)

Subsetting via a foreign table   If you look at the World Bank data, you will see a large number of countries that are small islands of a few million people. Say that we are unconcerned with these countries, and want only the GDP of countries where `population` > 270.

Q3.12 | Write a query to pull only the GDP of countries where the population is greater than 270 million, using the standard `where leftcol=rightcol` join syntax from the head of this section.

But the full join (as per the exercise) is not necessary: we are not particularly concerned with the population *per se*, but are just using it to eliminate rows. It would thus be logical to fit the query into the `where` clause, since that is the clause that is typically used to select a subset of the rows. Indeed, we can put a query directly into a `where` ... in clause:

```
select *
from gdp
where country in (select country from pop where population > 270)
```

The subquery will return a list of country names, and the main query can then use those as if you had directly typed them in.

This is typically much faster than a full join operation, because there was no need to make (left table row count) × (right table row count) comparisons.

The boost in efficiency implies some slight restrictions: because the `from` clause does not list the table used in the subquery, you can not refer to any of the subquery's columns in the output.

※ *Joining via a `for` loop*   The time it takes to do an especially large join is not linear in the number of rows, primarily for real-world reasons of hardware and software engineering. If your computer can not store all the data points needed for a query in fast memory, it will need to do plenty of swapping back and forth between different physical locations in the computer. But your computer may be able to store a hundredth or a thousandth of the data set in fast memory, and so you can perhaps get a painfully slow query to run in finite time by breaking it down into a series of shorter queries.

Here is an example from my own work (Baum *et al.*, 2008). We had gathered 550,000 genetic markers (SNPs) from a number of pools of subjects, and wanted the mean for each pool. Omitting a few details, the database included a `pools` table with the subject `id` and the `poolid` of its pool, with only about a hundred elements; and a table of individual `ids`, the `SNP` labels, and their values, which had tens of millions of values. Even after creating the appropriate indices, the straight join—

```
select pools.poolid as poolid, SNP, avg(val) as val, var(val) as var
from genes, pools
where genes.id=pools.id
group by pools.poolid, SNP
```

—was taking hours.

Our solution was to use a C-side `for` loop, plus subsetting via a foreign table, to avoid the join that was taking so long. There are three steps to the process: create a blank table to be filled, get a list of `poolids`, and then use `insert into ... select ...` to add each `poolid`'s data to the main table. The details of the functions will be discussed below, but these three steps should be evident in this code snippet.

```
apop_query("create table t (poolname, SNP, val, var);");
apop_data *names = apop_query_to_text("select distinct poolid from pools");
for (int i=0; i< names->textsize[0]; i++)
    apop_query("insert into t \n\
        select '%s', SNP, avg(val), var(val) \n\
        from genes \n\
        where id in (select id from pools where poolid = '%s') \n\
        group by SNP; \n\
            ", names[i][0], names[i][0]);
```

This allowed the full aggregation process to run in only a few minutes. The next week we bought better hardware.

As noted above, if there is no natural grouping like the pools in this example, a `for` loop using the `limit ...offset` form can also break a too-long table into smaller pieces.

STACKING TABLES   You can think of joining two tables as setting one table to the right of another table. But now and then, you need to stack one on top of the other. There are four keywords to do this.

- Union: Sandwiching union between two complete queries, such as

```
select id, age, zip
from data_set_1
union
select id, age, zip
from data_set_2
```

will produce the results of the first query stacked directly on top of the second query. Be careful that both tables have the same number of columns.

- Union all: If a row is duplicated in both tables, then the union operation throws out one copy of the duplicate lines, much like select distinct includes only one of the duplicates. Replacing union with union all will retain the duplicates.

- Intersect: As you can guess, putting intersect between two select statements returns a single copy of only those lines that appear in both tables.

- Except: This does subtraction, returning only elements from the first table that do not appear in the second. Notice the asymmetry: nothing in the second table will appear.

➤ You can put the output of a query into the from clause of a parent query.

➤ You can join tables by listing multiple tables in the from clause. When you do, you will need to specify a where clause, and possibly the distinct keyword, to prevent having an unreasonably long output table.

➤ If you intend to join elements, you can speed up the join immensely by creating an index first.

➤ If the join still takes too long, you can sidestep it via the select ... where col in (select ...) form, or via a C-side for loop.

➤ Tables can be stacked using union, union all, intersect, and except.

3.4   ON DATABASE DESIGN   Say that you are not reading in existing data, but are gathering your own, either from a simulation or data collected from the real world. Here are some considerations and suggestions for how you could design your database, summarizing the common wisdom about the best way to think about database tables.

The basic premise is that each type of object should have a single table, and each object should have a single row in that table.

Figure 3.2 shows a table of observations for a generic study involving several subjects and treatments, whose information was measured at several times. The simple one-table design is how the typical spreadsheet is designed. This version has one row per subject, so each row has two observations, and information about subjects, treatments, observations, and pools are mixed together.

Figure 3.3 shows a structure better suited for databases. For most statistical studies, the key object is the observation, and that gets its own table; we now see that there were twenty observations. The other objects in the study—subjects, pools, and treatments—all get their own tables as well. By giving each element of each table an ID number, each table can easily cross-reference others. This setup has many advantages.

Minimize redundancy   This is rule number one in database design, and many a book and article has been written about how one goes about reducing data to the redundancy-minimized *normal form* (Codd, 1970). If a human had to enter all of the redundant data, this creates more chances for error, and the same opportunities for failure come up when the data needs to be modified when somebody notices that there were actually nine subjects in the pool from 6/2/02. In the single-table form, information about the pool was repeated for every member of the pool, while having a separate table for pools means that each pool's information is listed exactly once.

Ask non-observation questions   There are reasons to ask questions based on treatments or pools, but a setup with only an observation-based table does not facilitate this. From the multiple tables, it is easy to ask questions that focus on data, treatments, or pools, via join operations on the observation, pool, subject, or treatment IDs.

Gelman & Hill (2007, p 239) point out that separating subjects and groups facilitates multilevel models, where each group has parameters for its own submodel estimated, and then those parameters are used to estimate an overall model. This sort of modeling will be covered in later chapters.

Use the power of row subsets   Figure 3.2 includes multiple observations on one line, for the morning and evening measurements. But what if we went from two observations to hourly observations for 24 hours? Remember, there is no way to arbitrarily select a subset of columns, so columns

subjid	value_morn	value_eve	poolcount	pooldate	t_type	t_dosage
1	23.28	NaN	12	2/2/02	control	NaN
2	14.07	NaN	12	2/2/02	control	NaN
3	20.98	NaN	12	2/2/02	control	NaN
4	12.12	NaN	12	2/2/02	control	NaN
5	30.28	28.11	11	4/2/02	case	0.2
6	22.15	14.05	11	4/2/02	case	0.2
7	19.78	12.54	8	4/2/02	case	0.4
8	21.53	9.01	8	4/2/02	case	0.4
9	27.42	23.20	19	6/2/02	case	0.2
10	18.57	12.29	19	6/2/02	case	0.2

Figure 3.2 Spreadsheet style: one monolithic table, with much redundancy.

obsid	subjid	value	time
1	1	23.28	morn
2	2	14.07	morn
3	3	20.98	morn
4	4	12.12	morn
5	5	30.28	morn
6	6	22.15	morn
7	7	19.78	morn
8	8	21.53	morn
9	9	27.42	morn
10	10	18.57	morn
11	1	NaN	eve
12	2	NaN	eve
13	3	NaN	eve
14	4	NaN	eve
15	5	28.11	eve
16	6	14.05	eve
17	7	12.54	eve
18	8	9.01	eve
19	9	23.20	eve
20	10	12.29	eve

subjid	poolid	treatmentid
1	1	1
2	1	1
3	1	1
4	1	1
5	2	2
6	2	2
7	3	3
8	3	3
9	4	2
10	4	2

poolid	poolcount	pooldate
1	12	2/2/02
2	11	4/2/02
3	8	4/2/02
4	19	6/2/02

treatmentid	t_type	t_dosage
1	control	NaN
2	case	0.2
3	case	0.4
4	case	0.6

Figure 3.3 Database style: one table for each object type, one row for each object.

named 1AM, 2AM, ..., would be difficult to use. If we needed the mean of all morning observations, we'd need to do something like select (12AM + 1AM + 2AM + 3AM + ...)/12, but if the table in Figure 3.3 had an hour column, we could simply use:

> **select avg(value)**
>   **from** observations
>   **where time like** '%am'

(or where time < 12, depending on the format we choose for the time).

If there is any chance that two observations will somehow be compared or aggregated, then they should probably be recorded in different rows of the same column. For 24 hours and ten subjects, the table would be 240 rows, which is not nearly as pleasing or human-digestible as a 10×24 spreadsheet. But you will rarely need to look at all the data at once, and can easily construct the crosstab if need be via apop_db_to_crosstab.

Even worse than having two data points of the same type in separate columns is having two data points of the same type in separate tables, such as a cases table and a controls table. Or, say that a political scientist wants to do a study of county-level data throughout the United States, including variables such as correlations between tax rates, votes by Senators, and educational outcomes. Because DC has no county subdivisions and its residents have no Congressional representation, the DC data does not fit the form of the data for the states and commonwealths of the United States. But the correct approach is nonetheless to put DC data in the same table as the counties of the fifty states, rather than creating a table for DC and a table for all other states—or still worse, a separate table for every state.

It is easy to select * from alldata where senate_vote is not null if DC's lack of representation will affect the analysis.[12]

> ➤ Databases are not spreadsheets. They are typically designed for many tables, which may have millions of rows if necessary.
>
> ➤ Each type of object (observations, treatments, groups) should have a single table, and each object should have a single row in that table.
>
> ➤ Bear in mind the tools you have when designing your table layouts. It is easy to join tables, find subsets of tables, and create spreadsheet-like crosstabs from data tables.

---

[12]By the way, select * from alldata where population > (select population from alldata where state = 'DC') won't work: it will return only 49 out of 50 states, because the population of DC (zero Senators, zero Representatives) is 572,000, while Wyoming (two Senators, one Representative) has a population of 494,000. [2000 census data]

3.5 FOLDING QUERIES INTO C CODE   This section covers the functions in the Apophenia library that will create and query a database. All of these functions are wrappers of functions in the SQLite or mySQL libraries that do the dirty work, but they are sufficiently complete that you should never need to use the functions in the SQLite/mySQL C libraries directly. The details of the main discussion will apply to SQLite; mySQL users, see page 106 for the list of differences.

IMPORTING   The first command you will need is `apop_open_db`. If you give it the name of a file, like `apop_open_db("study.db")`, then the database will live on your hard drive. This is slower than memory, but will exist after you stop and restart the program, and so other programs will be able to use the file, you have more information for debugging, and you can re-run the program without re-reading in the data. Conversely, if you give a null argument—`apop_open_db(NULL)`—then the database is kept in memory, and will run faster but disappear when the program exits. Apophenia uses only one database at a time, but see the `apop_merge_dbs` and SQLite's `attach` functions below.

Command-line utilities

Apophenia includes a handful of command-line utilities for handling SQLite databases where there is no need to write a full-blown C program. `apop_text_to_db` reads a text file into a database table, `apop_merge_dbs` will send tables from one database to another, `apop_plot_query` will send query output directly to Gnuplot, and `apop_db_to_crosstab` will take a table from the SQLite database and produce a crosstab. All of these are simply wrappers for the corresponding Apophenia functions. For all of the utilities, you can use the `-h` parameter to get detailed instructions (e.g., `apop_plot_query -h`).

Unless your program is generating its own data, you will probably first be importing data from a text file. The `apop_text_to_db` function will do this for you, or you can try it on the command line (see box). The first line of the text file can be column names, and the remaining rows are the data. If your data file is not quite in the right format (and it rarely is), see Appendix B for some text massaging techniques.

When you are done with all of your queries, run `apop_close_db` to close the database. If you send the function a one—`apop_close_db(1)`—then SQLite will take a minute to clean up the database before exiting, leaving you with a smaller file on disk; sending in a zero doesn't bother with this step. Of course, if your database is in memory, it's all moot and you can forget to close the database without consequence.

The queries   The simplest function is `apop_query`, which takes a single text argument: the query. This line runs the query and returns nothing, which is appropriate for `create` or `insert` queries:

```
int page_limit = 600;
apop_query(
    "create table tourist_traps as \
    select country \
    from lonely_planet \
    where (pp + 0.0) > %i ", page_limit);
```

- A string is easiest for you as a human to read if it is broken up over several lines; to do this, end every line with a backslash, until you reach the end of the string. The next example will use another alternative.

- As the example shows, all of Apophenia's query functions accept the `printf`-style arguments from page 26, so you can easily write queries based on C-side calculations.

There are also a series of functions to query the database and put the result in a C-side variable. This function will run the given query and return the resulting table for your analysis:

```
int page_limit = 600;
apop_data *tourist_traps = apop_query_to_text(
    "select country "
    "from lonely_planet "
    "where (0.0+pp) > %i ", page_limit);
```

- C merges consecutive strings, so `"select country "` `"from"` will be merged into `"select country from"`. We can use this to split a string over several lines. But be careful to include whitespace: `"select country"` `"from"` merges into `"select countryfrom"`.

After this snippet, `tourist_traps` is allocated, filled with data, and ready to use—unless the query returned no data, in which case it is NULL. It is worth checking for NULL output after any query that could return nothing. There are `apop_query_...` functions for all of the types you will meet in the next chapter, including `apop_query_to_matrix` to pull a query to a `gsl_matrix`, `apop_query_to_text` to pull a query into the text part of an `apop_data` set, `apop_query_to_data` to pull data into the matrix part, and `apop_query_to_vector` and `apop_query_to_float` to pull the first column or first number of the returned table into a `gsl_vector` or a `double`.

For immediate feedback, you can use `apop_data_show` to dump your data to screen or `apop_data_print` to print to a file (or even back to the database). If you want a quick on-screen picture of a table, try

```
apop_data_show(apop_query_to_data("select * from table"));
```

Listing 3.4 gives an idea of how quickly data can be brought from a database-side table to a C-side matrix. The use of these structures is handled in detail in Chapter 4, so the application of the `percap` function may mystify those reading this book sequentially. But the `main` function should make sense: it opens the database, sets the `apop_opts.db_name_column` to an appropriate value, and then uses `apop_query_to_data` to pull out a data set. Its last two steps do the math and show the results on screen.

```
1   #include <apop.h>
2
3   void percap(gsl_vector *in){
4       double gdp_per_cap = gsl_vector_get(in, 1)/gsl_vector_get(in, 0);
5       gsl_vector_set(in, 2, gdp_per_cap); //column 2 is gdp_per_cap.
6   }
7
8   int main(){
9       apop_opts.verbose ++;
10      apop_db_open("data−wb.db");
11      strcpy(apop_opts.db_name_column, "country");
12      apop_data *d = apop_query_to_data("select pop.country as country, \
13          pop.population as pop, gdp.GDP as GDP, 1 as GDP_per_cap\
14          from pop, gdp \
15          where pop.country == gdp.country");
16      apop_matrix_apply(d−>matrix, percap);
17      apop_data_show(d);
18      apop_opts.output_type = 'd';
19      apop_data_print(d, "wbtodata_output");
20  }
```

Listing 3.4 Query populations and GDP to an `apop_data` structure, and then calculate the GDP per capita using C routines. Online source: `wbtodata.c`.

- Line 11: As above, SQL tables have no special means of handling row names, while `apop_data` sets can have both row and column labels. You can set `apop_opts.db_name_column` to a column name that will be specially treated as holding row names for the sake of importing to an `apop_data` set.
- Lines 12–15: The final table will have three columns (pop, GDP, GDP/cap), so the query asks for three columns, one of which is filled with ones. This is known as *planning ahead*: it is difficult to resize `gsl_matrixes` and `apop_data` sets, so we query a table of the appropriate size, and then fill the column of dummy data with correct values in the C-side matrix.

Data to db   To go from C-side matrices to database-side tables, there are the plain old print functions like `apop_data_print` and `apop_matrix_print`. Lines 18–19 of Listing 3.4 will write the data table to a table named `wbtodata_output`. Say that tomorrow you decide you would prefer to have the data dumped to a file; then just change the `'d'` to an `'f'` and away you go.

Crosstabs   In the spreadsheet world, we often get tables in a form where both the X-
            and Y-dimensions are labeled, such as the case where the X-dimension
is the year, the Y-dimension is the location, and the (x, y) point is a measurement
taken that year at that location.

Conversely, the most convenient form for this data in a database is three columns:
year, location, statistic. After all, how would you write a query such as `select
statistic from tab where year < 1990` if there were a separate column for
each year? Converting between the two forms is an annoyance, and so Apophenia
provides functions to do conversions back and forth, `apop_db_to_crosstab` and
`apop_crosstab_to_db`.

Imagine a data table with two columns, `height` and `width`, where `height` may
take on values like up, `middle`, or down, and `width` takes on values like `left` and
`right`. Then the query

```
create table anovatab as
    select height, width, count(*) as ct
    group by height, width
```

will produce a table looking something like

height	width	ct
up	left	12
up	right	18
middle	left	10
middle	right	7
down	left	6
down	right	18

Then, the command

```
apop_data *anova_tab = apop_db_to_crosstab("anovatab", "height", "width", "ct");
```

will put into `anova_tab` data of the form

	Left	Right
Up	12	18
Middle	10	7
Down	6	18

You can print this table as a summary, or use it to run ANOVA tests, as in Section 9.4. The `apop_crosstab_to_db` function goes in the other direction; see the online reference for details.

$Q_{3.13}$ | Use the command-line program `apop_db_to_crosstab` (or the corresponding C function) and the `data-climate.db` database to produce a table of temperatures, where each row is a year and each column a month. Import the output into your favorite spreadsheet program.

Multiple databases   For both SQL and C, the dot means *subelement*. Just as a C struct named `person` might have a subelement named `person.height`, the full name of a column is `dbname.tablename.colname`.

The typical database system (including mySQL and SQLite) begins with one database open, which always has the alias `main`, but allows you to attach additional databases. For SQLite, the syntax is simply `attach database "newdb.db" as dbalias`; after this you can refer to tables via the `dbalias.tablename` form. For mySQL, you don't even need the `attach` command, and can refer to tables in other mySQL databases using the `dbname.tablename` form at any time.

Aliases again help to retain brevity. Instead of using the full `db.table.col` format for a column, this query assigns aliases for the `db.table` parts in the `from` clause, then uses those alises in the `select` clause:

```
attach database newdb as n;
select t1.c1, t2.c2
from main.firsttab t1, n.othertab t2
```

Given two attached databases, say `main` and `new`, you could easily copy tables between them via

```
create table new.tablecopy
    as select * from main.origial
```

Apophenia also provides two convenience functions, `apop_db_merge` and `apop_db_merge_table`, which facilitate such copying.

In-memory databases are faster, but at the close of the program, you may want the database on the hard drive. To get the best of both worlds, use an in-memory database for the bulk of the work, and then write the database to disk at the end of the program, e.g.:

```
int main(void){
    apop_db_open(NULL); //open a db in memory.
    do_hard_math(...);
    remove("on_disk.db");
    apop_db_merge("on_disk.db");
}
```

- `remove` is the standard C library function to delete a file.
- Removing the file before merging prevented the duplication of data (because duplicate tables are appended to, not overwritten).

➤ Open an SQLite database in memory using `apop_db_open(NULL)`, and on the hard drive using `apop_db_open("filename")`.

➤ Import data using `apop_text_to_db`.

➤ If you don't need output, use `apop_query` to send queries to the database engine.

➤ Use `apop_query_to_(data|matrix|vector|text|float)` to write a query result to various formats.

3.6 MADDENING DETAILS Data are never as clean as it seems in the textbooks, and our faster computers have done nothing to help the fact that everybody has different rules regarding how data should be written down. Here are a few tips on dealing with some common frustrations of data importation and use; Appendix B offers a few more tools.

Spaces in column names Column names should be short and have no punctuation but underscores. Instead of a column name like `Percent of male treatment 1 cases showing only signs of nausea`, give a brief name like `male_t1_moderate`, and then create a documentation table that describes exactly what that abbreviation means.

Not everybody follows this advice, however, which creates a small frustration. The query `select 'percent of males treatment 1' from data` will produce a table with the literal string `percent of males treatment 1` repeated for each row, which is far from what you meant. The solution is to use the dot notation to specify a table: `select data.'percent of males treatment 1'` as `males_t1 from data` will correctly return the data column, and give it an alias that is much easier to use.

Text and numbers   In some cases, you need both text and numeric data in the same data set. As you will see in the next chapter, the `apop_data` structure includes slots for both text and numbers, so you only need to specify which column goes where. The first argument to the `apop_query_to_mixed_-data` function is a specifier consisting of the letters n, v, m, t, indicating whether each column should be read in to the output `apop_data`'s name, vector, a matrix column, or a text column. For example, `apop_query_to_mixed_data("nmt", "select a, b*%i, c from data", counter)` would use column a as the row names, b*counter as the first column of the matrix, and c as a column of text elements. This provides maximal flexibility, but requires knowing exactly what the query will output.[13]

Now that you have text in an `apop_data` set, what can you do with it? In most cases, the data will be unordered discrete data, and the only thing you can do with it is to turn it into a series of dummy variables. See page 123 for an example.

Missing data   Everybody represents missing data differently. SQLite uses `NULL` to indicate missing data; Section 4.5 will show that real numbers in C can take `NAN` values, whose use is facilitated by the GSL's `GSL_NAN` macro. The typical input data set indicates a missing value with a text marker like NaN, .., -, -1, NA, or some other arbitrary indicator.

When reading in text, you can set `apop_opts.db_nan` to a regular expression that matches the missing data marker. If you are unfamiliar with regular expressions, see Appendix B for a tutorial. For now, here are some examples:

```
//Apophenia's default NaN string, matching NaN, nan, or NAN:
strcpy(apop_opts.db_nan, "NaN");
//Literal text:
strcpy(apop_opts.db_nan, "Missing");
//Matches two periods. Periods are special in regexes, so they need backslashes.
strcpy(apop_opts.db_nan, "\\.\\.");
```

---

[13]Why doesn't Apophenia automatically detect the type of each column? Because it stresses replicability, and it is impossible to replicably guess column types. One common approach used by some stats packages is to look at the first row of data and use that to cast the entire column, but if the first element in a column is NAN, then numeric data may wind up as text or vice versa, depending on arbitrary rules. The system could search the entire column for text and presume that some count of text elements means the entire column is text, but this too is error-prone. Next month, when the new data set comes in, columns that used to be auto-typed as text may now be auto-typed as numbers, so scripts written around the first data set break. Explicitly specifying types may take work, but outguessing the system's attempts at cleaning real-world data frequently takes more work.

The searched-for text must be the entire string, plus or minus surrounding quotation marks or white space. None of these will match NANCY or missing persons.

Once the database has a NULL in the right place, Apophenia's functions to read between databases on one side and gsl_matrixes, apop_data, and other C structures on the other will translate between database NULLs and floating-point GSL_-NANs.

Mathematically, any operation on unknown data produces an unknown result, so you will need to do something to ensure that your data set is complete before making estimations based on the data. The naïve approach is to simply delete every observation that is not complete. Allison (2002) points out that this naïve approach, known in the jargon as *listwise deletion*, is a somewhat reasonable approach, especially if there is no reason to suspect that the pattern of missing data is correlated to the dependent variable in your study.[14] Missing data will be covered in detail on page 345.

Implementing listwise deletion in SQL is simple: given *datacol1* and *datacol2*, add a where *datacol1* is not null and *datacol2* is not null clause to your query. If both are numeric data, then you can even summarize this to where (*datacol1* + *datacol2*) is not null.

Q3.14 Using the above notes and the data-tattoo.db file, query to an apop_data set the number of tattoos, number of piercings, and the political affiliation of each subject. Make sure that all NaNs are converted to zeros at some point along the chain. Print the table to screen (via apop_data_show) to make sure that all is correctly in place. Then, query out a list of the political parties in the data set. (*Hint*: select distinct.) Write a for loop to run through the list, finding the mean number of tattoos and piercings for Democrats, Republicans, ... Would you keep the last person in the survey (who has far more tattoos than anybody else) or eliminate the person as an outlier, via a where clause restricting the tattoo count to under 30?

Outer join Another possibility is that a row of data is entirely missing from one table. The World Bank database includes a lonely_planet table listing the number of pages in the given country's Lonely Planet tourist guidebook. Antarctica has a 328-page guidebook, but no GDP and a negligible population, so the query

---

[14]Systematic relationships between missingness and the independent variables is much less of a concern.

```
select pp, gdp
    from lonely_planet lp, gdp
    where lp.country=gdp.country
```

will not return an Antarctica line, because there is no corresponding line in the
gdp table. The solution is the outer join, which includes all data in the first
table, plus data from the second table or a blank if necessary. Here is a join that
will include Antarctica in its output. The condition for joining the two tables (join
on l.country=gdp.country) now appears in a different location from the norm,
because the entire left outer join clause describes a single table to be used as a data
source.

```
select pp, gdp
    from lonely_planet lp left outer join gdp
        on l.country=gdp.country
    where l.country like 'A%'
```

Q3.15 The query above is a *left outer join*, which includes all data from the left
table, but may exclude data from the right table. As of this writing, this
is all that SQLite supports, but other systems also support the *right outer
join* (include all entries in the right table) and the *full outer join* (include all
entries from both tables).
Using the union keyword, generate a reference table with all of the country
names from both the Lonely Planet and GDP tables. Then use a few left
outer joins beginning with the reference table to produce a complete data
set.

※ *mySQL*   As well as SQLite, Apophenia supports mySQL. mySQL is somewhat
better for massive data sets, but will work only if you already have a
mySQL server running, have permission to access it, and have a database in place.
Your package manager will make installing the mySQL server, client, and development libraries easy, and mySQL's maintainers have placed online a comprehensive
manual with tutorial.

Once mySQL is set up on your system, you will need to make one of two changes:
either set your shell's APOP_DB_ENGINE environment variable to mysql,[15] or in
your code, set apop_opts.db_engine='m'. You can thus switch back and forth
between SQLite and mySQL; if the variable is 'm' then any database operations
will go to the mySQL engine and if it is not, then database operations will be sent

---

[15]As discussed in Appendix A, you will probably want to add export APOP_DB_ENGINE=mysql to your
.bashrc on systems using mySQL.

to the SQLite engine. This could be useful for transferring data between the two. For example:

```
apop_opts.db_engine = 'm';
apop_db_open("mysqldb");
apop_data *d = apop_query_to_data("select * from get_me");
apop_opts.db_engine = 'l';
apop_db_open("sqlitedb");
apop_opts.output_type = 'd'; //print to database.
apop_data_print(d, "put_me");
```

SQLite's concept of a database is a single file on the hard drive, or a database in memory. Conversely mySQL has a server that stores all databases in a central repository (whose location is of no concern to end-users). It has no concept of an in-memory database.

As noted above, every SQL system has its own rules for metatadata. From the `mysql` prompt, you can query the mySQL server for a complete list of databases with `show databases`, and then attach to one using `use dbname`; (or type `mysql dbname` at the command prompt to attach to `dbname` from the outset). You can use `show tables`; to get the list of tables in the current database (like the SQLite prompt's `.tables` command), or use `show tables from your_db`; to see the tables in `your_db` without first attaching to it. Given a table, you can use `show columns from your_table` to see the column names of `your_table`.[16]

mySQL digresses from the SQL standard in different manners from SQLite's means of digressing from the standard:

- SQLite is somewhat forgiving about details of punctuation, such as taking == and = as equivalent, and "double-ticks" and 'single-ticks' as equivalent. mySQL demands a single = and 'single-ticks'.

- After every `select`, `create`, and so on, mySQL's results need to be internally processed, lest you get an error about commands executed out of order. Apophenia's functions handle the processing for you, but you may still see odd effects when sending a string holding multiple semicolon-separated queries to the `apop_query...` functions. Similarly, you may have trouble using `begin`/`commit` wrappers to bundle queries, though mySQL's internal cache management may make such wrappers unnecessary.

- mySQL includes many more functions beyond the SQL standard, and has a number of additional utilities. For example, there is a `LOAD` command that will read in a text file much more quickly than `apop_text_to_db`.

---

[16]Or, use the command-line program `mysqlshow` to do all of these things in a slightly more pleasant format.

> ➤ SQL represents missing data via a NULL marker, so queries may include conditions like `where col is not null`.

> ➤ Data files use whatever came to mind to mark missing data, so set `apop_opts.db_nan` to a regular expression appropriate for your data.

> ➤ If a name appears in one table but not another, and you would like to joint tables by name, use the `outer join` to ensure that all names appear.

3.7 SOME EXAMPLES   Here are a few examples of how C code and SQL calls can neatly interact.

TAKING SIMULATION NOTES   Say that you are running a simulation and would like to take notes on its state each period. The following code will open a file on the hard drive, create a table, and add an entry each period. The begin-commit wrapper puts data in chunks of 10,000 elements, so if you get tired of waiting, you can halt the program and walk away with your data to that point.[17]

```
double sim_output;
apop_db_open("sim.db");
apop_table_exists("results", 1); //See below.
apop_query("create table results (period, output); begin;");
for (int i=0; i< max_periods; i++){
    sim_output = run_sim(i);
    apop_query("insert into results values(%i, %g);", i, sim_output);
    if (!(i%1e4))
        apop_query("commit; begin;");
}
apop_query("commit;");
apop_db_close(0);
```

• The `apop_table_exists` command checks for whether a table already exists. If the second argument is one, as in the example above, then the table is deleted so that it can be created anew subsequently; if the second argument is zero, then the function simply returns the answer to the question "does the table exist?" but leaves the table intact if it is there. It is especially useful in `if` statements.

• Every 1e4 entries, the system commits what has been entered so far and begins a new batch. With some SQLite systems, this can add significant speed. mySQL

---

[17]Sometimes such behavior will leave the database in an unclean state. If so, try the SQLite command `vacuum`.

does its own batch management, so the `begins` and `commits` should be omitted for mySQL databases.

EASY T-TESTS   People on the East and West coasts of the United States sometimes joke that they can't tell the difference between all those states in the middle. This is a perfect chance for a t test: are incomes in North Dakota significantly different from incomes in South Dakota? First, we will go through the test algorithm in English, and then see how it is done in code.

Let the first data set be the income of counties in North Dakota, and let the second be the income of counties in South Dakota. If $\hat{\mu}$, $\hat{\sigma}^2$, and n are the estimated mean, variance, and actual count of elements of the North and South data sets,

$$\texttt{stat} = \frac{\hat{\mu}_N - \hat{\mu}_S}{\sqrt{\hat{\sigma}_N^2/n_N + \hat{\sigma}_S^2/n_S}} \sim t_{n_N+n_S-2}. \qquad (3.7.1)$$

[That is, the given ratio has a t distribution with $n_N + n_S - 2$ degrees of freedom.]

The final step is to look up this statistic in the standard t tables as found in the back of any standard statistics textbook. Of course, looking up data is the job of a computer, so we instead ask the GSL for the two-tailed confidence level (see page 305 for details):

```
double confidence = (1 − 2* gsl_cdf_tdist_Q(|stat|, nN + nS −2));
```

If `confidence` is large, say $> 95\%$, then we can reject the null hypothesis that North and South Dakotan incomes (by county) are different. Otherwise, there isn't enough information to say much with confidence.

Listing 3.5 translates the process into C.

- Lines 4–8 comprise two queries, that are read into a `gsl_vector`. Both ask for the same data, but one has a `where` clause restricting the query to pull only North Dakotan counties, and the other has a `where` clause restricting the query to South Dakota.
- Lines 10–15 get the vital statistics from the vectors: count, mean, and variance.
- Given this, line 17 is the translation of Equation 3.7.1.
- Finally, line 18 is the confidence calculation from above, which line 19 prints as a percentage.

```
 1   #include <apop.h>
 2
 3   int main(){
 4       apop_db_open("data−census.db");
 5       gsl_vector *n = apop_query_to_vector("select in_per_capita from income "
 6               "where state= (select state from geography where name ='North Dakota')");
 7       gsl_vector *s = apop_query_to_vector("select in_per_capita from income "
 8               "where state= (select state from geography where name ='South Dakota')");
 9
10       double n_count = n−>size,
11               n_mean = apop_vector_mean(n),
12               n_var = apop_vector_var(n),
13               s_count = s−>size,
14               s_mean = apop_vector_mean(s),
15               s_var = apop_vector_var(s);
16
17       double stat = fabs(n_mean − s_mean)/ sqrt(n_var/ (n_count−1) + s_var/(s_count−1));
18       double confidence = 1 − (2 * gsl_cdf_tdist_Q(stat, n_count + s_count −2));
19       printf("Reject the null with %g%% confidence\n", confidence*100);
20   }
```

Listing 3.5 Are North Dakota incomes different from South Dakota incomes? Answering the long
way. Online source: `ttest.long.c`.

No, easier   But this is not quite as easy as it could be, because Apophenia provides
a high-level function to do the math for you, as per Listing 3.6. The
code is identical until line eight, but then line nine calls the `apop_t_test` function,
which takes the two vectors as input, and returns an `apop_data` structure as output,
listing the relevant statistics. Line ten prints the entire output structure, and line
eleven selects the single confidence statistic regarding the two-tailed hypothesis
that income$_{ND} \neq$ income$_{SD}$.

DUMMY VARIABLES   The `case` command is the if-then-else of SQL. Say that you
have data that are true/false or yes/no. One way to turn this
into a one-zero variable would be via the `apop_data_to_dummies` function on
the matrix side. This works partly because of our luck that $y > n$ and $T > F$ in
English, so y and T will map to one and n and F will map to zero. But say that our
survey used *affirmative* and *negative*, so the mapping would be backward from our
intuition. Then we can put a `case` statement in with the other column definitions
to produce a column that is one when `binaryq` is affirmative and zero otherwise:

```
select id,
    case binaryq when "affirmative" then 1 else 0 end,
    other_vars
from datatable;
```

```
 1   #include <apop.h>
 2
 3   int main(){
 4       apop_db_open("data−census.db");
 5       gsl_vector *n = apop_query_to_vector("select in_per_capita from income "
 6               "where state= (select state from geography where name ='North Dakota')");
 7       gsl_vector *s = apop_query_to_vector("select in_per_capita from income "
 8               "where state= (select state from geography where name ='South Dakota')");
 9       apop_data *t = apop_t_test(n,s);
10       apop_data_show(t); //show the whole output set...
11       printf ("\n confidence: %g\n", apop_data_get_ti(t, "conf.*2 tail", −1)); //...or just one value.
12   }
```

Listing 3.6 Are North Dakota incomes different from South Dakota incomes? Online source: `ttest.c`.

To take this to the extreme, we can turn a variable that is discrete but not or-dered (such as district numbers in the following example) into a series of dummy variables. It requires writing down a separate `case` statement for each value the variable could take, but that's what `for` loops are for. [Again, this is demonstration code. Use `apop_data_to_dummies` to do this in practice.] Listing 3.7 creates a series of dummy variables using this technique.

- On lines 5–6, the `build_a_query` function queries out the list of districts.
- Then the query writes a select statement with a line `case State when` *state_-name* `then 1 else 0` for every *state_name*.
- Line 11 uses the obfuscatory if (page 211) to print a comma between items, but not at the end of the `select` clause.
- Line 18 pulls the data from this massive query, and line 19 runs an OLS regression on the returned data.
- You can set `apop_opts.verbose=1` at the head of `main` to have the function dis-play the full query as it executes.
- Lines 20–21 show the parameter estimates, but suppress the gigantic variance–covariance matrix.

Note well that the `for` loop starting on line eight goes from i=1, not i=0. When including dummy variables, you always have to exclude one baseline value to pre-vent **X** from being singular; excluding i=0 means Alabama will be the baseline. ℚ: Rewrite the `for` loop to use another state as a baseline. Or, set the `for` loop to run the full range from zero to the end of the array, and watch disaster befall the analysis.

```
1   #include <apop.h>
2
3   char *build_a_query(){
4       char *q = NULL;
5       apop_data *state = apop_query_to_text("select Name as state, State as id \
6                       from geography where sumlevel+0.0 = 40");
7       asprintf(&q, "select in_per_capita as income, ");
8       for (int i=1; i< state->textsize[0]; i++)
9           asprintf(&q, "%s (case state when '%s' then 1 else 0 end) '%s' %c \n",
10                      q, state->text[i][1], state->text[i][0],
11                      (i< state->textsize[0]-1) ? ',':' ');
12      asprintf(&q,"%s from income\n", q);
13      return q;
14  }
15
16  int main(){
17      apop_db_open("data-census.db");
18      apop_data *d = apop_query_to_data(build_a_query());
19      apop_model *e = apop_estimate(d, apop_ols);
20      e->covariance = NULL; //don't show it.
21      apop_model_show(e);
22  }
```

Listing 3.7 A sample of a `for` loop that creates SQL that creates dummy variables. Online source: `statedummies.c`.

➤ There is no standard for `for` loops, assigning variables, or matrix-style manipulation within SQL, so you need to do these things on the C-side of your analysis.

➤ Functions exist to transfer data between databases and matrices, so you can incorporate database-side queries directly into C code.

MATRICES AND MODELS

My freedom thus consists in moving about within the narrow frame that I have assigned myself for each one of my undertakings.... Whatever diminishes constraint diminishes strength. The more constraints one imposes, the more one frees one's self of the chains that shackle the spirit.

—Stravinsky (1942, p 65)

Recall that the C language provides only the most basic of basics, such as addition and division, and everything else is provided by a library. So before you can do data-oriented mathematics, you will need a library to handle matrices and vectors.

There are many available; this book uses the GNU Scientific Library (*GSL*). The GSL is recommended because it is actively supported and will work on about as many platforms as C itself. Beyond functions useful for statistics, it also includes a few hundred functions useful in engineering and physics, which this book will not mention. The full reference documentation is readily available online or in book form (Gough, 2003). Also, this book co-evolved with the Apophenia library, which builds upon the GSL for more statistics-oriented work.

This chapter goes over the basics of dealing with the GSL's matrices and vectors. Although insisting that matrices and vectors take on a specific, rigid form can be a constraint, it is the constraint that makes productive work possible. The predictable form of the various structures makes it is easy to write functions that allocate and fill them, multiply and invert them, and convert between them.

4.1   THE GSL'S MATRICES AND VECTORS   Quick—what's 14 times 17?
Thanks to calculators, we are
all a bit rusty on our multiplication, so Listing 4.1 produces a multiplication table.

```
1   #include <apop.h>
2
3   int main(){
4      gsl_matrix *m = gsl_matrix_alloc(20,15);
5      gsl_matrix_set_all(m, 1);
6      for (int i=0; i< m−>size1; i++){
7           Apop_matrix_row(m, i, one_row);
8           gsl_vector_scale(one_row, i+1);
9      }
10     for (int i=0; i< m−>size2; i++){
11          Apop_matrix_col(m, i, one_col);
12          gsl_vector_scale(one_col, i+1);
13     }
14     apop_matrix_show(m);
15     gsl_matrix_free(m);
16  }
```

Listing 4.1 Allocate a matrix, then multiply each row and each column by a different value to
produce a multiplication table. Online source: `multiplicationtable.c`.

- The matrix is allocated in the introductory section, on line four. It is no surprise that it has `alloc` in the name, giving indication that memory is being allocated for the matrix. In this case, the matrix has 20 rows and 15 columns. Row always comes first, then Column, just like the order in Roman Catholic, Randy Choirboy, or RC Cola.

- Line five is the first matrix-level operation: set every element in the matrix to one.

- The rest of the file works one row or column at a time. The first loop, from lines six to nine, begins with the `Apop_matrix_row` macro to pull a single row, which it puts into a vector named `one_row`.

- Given the vector `one_row`, line eight multiplies every element by `i+1`. When this happens again by columns on line 12, we have a multiplication table.

- Line 14 displays the constructed matrix to the screen.

- Line 15 frees the matrix.[1] The system automatically frees all matrices at the end of the program. Some consider it good style to free matrices and other allocated memory anyway; others consider freeing at the end of `main` to be a waste of time.

---

[1] Due to magic discussed below, vectors allocated by `Apop_matrix_row` and `_col` do not really exist and do not need to be freed.

Naming conventions   Every function in the GSL library will begin with `gsl_`, and the first argument of all of these functions will be the object to be acted upon. Most GSL functions that affect a matrix will begin with `gsl_matrix_` and most that operate on vectors begin with `gsl_vector_`. The other libraries used in this book stick to such a standard as well: 100% of Apophenia's functions begin with `apop_` and a great majority of them begin with a data type such as `apop_data_` or `apop_model_`, and GLib's functions all begin with `g_object`: `g_tree_`, `g_list_`, et cetera.[2]

This custom is important because C is a general-purpose language, and the designers of any one library have no idea what other libraries authors may be calling in the same program. If two libraries both have a function named `data_alloc`, then one will break.

C's library-loaded matrix and vector operations are clearly more verbose and redundant than comparable operations in languages that are purpose-built for matrix manipulation. But C's syntax does provide a few advantages—notably that it is verbose and redundant. As per the discussion of debugging strategy on page 46, spacing out the operations can make debugging numerical algorithms less painful. When there is a type name in the function name, there is one more clue in the function call itself whether you are using the function correctly.

The authors of the Mathematica package chose not to use abbreviations; here is their answer to the question of why, which applies here as well:

> The answer... is consistency. There is a general convention ... that all function names are spelled out as full English words, unless there is a standard mathematical abbreviation for them. The great advantage of this scheme is that it is *predictable*. Once you know what a function does, you will usually be able to guess exactly what its name is. If the names were abbreviated, you would always have to remember which shortening of the standard English words was used. (Wolfram, 2003, p 35)

The naming convention also makes indices very helpful. For example, the index of the GSL's online reference gives a complete list of functions that operate on vectors alphabetized under `gsl_vector_...`, and the index of this book gives a partial list of the most useful functions.

---

[2]There is one awkward detail to the naming scheme: some functions in the Apophenia library act on `gsl_matrixes` and `gsl_vectors`. Those have names beginning with `apop_matrix` and `apop_vector`, compromising between the library name and the name of the main input.

Don't delay—have a look at the `gsl_vector_...` and `gsl_matrix_...` sections of the index to this book or the GSL's online reference and skim over the sort of operations you can do. The Apophenia package has a number of higher-level operations that are also worth getting to know, so have a look at the `apop_vector_...`, `apop_matrix_...`, and `apop_data_...` sections as well.

If you find the naming scheme to be too verbose, you can write your own wrapper functions that require less typing. For example, you could write a file `my_-convenience_fns.c`, which could include:

```
void mset(gsl_matrix *m, int row, int col, double data){
    gsl_matrix_set(m, row, col, data);
}

void vset(gsl_vector *v, int row, double data){
    gsl_vector_set(v, row, data);
}
```

You would also need a header file, `my_convenience_fns.h`:

```
#include <gsl/gsl_matrix.h>
#include <gsl/gsl_vector.h>
void mset(gsl_matrix *m, int row, int col, double data);
void vset(gsl_vector *v, int row, double data);

#define VECTOR_ALLOC(vname, length) gsl_vector *vname = gsl_vector_alloc(length);

// For simple functions, you can rename them via #define; see page 212:
#define vget(v, row) gsl_vector_get(v, row)
#define mget(m, row, col) gsl_matrix_get(m, row, col)
```

After throwing an `#include "my_convenience_fns.h"` at the top of your program, you will be able to use your abbreviated syntax such as `vget(v,3)`. It's up to your æsthetic as to whether your code will be more or less legible after you make these changes. But the option is always there: if you find a function's name or form annoying, just write a more pleasant wrapper function for your personal library that hides the annoying parts.

BASIC MATRIX AND VECTOR OPERATIONS   The simplest operations on matrices and vectors are element-by-element operations such as adding the elements of one matrix to those of another. The GSL provides the functions you would expect to do such things. Each modifies its first argument.

gsl_matrix_add (a,b);   // $a_{ij} \leftarrow a_{ij} + b_{ij}, \forall\, i, j$
gsl_matrix_sub (a,b);   // $a_{ij} \leftarrow a_{ij} - b_{ij}, \forall\, i, j$
gsl_matrix_mul_elements (a,b);   // $a_{ij} \leftarrow a_{ij} \cdot b_{ij}, \forall\, i, j$
gsl_matrix_div_elements (a,b);   // $a_{ij} \leftarrow a_{ij}/b_{ij}, \forall\, i, j$
gsl_matrix_scale (a,x);   // $a_{ij} \leftarrow a_{ij} \cdot x, \forall\, i, j \in \mathbb{N},\, x \in \mathbb{R}$
gsl_matrix_add_constant (a,x);   // $a_{ij} \leftarrow a_{ij} + x, \forall\, i, j \in \mathbb{N},\, x \in \mathbb{R}$

gsl_vector_add (a,b);   // $a_i \leftarrow a_i + b_i, \forall\, i$
gsl_vector_sub (a,b);   // $a_i \leftarrow a_i - b_{ij}, \forall\, i$
gsl_vector_mul (a,b);   // $a_i \leftarrow a_i \cdot b_i, \forall\, i$
gsl_vector_div (a,b);   // $a_i \leftarrow a_i/b_i, \forall\, i$
gsl_vector_scale (a,x);   // $a_i \leftarrow a_i \cdot x, \forall\, i \in \mathbb{N},\, x \in \mathbb{R}$
gsl_vector_add_constant (a,x);   // $a_i \leftarrow a_i + x, \forall\, i \in \mathbb{N},\, x \in \mathbb{R}$
apop_vector_log(a);   // $a_i \leftarrow \ln(a_i), \forall\, i$
apop_vector_log10(a);   // $a_i \leftarrow \log_{10}(a_i), \forall\, i$
apop_vector_exp(a);   // $a_i \leftarrow e^{a_i}, \forall\, i$

The functions to multiply and divide matrix elements are given slightly lengthier names to minimize the potential that they will be confused with the process of multiplying a matrix with another matrix, $\mathbf{AB}$, or its inverse, $\mathbf{AB}^{-1}$. Those operations require functions with more computational firepower, introduced below.

Q4.2

Rewrite the structured birthday paradox program from page 35 using a `gsl_matrix` instead of the `struct` that it currently uses.

- `alloc` or `calloc` the matrix in `main`; pass it to both functions.

- Replace the `#include` directives to call in `apop.h`.

- Replace everything after the title-printing line in `print_days` with `apop_matrix_show(`*`data_matrix`*`)`.

- Put three `gsl_matrix_set` commands in the `for` loop of `calculate_days` to set the number of people, likelihood of matching the first, and likelihood of any match (as opposed to one minus that likelihood, as in `bdayfns.c`).

Apply and map   Beyond the simple operations above, you will no doubt want to transform your data in more creative ways. For example, the function in Listing 4.2 will take in a `double` indicating taxable income and will return US income taxes owed, assuming a head of household with two dependents taking the standard deduction (as of 2006; see Internal Revenue Service (2007)). This function can be applied to a vector of incomes to produce a vector of taxes owed.

```
1    #include <apop.h>
2
3    double calc_taxes(double income){
4        double cutoffs[] = {0, 11200, 42650, 110100, 178350, 349700, INFINITY};
5        double rates[] = {0, 0.10, .15, .25, .28, .33, .35};
6        double tax = 0;
7        int bracket = 1;
8        income −= 7850; //Head of household standard deduction
9        income −= 3400*3; //exemption: self plus two dependents.
10       while (income > 0){
11           tax += rates[bracket] * GSL_MIN(income, cutoffs[bracket]−cutoffs[bracket−1]);
12           income −= cutoffs[bracket];
13           bracket ++;
14       }
15       return tax;
16   }
17
18   int main(){
19       apop_db_open("data−census.db");
20       strncpy(apop_opts.db_name_column, "geo_name", 100);
21       apop_data *d = apop_query_to_data("select geo_name, Household_median_in as income\
22                           from income where sumlevel = '040'\
23                           order by household_median_in desc");
24       Apop_col_t(d, "income", income_vector);
25       d−>vector = apop_vector_map(income_vector, calc_taxes);
26       apop_name_add(d−>names, "tax owed", 'v');
27       apop_data_show(d);
28   }
```

Listing 4.2 Read in the median income for each US state and find the taxes a family at the median
would owe. Online source: `taxes.c`.

- Lines 24–27 of Listing 4.2 demonstrate the use of the `apop_data` structure, and will be explained in detail below. For now, it suffices to know that line 24 produces a `gsl_vector` named `income_vector`, holding the median household income for each state.

- The bulk of the program is the specification of the tax rates in the `calc_taxes` function. In the exercise on page 192, you will plot this function.

- The program does not bother to find out the length of the arrays declared in lines four and five. The `cutoffs` array has a final value that guarantees that the `while` loop on lines 10–14 will exit at some point. Similarly, you can always add a final value like NULL or NAN[3] to the end of a list and rewrite your `for` loop's header to `for (int i=0; data[i] != NAN; i++)`. This means you have to remember to put the sentinel value at the end of the list, but do not need to remember to fix a counter every time you fix the array.

---

[3]NAN is read as not-a-number, and will be introduced introduced on page 135.

You could write a `for` loop to apply the `calc_tax` function to each element of the income vector in turn. But the `apop_vector_map` function will do this for you. Let $c()$ be the `calc_taxes` function, and `i` be the `income_vector`; then the call to `apop_vector_map` on line 25 returns $c(i)$, which is then assigned to the vector element of the data set, `d->vector`. Line 27 displays the output.

But `apop_vector_map` is just the beginning: Apophenia provides a small family of functions to map and apply a function to a data set. The full index of functions is relegated to the manual pages, but here is a list of examples to give you an idea.

> **Threading**
>
> Even low-end laptops ship with processors that are capable of simultaneously operating on two or more stacks of frames, so the `map` and `apply` functions can split their work among multiple processors. Set `apop_opts.thread_count` to the desired number of threads (probably the number of processor *cores* in your system), and these functions apportion work to processors appropriately.
>
> When threading, be careful writing to global variables: if a thousand threads could be modifying a global variable in any order, the outcome is likely undefined. When writing functions for threading, your best bet is to take all variables that were not passed in explicitly as read-only.

- You saw that `apop_vector_map(income_vector, calc_taxes)` will take in a `gsl_vector` and returns another vector. Or, `apop_vector_apply(income_vector, calc_taxes)` would replace every element of `income_vector` with `calc_taxes(element)`.

- One often sees functions with a header like `double log_likelihood(gsl_vector indata)`, which takes in a data vector and returns a log likelihood. Then if every row of *dataset* is a vector representing a separate observation, then `apop_matrix_map(dataset, log_likelihood)` would return the vector of log likelihoods of each observation.

- Functions with `..._map_..._sum`, like `apop_matrix_map_all_sum`, will return the sum of $f(\text{item})$ for every item in the matrix or vector. For example, `apop_matrix_map_all_sum(m, gsl_isnan)` will return the total number of elements of *m* that are NAN. Continuing the log likelihood example from above, `apop_matrix_map_sum(dataset, log_likelihood)` would be the total log likelihood of all rows.

- Another example from the family appeared earlier: Listing 3.4 (page 100) used `apop_matrix_apply` to generate a vector of GDP per capita from a matrix with GDP and population.

➤ You can express matrices and vectors via `gsl_matrix` and `gsl_vector` structures.

➤ Refer to elements using `gsl_matrix_set` and `gsl_matrix_get` (and similarly for `apop_data` sets and `gsl_vectors`).   ≫

	"v"	"c0"	"c1"	"c2"	"t0"	"t1"	"t2"
"r0"	$(0, -1)$	$(0, 0)$	$(0, 1)$	$(0, 2)$	$(0, 0)$	$(0, 1)$	$(0, 2)$
"r1"	$(1, -1)$	$(1, 0)$	$(1, 1)$	$(1, 2)$	$(1, 0)$	$(1, 1)$	$(1, 2)$
"r2"	$(2, -1)$	$(2, 0)$	$(2, 1)$	$(2, 2)$	$(2, 0)$	$(2, 2)$	$(2, 2)$

Figure 4.3 The vector is column -1 of the matrix, while the text gets its own numbering system. Row names are shared by all three elements.

➤ Once your data set is in these forms, you can operate on the matrix or vector as a whole using functions like `gsl_matrix_add(a,b)` or `gsl_vector_scale(a,x)`.

➤ Use the `apop_(matrix|vector)_(map|apply)` family of functions to send every row of a vector/matrix to a function in turn.

4.2 apop_data   The `apop_data` structure is the joining-together of four data types: the `gsl_vector`, `gsl_matrix`, a table of strings, and an `apop_name` structure.

The conceptual layout is given in Figure 4.3. The vector, columns of the matrix, and columns of text are all named. Also, all rows are named, but there is only one set of row names, because the presumption is that each row of the structure holds information about a single observation.

- Think of the vector as the -1st element of the matrix, and the text elements as having their own addresses.
- There are various means of creating an `apop_data` set, including `apop_query_-to_data`, `apop_matrix_to_data`, `apop_vector_to_data`, or creating a blank slate with `apop_data_alloc`; see below.
- For example, Listing 4.2 used `apop_query_to_data` to read the table into an `apop_data` set, and by setting the `apop_opts.db_name_column` option to a column name on line 20, the query set row names for the data. Line 25 sets the `vector` element of the data set, and line 26 adds an element to the `vector` slot of the `names` element of the set.
- You can easily operate on the subelements of the structure. If your *matrix_manipulate* function requires a `gsl_matrix`, but *your_data* is an apop_data structure, then you can call *matrix_manipulate(your_data->*matrix). Similarly,

you can manipulate the names and the table of text data directly. The size of the text data is stored in the `textsize` element. Sample usage:

```
apop_data *set = apop_query_to_text(...);
for (int r=0; r< set->textsize[0]; r++){
    for (int c=0; c< set->textsize[1]; c++)
        printf("%s\t", set->text[r][c]);
    printf("\n");
}
```

- There is no consistency-checking to make sure the number of row names, the `vector->size`, or `matrix->size1` are equal. If you want to put a vector of fifteen elements and a 10×10 matrix in the same structure, and name only the first two columns, you are free to do so. In fact, the typical case is that not all elements are assigned values at all. If *vector_size* is zero, then

```
apop_data *newdata_m = apop_data_alloc(vector_size, n_rows, n_cols);
```

will initialize most elements of `newdata` to NULL, but produce a `n_rows`×`n_cols` matrix with an empty set of names. Alternatively, if `n_rows == 0` but `vector_-size` is positive, then the vector element is initialized and the matrix set to NULL.[4]

Get, set, and point   You can use any of the GSL tools above to dissect the `gsl_-matrix` element of the `apop_data` struct, and similarly for the `vector` element. In addition, there is a suite of functions for setting and getting an element from an `apop_data` set using the names. Let t be a title and i be a numeric index; then you may refer to the row–column coordinate using the (i, i), (t, i), (i, t), or (t, t) form:

```
apop_data_get(your_data, i, j);
apop_data_get_ti(your_data, "rowname", j);
apop_data_get_it(your_data, i, "colname");
apop_data_get_tt(your_data, "rowname", "colname");
apop_data_set(your_data, i, j, new_value);
apop_data_set_ti(your_data, "rowname", j, new_value);
...
apop_data_ptr(your_data, i, j);
apop_data_ptr_ti(your_data, "rowname", j);
...
```

---

[4]Seasoned C programmers will recognize such usage as similar to a `union` between a `gsl_vector`, a `gsl_-matrix`, and a `char` array, though the `apop_data` set can hold both simultaneously. C++ programmers will observe that the structure allows a form of polymorphism, because you can write one function that takes an `apop_data` as input, but operates on one or both of a `gsl_vector` or a `gsl_matrix`, depending on which is not NULL in the input.

- The `apop_data_ptr...` form returns a pointer to the given data point, which you may read from, write to, increment, et cetera. It mimics the `gsl_matrix_ptr` and `gsl_vector_ptr` functions, which do the same thing for their respective data structures.

- As above, you can think about the vector as the -1st element of the matrix, so for example, `apop_data_set_ti(your_data, "rowname", -1)` will operates on the `apop_data` structure's vector rather than the matrix. This facilitates forms like `for (int i=-1; i< data->matrix->size2; i++)`, that runs across an entire row, including both vector and matrix.

- These functions use case-insensitive regular-expression matching to find the right name, so you can even be imprecise in your column request. Appendix B discusses regular expressions in greater detail; for now it suffices to know that you can be approximate about the name: `"p.val.*"` will match `P value`, `p-val` and `p.values`.

For an example, flip back to `ttest.c`, listed on page 111. Line ten showed the full output of the t test, which was a list of named elements, meaning that the output used the set's rownames and `vector` elements. Line eleven pulled a single named element from the vector.

⁂ *Forming partitioned matrices*   You can copy the entire data set, stack two data matrices one on top of the other (stack rows), stack two data matrices one to the right of the other (stack columns), or stack two data vectors:

```
apop_data *newcopy = apop_data_copy(oldset);
apop_data *newcopy_tall = apop_data_stack(oldset_one, oldset_two, 'r');
apop_data *newcopy_wide = apop_data_stack(oldset_one, oldset_two, 'c');
apop_data *newcopy_vector = apop_data_stack(oldset_one, oldset_two, 'v');
```

Again, you are generally better off doing data manipulation in the database. If the tables are in the database instead of `apop_data` sets the vertical and horizontal stacking commands above are equivalent to

```
select * from oldset_one
union
select * from oldset_two

/* and */

select t1.*, t2.*
   from oldset_one t1, oldset_two t2
```

The output of the exercise on page 105 is a table with tattoos, piercings, and political affiliation. Run a Probit regression to determine whether political affiliation affects the count of piercings.

- The function `apop_data_to_dummies` will produce a new data matrix with a column for all but the first category.

- Stack that matrix to the right of the original table.

- Send the augmented table to the `apop_probit.estimate` function. The output for the categorical variables indicates the effect relative to the omitted category.

- Encapsulate the routine in a function: using the code you just wrote, put together a function that takes in data and a text column name or number and returns an augmented data set with dummy variables.

➤ The `apop_data` structure combines a vector, matrix, text array, and names for all of these elements.

➤ You can pull named items from a data set (such as an estimation output) using `apop_data_get_ti` and family.

4.3   SHUNTING DATA

Igor Stravinsky, who advocated constraints at the head of this chapter, also points out that "Rigidity that slightly yields, like Justice swayed by mercy, is all the beauty of earth."[5] None of function to this point would make any sense if they did not operate on a specific structure like the `gsl_matrix` or `gsl_vector`, but coding is much easier when there is the flexibility of easily switching among the various constrained forms. To that end, this section presents suggestions for converting among the various data formats used in this book. It is not an exciting read (to say the least); you may prefer to take this section as a reference for use as necessary.

Table 4.4 provides the key to the method most appropriate for each given conversion. From/to pairs marked with a dot are left as an exercise for the reader; none are particularly difficult, but may require going through another format; for example, you can go from a `double[ ]` to an `apop_data` set via `double[ ]` $\Rightarrow$ `gsl_matrix` $\Rightarrow$ `apop_data`. As will be proven below, it is only two steps from any format to any other.

---

[5]Stravinsky (1942), p 54, citing GK Chesterton, "The furrows," in *Alarms and discursions*.

	To					
From	Text file	Db table	double[ ]	gsl_vector	gsl_matrix	apop_data
Text file	C	F	·	·	·	F
Db table	·	Q	·	Q	Q	Q
double[ ]	·	·	C	F	F	·
gsl_vector	P	P	F	C	F	F
gsl_matrix	P	P	F	V	C	F
apop_data	P	P	F	S	S	C

Table 4.4 A key to methods of conversion.

❊ Copying structures   The computer can very quickly copy blocks without bothering to comprehend what that data contains; the function to do this is memmove, which is a safe variant of memcpy. For example, borrowing the complex structure from Chapter 2:

```
complex first = {.real = 3, .imaginary = −1};
complex second;
memmove(&second, &first, sizeof(complex));
```

The computer will go to the location of first and blindly copy what it finds to the location of second, up to the size of one complex struct. Since first and second now have identical data, their constituent parts are guaranteed to also be identical.[6]

But there is one small caveat: if one element of the struct is a pointer, then it is the pointer that is copied, not the data itself (which is elsewhere in memory). For example, the gsl_vector includes a data pointer, so using memmove would result in two identical structs that both point to the same data. If you want this, use a view, as per Method V below; if you want a copy, then you need to memmove both the base gsl_vector and the data array. This sets the stage for the series of functions below with memcpy in the name that are modeled on C's basic memmove/memcpy functions but handle internal pointers correctly.

Method C: Copying   The gsl_..._memcpy functions assume that the destination to which you are copying has already been allocated; this allows you to reuse the same space and otherwise carefully oversee memory. The

---

[6]How to remember the order of arguments: computer scientists think in terms of data flowing from left to right: in C, dest = source; in R, dest <- source; in pseudocode, dest ← source. Similarly, most copying functions have the data flow from end of line to beginning: memmove(dest, source).

`apop_...._copy` functions allocate and copy in one step, so you can declare and copy on the same line, and more easily embed a copy into a filtering operation.

```
//Text file ⇒ Text file
//Just use the system's file copy command. The apop_system function acts like
//the standard C system command, but accepts printf-style arguments:
    apop_system("cp %s %s", from_file_name, to_file_name);
//gsl_vector ⇒ gsl_vector
    gsl_vector *copy = gsl_vector_alloc(original->size);
    gsl_vector_memcpy(copy, original);
    gsl_vector *copy2 = apop_vector_copy(original);
//double[ ] ⇒ double[ ]
//Let original_size be the length of the original array.[7]
    double *copy1 = malloc(sizeof(double) * original_size);
    memmove(copy1, original, sizeof(double) * original_size);
    double copy2[original_size];
    memmove(&copy2, original, sizeof(original));
//gsl_matrix ⇒ gsl_matrix
    gsl_matrix *copy = gsl_matrix_alloc(original->size1, original->size2);
    gsl_matrix_memcpy(copy, original);
    gsl_matrix *copy2 = apop_matrix_copy(original);
//apop_data ⇒ apop_data
    apop_data *copy1 = apop_data_alloc(original->vector->size, original->matrix->size1,
        original->matrix->size2);
    apop_data_memcpy(copy1, original);
    apop_data *copy2 = apop_data_copy(original);
```

Method F: Function call These are functions designed to convert one format to another.

There are two ways to express a matrix of `doubles`. The analog to using a pointer is to declare a list of pointers-to-pointers, and the analog to an automatically allocated array is to use double-subscripts:

```
double **method_one = malloc(sizeof(double*)*size_1);
for (int i=0; i< size_1; i++)
    method_one[i] = malloc(sizeof(double) * size_2);
double method_two[size_1][size_2] = {{2,3,4},{5,6,7}};
```

The first method is rather inconvenient. The second method seems convenient, because it lets you allocate the matrix at once. But due to minutiæ that will not be

---

[7]The `sizeof` function is not just for types: you can also send an array or other element to `sizeof`. If `original` is an array of 100 doubles, then `sizeof(original)=100*sizeof(double)`, while `sizeof(*original)=sizeof(double)`, and so you could use `sizeof(original)` as the third argument for `memmove`. However, this is incredibly error prone, because this is one of the few places in C where you could send either an object or a pointer to an object to the same function without a warning or error. In cases with modest complexity, the difference between an array and its first element can be easy to confuse and hard to debug.

discussed here (see Kernighan & Ritchie (1988, p 113)), that method is too much
of a hassle to be worth anything.

Instead, declare your data as a single line, listing the entire first row, then the second, et cetera, with no intervening brackets. Then, use the `apop_line...` functions to convert to a matrix. For another example, see page 9.

```
//text ⇒ db table
//The first number states whether the file has row names; the second
//whether it has column names. Finally, if no colnames are present,
//you can provide them in the last argument as a char **
    apop_text_to_db("original.txt", "tablename", 0 , 1, NULL);
//text ⇒ apop_data
    apop_data *copyd = apop_text_to_data("original.txt", 0 , 1);
//double[ ][ ] ⇒ gsl_vector,gsl_matrix
    double original[] = {{2,3,4}, {5,6,7}};
    gsl_vector *copv = apop_array_to_vector(original, original_size);
    gsl_matrix *copm = apop_array_to_matrix(original, original_size1, original_size2);

//double[ ] ⇒ gsl_matrix
    double original[] = {2,3,4,5,6,7};
    int orig_vsize = 0, orig_size1 = 2, orig_size2 = 3;
    gsl_matrix *copym = apop_line_to_matrix(original, orig_size1, orig_size2);
//double[ ] ⇒ apop_data
    apop_data *copyd = apop_line_to_data(original, orig_vsize, orig_size1, orig_size2);

//gsl_vector ⇒ double[ ]
    double *copyd = apop_vector_to_array(original_vec);
//gsl_vector ⇒ n × 1 gsl_matrix
    gsl_matrix *copym = apop_vector_to_matrix(original_vec);
//gsl_vector, gsl_matrix ⇒ apop_data
    apop_data *copydv = apop_vector_to_data(original_vec);
    apop_data *copydm = apop_matrix_to_data(original_matrix);
```

Method P: Printing   Apophenia's printing functions are actually four-in-one functions: you can dump your data to either the screen, a file, a database, or a system pipe [see Appendix B for an overview of pipes]. Early in putting together an analysis, you will want to print all of your results to screen, and then later, you will want to save temporary results to the database, and then next month, a colleague will ask for a text file of the output; you can make all of these major changes in output by changing one character in your code.

The four choices for the `apop_opts.output_type` variable are

```
apop_opts.output_type = 's'; //default: print to screen.
apop_opts.output_type = 'f'; //print to file.
apop_opts.output_type = 'd'; //store in a database table.
apop_opts.output_type = 'p'; //write to the pipe in apop_opts.output_pipe.
```

- The screen output will generally be human-readable, meaning different column sizes and other notes and conveniences for you at the terminal to understand what is going on. The file output will generally be oriented toward allowing a machine to read the output, meaning stricter formatting.

- The second argument to the output functions is a string. Output to screen or pipe ignores this; if outputting to file, this is the file name; if writing to the database, then this will be the table name.[8]

```
//gsl_vector, gsl_matrix, apop_data ⇒ text file
    apop_opts.output_type = 't'
    apop_vector_print(original_vector, "text_file_copy");
    apop_matrix_print(original_matrix, "text_file_copy");
    apop_data_print(original_data, "text_file_copy");
//gsl_vector, gsl_matrix, apop_data ⇒ db table
    apop_opts.output_type = 'd'
    apop_vector_print(original_vector, "db_copy");
    apop_matrix_print(original_matrix, "db_copy");
    apop_data_print(original_data, "db_copy");
```

Method Q: Querying   The only way to get data out of a database is to query it out.

```
//db table ⇒ db table
    apop_query("create table copy as \
    select * from original");
//db table ⇒ double, gsl_vector, gsl_matrix, or apop_data
    double d = apop_query_to_float("select value from original");
    gsl_vector *v = apop_query_to_vector("select * from original");
    gsl_matrix *m = apop_query_to_matrix("select * from original");
    apop_data *d = apop_query_to_data("select * from original");
```

Method S: Subelements   Sometimes, even a function is just overkill; you can just pull a subelement from the main data item.

Notice, by the way, that the `data` subelement of a `gsl_vector` can not necessarily be copied directly to a `double[ ]`—the *stride* may be wrong; see Section 4.6 for details. Instead, use the copying functions from Method F above.

```
//apop_data ⇒ gsl_matrix, gsl_vector
    my_data_set -> matrix
    my_data_set -> vector
```

---

[8]File names tend to have periods in them, but periods in table names produce difficulties. When printing to a database, the file name thus has its dots stripped: `out.put.csv` becomes the table name `out_put`.

Method V: Views   Pointers make it reasonably easy and natural to look at subsets of a matrix. Do you want a matrix that represents **X** with the first row lopped off? Then just set up a matrix whose `data` pointer points to the second row. Since the new matrix is pointing to the same data as the original, any changes will affect both matrices, which is often what you want; if not, then you can copy the submatrix's data to a new location.

However, it is not quite as easy as just finding the second row and pointing to it, since a `gsl_matrix` includes information about your data (i.e., *metadata*), such as the number of rows and columns. Thus, there are a few macros to help you pull a row, column, or submatrix from a larger matrix. For example, say that m is a `gsl_matrix*`, then

> Apop_matrix_row(m, 3, row_v);
> Apop_matrix_col(m, 5, col_v);
> Apop_submatrix(m, 2, 4, 6, 8, submatrix);

will produce a `gsl_vector*` named `row_v` holding the third row, another named `col_v` holding the fifth column, and a 6×8 `gsl_matrix*` named `submatrix` whose $(0, 0)$th element is at $(2, 4)$ in the original.

For an `apop_data` set, we have the names at our disposal, and so you could use either `Apop_row(m, 3, row_v)` and `Apop_col(m, 5, col_v)` to pull the given vectors from the matrix element of an `apop_data` structure using row/column number, or `Apop_row_t(m, "fourth row", row_v)` and `Apop_col_t(m, "sixth column", col_v)` to pull these rows and columns by their titles.

The macros work a bit of magic: they internally declare an automatically-allocated `gsl_matrix` or `vector` with the requisite metadata, and then declare a pointer with the name you selected, that can be used like any other pointer to a matrix or vector. However, because these macros used only automatically allocated memory, you do not need to free the matrix or vector generated by the macro. Thus, they provide a quick, disposable view of a portion of the matrix.[9] If you need a more permanent record, them copy the view to a regular vector or matrix using any of the methods from prior pages (e.g., `gsl_vector *permanent_copy = apop_-vector_copy(temp_view);`).

---

[9]These macros are based on GSL functions that are slightly less convenient. For example:
`gsl_vector v = gsl_matrix_col(a_matrix, 4).vector;`
`apop_vector_show(&v);`
If the macro seems to be misbehaving, as macros sometimes do, you can fall back on this form.

4.4 **LINEAR ALGEBRA**   Say that we have a transition matrix, showing whether
the system can go from a row state to a column state.
For example, Figure 4.4 was such a transition matrix, showing which formats can
be converted to which other formats.

Omitting the labels and marking each transition with a one and each dot in Figure
4.4 with a zero, we get the following transition matrix:

$$
\begin{bmatrix}
1 & 1 & 0 & 0 & 0 & 1 \\
0 & 1 & 0 & 1 & 1 & 1 \\
0 & 0 & 1 & 1 & 1 & 0 \\
1 & 1 & 1 & 1 & 1 & 1 \\
1 & 1 & 1 & 1 & 1 & 1 \\
1 & 1 & 1 & 1 & 1 & 1
\end{bmatrix}
$$

Listing 4.5 shows a brief program to read the data set from a text file, take the dot
product of t with itself, and display the result.

```
#include <apop.h>
int main(){
    apop_data *t = apop_text_to_data("data-markov", 0, 0);
    apop_data *out = apop_dot(t, t, 0, 0);
    apop_data_show(out);
}
```

Listing 4.5 Two transitions along a transition matrix. Online source: `markov.c`.

Before discussing the syntax of `apop_dot` in detail, here is the program's output:

$$
\begin{bmatrix}
2 & 3 & 1 & 2 & 2 & 3 \\
3 & 4 & 3 & 4 & 4 & 4 \\
2 & 2 & 3 & 3 & 3 & 2 \\
4 & 5 & 4 & 5 & 5 & 5 \\
4 & 5 & 4 & 5 & 5 & 5 \\
4 & 5 & 4 & 5 & 5 & 5
\end{bmatrix}
$$

This tells us, for example, that there are three ways to transition from the first state
to the second in two steps (you can verify that they are: $1 \Rightarrow 1 \Rightarrow 2$, $1 \Rightarrow 2 \Rightarrow 2$,
and $1 \Rightarrow 6 \Rightarrow 2$).

The `apop_dot` function takes up to four arguments: two `apop_data` structures, and
one flag for each matrix indicating what to do with it ('t'=transpose the matrix,
'v'=use the vector element, 0=use the matrix as-is). For example, if X is a matrix,
then

```
apop_dot(X, X, 't', 0);
```

will find $\mathbf{X}'\mathbf{X}$: the function takes the dot product of X with itself, and the first version is transposed and the second is not.

- If a data set has a `matrix` component, then it will be used for the dot product, and if the `matrix` element is NULL then the `vector` component is used.

- There should be exactly as many transposition flags as matrices. If the first element is a vector, it is always taken to be a row; if the second element is a vector, it is always a column. In both cases, if the other element is a matrix, you will need one flag to indicate whether to use the apop_data set's `vector` element (`'v'`), use the transposed matrix (`'t'`), or use the matrix as written (any other character).[10]

- If both elements are vectors, then you are probably better off just using `gsl_blas_ddot`, below, but if you use `apop_dot`, the output will be an apop_data set that has a vector element of length one.

$\mathbb{Q}_{4.4}$ The quadratic form $\mathbf{X}'\mathbf{Y}\mathbf{X}$ appears very frequently in statistical work. Write a function with the header apop_data *quadratic_form(apop_- data *x, apop_data *y); that takes two gsl_matrixes and returns the quadratic form as above. Be sure to check that y is square and has the same dimension as x->size1.

Vector · vector   Given two vectors **x** and **y**, `gsl_blas_ddot` returns $x_1 y_1 + x_2 y_2 + \cdots + x_n y_n$. Rather than outputting the value of **x·y** as the function's return value, it takes the location of a `double`, and places the output there. E.g., if x and y are `gsl_vector*`s, use

```
double dotproduct;
gsl_blas_ddot (x, y, &dotproduct);
```

---

[10]Why do you have to tell a computer whether to transpose or not? Some feel that if you send a 1 to indicate transposition when you meant 0 (or vice versa), the system should be able to determine this. Say that you have a 1×10 vector that you will multiply against three data sets, where the first is 8×10, the second is 15×10, and the third is 10×10. You write a simple `for` loop:
```
for(int i=0; i<3; i++)
        out[i] = apop_dot(data[i], v, 1);
```
At i=0, a 'smart' system realizes that you committed a faux pas: an 8×10 matrix dot a 10×1 column vector works without transposition. So it corrects you without telling you, and does the same with `data[1]`. With `data[2]`, the transposition works, since there are both ten rows and ten columns. So `out[0]` and `out[1]` are correct and `out[2]` is not. Good luck catching and debugging that.

Write a table displaying the sum of squares $1^2 + 2^2 + 3^2 + \cdots + n^2$ for $n = 1$ through 10.

$\mathbb{Q}_{4.5}$

- Write a function that takes in n and

    - allocates a `gsl_vector*` of size n,
    - fills the vector with $1, \ldots, n$,
    - calculates and returns $\mathbf{v} \cdot \mathbf{v}$ using `gsl_blas_ddot`,
    - and finally frees v.

- Write a loop for $n = 1$ through 10 that calls the above function and then prints n and the returned value.

- Verify your work, by printing $n(n + 1)(2n + 1)/6$ alongside your calculation of the sum of squares up to n.

An example: Cook's distance   Cook's distance is an estimate of how much each data point affects a regression (Cook, 1977). The formula is

$$C_i = \frac{\sum_j (\hat{y}_j^{\mathrm{r}} - \hat{y}_j^{\mathrm{ri}})^2}{p \cdot MSE}, \tag{4.4.1}$$

where p is the number of parameters, MSE is mean squared error for the overall regression, $\hat{y}_j^{\mathrm{r}}$ is the jth element of the predicted value of $\hat{y}$ based on the overall regression, and $\hat{y}_j^{\mathrm{ri}}$ is the jth element of the predicted value of $\hat{y}$ based on a regression excluding data point i. That is, to find Cook's distance for 3,000 data points, we would need to do a separate regression on 3,000 data sets, each excluding a different data point. The formula simply quantifies whether the predictions made by the main regression change significantly when excluding a given data point.

The procedure provides us a good opportunity to do some matrix-shunting and linear algebra, since we will need functions to produce the subsets, functions to calculate $\hat{y} = \mathbf{X}\hat{\beta}$, and to find the squared differences and MSE.

The first function is in Listing 4.6. It produces the series of data sets, each with one row missing. The function is named after the *jackknife* procedure, which uses the same delete-one loop for calculating covariances or correcting bias.[11]

- Lines 9–10 use a submatrix to produce a view of the main matrix starting at the

---

[11]The Jackknife is not discussed in this book; see the online documentation for Apophenia's `apop_-jackknife_cov`.

```
1   #include <apop.h>
2
3   typedef double (*math_fn)(apop_data *);
4
5   gsl_vector *jack_iteration(gsl_matrix *m, math_fn do_math){
6       int height = m->size1;
7       gsl_vector *out = gsl_vector_alloc(height);
8       apop_data *reduced = apop_data_alloc(0, height - 1, m->size2);
9       APOP_SUBMATRIX(m, 1, 0, height - 1, m->size2, mv);
10      gsl_matrix_memcpy(reduced->matrix, mv);
11      for (int i=0; i< height; i++){
12          gsl_vector_set(out, i, do_math(reduced));
13          if (i < height - 1){
14              APOP_MATRIX_ROW(m, i, onerow);
15              gsl_matrix_set_row(reduced->matrix, i, onerow);
16          }
17      }
18      return out;
19  }
```

Listing 4.6 Iteratively produce `in->size1` submatrices, each with one omitted row of data. Online source: `jackiteration.c`.

position (1,0), and with size (`m->size1 - 1`, `m->size2`)—that is, the original matrix with the first row missing—and then uses `gsl_matrix_memcpy` to copy that to a new matrix.

• The `for` loop then repeats the view-and-copy procedure row by row. It begins with row zero, which was omitted before, and overwrites row zero in the copy, aka row one of the original. It then copies over original row one, overwriting the copy of original row two, and so on to the end of the matrix.

• Line 12 calls the function which was sent in as an argument. See page 190 for notes on writing functions that take functions as inputs, including the meaning of the `typedef` on line 3.

Now that we have the matrix-shunting out of the way, Listing 4.7 provides additional functions to do the linear alegbra.

• The `sum_squared_diff` function calculates $\sum_i (L_i - R_i)^2$. The first line finds $L - R$, and the second line applies the function `gsl_pow_2` to each element of $L - R$ (that is, it squares each element) and returns the post-squaring sum.[12]

• The `project` function is taking the dot product $\mathbf{y}_{\text{est}} = \mathbf{X}\beta$. By giving this single

---

[12]The GSL provides efficient power calculators from `gsl_pow_2` up to `gsl_pow_9`, and the catch-all function `gsl_pow_int(value, exponent)`, that will raise `value` to any integer exponent in a more efficient manner than the general-purpose `pow`.

```
#include <apop.h>

typedef double (*math_fn)(apop_data *);
gsl_vector *jack_iteration(gsl_matrix *, math_fn);
apop_data *ols_data;
gsl_vector * predicted;
double p_dot_mse;

double sum_squared_diff(gsl_vector *left, gsl_vector *right){
    gsl_vector_sub(left, right); //destroys the left vector
    return apop_vector_map_sum(left, gsl_pow_2);
}

gsl_vector *project(apop_data *d, apop_model *m){
    return apop_dot(d, m->parameters, 0, 'v')->vector;
}

double cook_math(apop_data *reduced){
    apop_model *r = apop_estimate(reduced, apop_ols);
    double out =sum_squared_diff(project(ols_data, r), predicted)/p_dot_mse;
    apop_model_free(r);
    return out;
}

gsl_vector *cooks_distance(apop_model *in){
  apop_data *c = apop_data_copy(in->data);
  apop_ols.prep(in->data, in);
  ols_data = in->data;
  predicted = project(in->data, in);
  p_dot_mse = c->matrix->size2 * sum_squared_diff(in->data->vector, predicted);
  return jack_iteration(c->matrix, cook_math);
}

int main(){
    apop_data *dataset = apop_text_to_data("data-regressme", 0, 1);
    apop_model *est = apop_estimate(dataset, apop_ols);
    printf("plot '-'\n");
    strcpy(apop_opts.output_delimiter, "\n");
    apop_vector_show(cooks_distance(est));
}
```

Listing 4.7 Calcluate the Cook's distance, by running 3,200 regressions. Compile with `jackiteration.c`. Online source: `cooks.c`.

line a function of its own, we hide some of the details and get a self-documenting indication of the code's intent.

• The `cook_math` function calculates Equation 4.4.1 for each value of `i`. It is not called directly, but is passed to `jack_iteration`, which will apply the function to each of its 3,200 submatrices.

- The `cooks_distance` function produces two copies of the data set: an untainted copy c, and a regression-style version with the dependent variable in the `vector` element of the data set, and the first column of the `matrix` all ones.

- After `main` calls `cooks_distance`, which calls the various linear algebra procedures and `jack_iteration`, which calls `cook_math` for each submatrix, we have a list of the Cook's distance for every point in the set, which we can use to search for outliers.

- The `main` function produces Gnuplot-ready output, so run this using, e.g., `./cook | gnuplot -persist`. Some researchers prefer to sort the data points before plotting; i.e., try sending the output vector to `gsl_vector_sort` before plotting.

$Q_{4.6}$ Add some bad data points to the `data-regressme` file, like 1|1|1, to simulate outliers or erroneous data. Does the Cook's distance of the bad data stand out as especially large?

MATRIX INVERSION AND EQUATION SOLVING Inverting a matrix requires significantly more computation than the element-by-element operations above. But here in the modern day, it is not such a big deal: my old laptop will invert a $1,000 \times 1,000$ matrix in about ten seconds, and does the inversion step for the typical OLS regression, around a 10×10 matrix at the most, in well under the blink of an eye.

Apophenia provides functions to find determinants and inverses (via the GSL and BLAS's triangular decomposition functions), named `apop_matrix_inverse`, `apop_matrix_determinant`, and for both at once, `apop_det_and_inv`. Examples for using this function are located throughout the book; e.g., see the calculation of OLS coefficients on page 280.

Sometimes, you do not have to bother with inversion. For example, we often write the OLS parameters as $\beta = (\mathbf{X'X})^{-1}(\mathbf{X'Y})$, but you could implement this as solving $(\mathbf{X'X})\beta = \mathbf{X'Y}$, which involves no inversion. If xpx is the matrix $\mathbf{X'X}$ and xpy is the vector $\mathbf{X'Y}$, then `gsl_linalg_HH_solve(xpx, xpy, betav)` will return the vector β.

Σ
- ➤ `apop_data · apop_data: apop_dot`.
- ➤ Vector · vector: `gsl_blas_ddot`.
- ➤ Inversion: `apop_matrix_inverse`, `apop_matrix_determinant`, or `apop_det_and_inv`.

4.5 NUMBERS   Floating-point numbers can take several special values, the most important of which are INFINITY, -INFINITY, and NAN.[13] Data-oriented readers will mostly be interested in NAN (read: not a number), which is an appropriate way to represent missing data. Listing 4.8 shows you the necessary vocabulary. All four if statements will be true, and print their associated statements.

```
#include <math.h> //NaN handlers
#include <stdio.h> //printf

int main(){
    double missing_data = NAN;
    double big_number = INFINITY;
    double negative_big_number = −INFINITY;
        if (isnan(missing_data))
            printf("missing_data is missing a data point.\n");
        if (isfinite(big_number)== 0)
            printf("big_number is not finite.\n");
        if (isfinite(missing_data)== 0)
            printf("missing_data isn't finite either.\n");
        if (isinf(negative_big_number)== −1)
            printf("negative_big_number is negative infinity.\n");
}
```

Listing 4.8 Some functions to test for infinity or NaNs. Online source: notanumber.c.

Because floating-point numbers can take these values, division by zero won't crash your program. Assigning double d = 1.0/0.0 will result in d == INFINITY, and d = 0.0/0.0 will result in d being set to NAN. However, integers have none of these luxuries: try int i = 1/0 and you will get something in the way of Arithmetic exception (core dumped).

Comparison to an NAN value always fails:

```
double blank = NAN;
blank  == NAN;      // This evaluates to false.
blank  == blank;    // This evaluates to false. (!)
isnan(blank);       // Returns 1: the correct way to check for an NaN value.
```

---

[13]Pedantic note on standards: These values are defined in the C99 standard (§7.12) only on machines that support the IEEE 754/IEC 60559 floating-point standards, but since those standards are from 1985 and 1989, respectively, they may be taken as given: to the best of my knowledge, all current hardware supports INFINITY and NAN. Recall that gcc requires -std=gnu99 for C99 features; otherwise, the GSL provides GSL_POSINF, GSL_NEGINF, and GSL_NAN that work in non-C99 and non-IEEE 754 systems.

	GSL constant	approx.		GSL constant	approx.
e	M_E	2.71828	π	M_PI	3.14159
$\log_2 e$	M_LOG2E	1.44270	$\pi/2$	M_PI_2	1.57080
$\log_{10} e$	M_LOG10E	0.43429	$\pi/4$	M_PI_4	0.78540
$\ln 2$	M_LN2	0.69315	$1/\pi$	M_1_PI	0.31831
$\ln 10$	M_LN10	2.30259	$2/\pi$	M_2_PI	0.63662
$\sqrt{2}$	M_SQRT2	1.41421	$\sqrt{\pi}$	M_SQRTPI	1.77245
$\sqrt{1/2} = 1/\sqrt{2}$	M_SQRT1_2	0.70711	$2/\sqrt{\pi}$	M_2_SQRTPI	1.12838
$\sqrt{3}$	M_SQRT3	1.73205	$\ln \pi$	M_LNPI	1.14473
Euler constant (γ)	M_EULER	0.57722			

Table 4.9 The GSL defines a number of useful constants.

※ *Predefined constants* There are a number of useful constants that are defined by the GSL (via preprocessor #defines); they are listed in Table 4.9.[14] It is generally better form to use these constants, because they are more descriptive than raw decimal values, and they are defined to greater precision than you will want to type in yourself.

PRECISION As with any finite means of writing real numbers, there is a roundoff error to the computer's internal representation of numbers. The computer basically stores non-integer numbers using scientific notation. For those who have forgotten this notation, π is written as 3.14159×10^0, or 3.14159e0, and 100π as 3.14159×10^2, or 3.14159e2. Numbers always have exactly one digit before the decimal point, and the exponent is chosen to ensure that this is the case.

Your computer works in binary, so floating-point numbers (of type float and double) are of the form $d \times 2^n$, where d is a string of ones and zeros and n is an exponent.[15]

The *scale* of a number is its overall magnitude, and is expressed by the exponent n. The floating-point system can express a wide range of scales with equal ease: it is as easy to express three picometers ($3e-12$) as three million kilometers ($3e9$). The *precision* is how many significant digits of information are held in d: $3.14e-12$ and $3.14e9$ both have three significant decimal digits of information.

There is a fixed space for d, and when that space is exceeded, n is adjusted to suit, but that change probably means a loss in precision for d. To give a small base-ten

---

[14]The GSL gets most of these constants from BSD and UNIX, so you may be able to find them even when the GSL is not available. The exceptions are M_SQRT3 and M_SQRTPI, which are GSL-specific.

[15]This oversimplifies some details that are basically irrelevant for users. For example, the first digit of d is always one, so the computer normally doesn't bother storing it.

i	2^i	2^{-i}
100	1.26765e+30	7.88861e−31
200	1.60694e+60	6.22302e−61
300	2.03704e+90	4.90909e−91
400	2.58225e+120	3.87259e−121
500	3.27339e+150	3.05494e−151
600	4.14952e+180	2.40992e−181
700	5.26014e+210	1.90109e−211
800	6.66801e+240	1.4997e−241
900	8.45271e+270	1.18305e−271
1000	1.07151e+301	9.33264e−302
1100	inf	0
1200	inf	0

Table 4.10 Multiplying together large columns of numbers will eventually fail.

example, say that the space for d is only three digits; then $5.89\text{e}0 \times 892\text{e}0 = 525\text{e}1$, though $5.89 \times 892 = 5,254$. The final four was truncated to zero to fit d into its given space.

Precision can easily be lost in this manner, and once lost can never be regained. One general rule of thumb implied by this is that if you are writing a precision-sensitive function to act on a `float`, use `double` variables internally, and if you are taking in `doubles`, use `long double` internally.

The loss of precision becomes especially acute when multiplying together a long list of numbers. This will be discussed further in the chapter on maximum likelihood estimation (page 330), because the likelihood function involves exactly such multiplication. Say that we have a column of a thousand values each around a half. Then the product of the thousand elements is about 2^{-1000}, which strains what a `double` can represent. Table 4.10 shows a table of powers of two as represented by a `double`. For $i > 1,000$—a modest number of data points—a `double` throws in the towel and calls $2^i = \infty$ and $2^{-i} = 0$. These are referred to as an *overflow error* and *underflow error*, respectively.

For those who would like to try this at home, Listing 4.11 shows the code used to produce this table, and also repeats the experiment using a `long double`, which doesn't give up until over 16,000 doublings and halvings.

```
#include <math.h>
#include <stdio.h>

int main(){
    printf("Powers of two held in a double:\n");
    for(int i=0; i< 1400; i+=100)
        printf("%i\t %g \t %g\n", i, ldexp(1,i), ldexp(1,−i));
    printf("Powers of two held in a long double:\n");
    for(int i=0; i< 18000; i+=1000)
        printf("%i\t %Lg \t %Lg\n", i, ldexpl(1,i), ldexpl(1,−i));
}
```

Listing 4.11 Find the computer's representation of 2^i and 2^{-i} for large i. Online source: `powersoftwo.c`.

- The program uses the `ldexp` family of functions, which manipulate the floating-point representation of a number directly (and are thus probably bad form).
- The `printf` format specifier for the `long double` type is `%Lg`.[16]

The solution to the problem of finding the product of a large number of elements is to calculate the log of the product rather than the product itself; see page 330.

If you need to calculate π to a million decimal points, you will need to find a library that can work with numbers to arbitrary precision. Such libraries typically work by representing all numbers as a data structure listing the ones place, tens place, hundreds place, . . . , and then extending the list in either direction as necessary. Another alternative is rational arithmetic, which leaves all numbers in `(int)/(int)` form for as long as possible. Either system will need to provide its own add/subtract/multiply/divide routines to act on its data structures, rather than using C's built-in operators. Unfortunately, the added layer of complexity means that the arithmetic operations that had been fast procedures (often implemented via special-purpose registers on the processor hardware itself) are now a long series of library calls.

So to do math efficiently on large matrices, we are stuck with finite precision, and therefore must not rely too heavily on numbers after around maybe four significant digits. For the purposes of estimating and testing the parameters of a model using real-world data, this is OK. If two numbers differ only after eight significant digits (say, 3.14159265 versus 3.14159268), there is rarely any reason to take these numbers as significantly different. Even if the hypothesis test indicates that they are different, it will be difficult to convince a referee of this.

---

[16]The `%g` and `%Lg` format specifiers round off large values, so change them to to `%f` and `%Lf` to see the precise value of the calculations without exponential notation.

CONDITIONING   Most matrix routines do badly when the determinant is near zero,
                or when eigenvalues are different orders of magnitude. One way
to cause such problems with your own data is to have one column that is of the
order of 1×10^{10} and another that is on the order of 1×10^{-10}. In finite-precision
arithmetic on two numbers of such wide range, the smaller number is often simply
swallowed: `3.14e10 + 5.92e-10 = 3.14e10`.

Thus, try to ensure that each column of the data is approximately of the same
order of magnitude before doing calculations. Say that you have a theory that mean
fingernail thickness is influenced by a location's population. You could modify the
scale when pulling data from the database,

```
select population/1000., nail_thickness∗1000.
from health_data;
```

or you could modify it in the `gsl_matrix`:

```
APOP_COL(data, 0, pop)
gsl_vector_scale(pop, 1/1000.);
APOP_COL(data, 1, nails)
gsl_vector_scale(nails, 1000.);
```

These notes about conditioning are not C-specific. Any mathematics package that
hopes to work efficiently with large matrices must use finite-precision arithmetic,
and will therefore have the same problems with ill-conditioned data matrices. For
much more about precision issues and the standard machine representation of numbers, see Goldberg (1991).

COMPARISON   Floating-point numbers are exact representations of a real number
              with probability zero. Simply put, there is a bit of fuzz, so expect
every number to be a little bit off from where it should be.

It is normally not a problem that $4 + 1e-20 \neq 4$, and such fuzz can be safely
ignored, but Polhill *et al.* (2005) point out that the fuzziness of numbers can be
a problem for comparisons, and those problems can create odd effects in simulations or agent-based models. After a long series of floating-point operations, a
comparison of the form (`t == 0`) will probably not work. For example, Listing
4.12 calculates $1.2 - 3 \cdot 0.4$ using standard IEEE arithmetic, and finds that it is less
than zero.[17]

There are labor-intensive solutions to the problem, like always using `long ints`

---

[17]This example is from a web site affiliated with the authors of the above paper, at `http://www.macaulay.ac.uk/fearlus/floating-point/`.

```
#include <stdio.h>

int main(){
    double t = 1.2;
        t −= 0.4;
        t −= 0.4;
        t −= 0.4;
    if (t<0)
            printf ("By the IEEE floating−point standard, 1.2 − 3*.04 < 0.\n");
}
```

Listing 4.12 The IEEE standard really does imply that $1.2 - 3 \cdot 0.4 < 0$. Online source: `fuzz.c`.

for everything[18], but the most sensible solution is to just bear in mind that no comparison is precise so, for example, agents should not die when their wealth is zero, but when it is less than maybe $1e-6$. Otherwise, the model should be robust to agents who have an iota of negative wealth.

➤ Floating-point numbers can take on values of `-INFINITY`, `INFINITY`, and `NAN`.

➤ Multiplying together a column of a thousand numbers will break, so get the log of the product by summing the logs of the column's elements.

➤ Reporting results based on the fifth significant digit (or so) is spurious.

➤ Try to keep the scale of your variables within a factor of about a thousand of each other.

➤ Exact comparisons of floating-point numbers can fail, so do not test `f == 0`, but `fabs(f) < 1e-6` (or so).

4.6 ※ `gsl_matrix` **AND** `gsl_vector` **INTERNALS**   First, a warning: the intent of this section is not to show you how to circumvent the GSL's access functions such as `gsl_matrix_get` and `gsl_vector_set`. Doing so is bad form, inviting errors and making code more difficult to read.[19] Instead, these notes will be useful to you for

[18]Recall that `(a/b)*b + a%b` is exactly `a` for `long int a, b`, $b \neq 0$.
[19]If you are really concerned about the overhead from these functions, then `#define GSL_RANGE_CHECK_-OFF`, either via that preprocessor directive or adding -DGSL_RANGE_CHECK_OFF to your compilation com-

understanding why the data structures are the way they are, and giving you a handle on what operations are easy and what operations are difficult.

Here is the relevant section of the declaration of the `gsl_matrix` structure:

```
typedef struct {
    size_t size1;
    size_t size2;
    size_t tda;
    double * data;
    [...]
    int owner;
} gsl_matrix;
```

(0,0) (0, 1) ...(0, 9) (1,0) (1, 1) ...(1, 9) (2,0) (2, 1) ...(2, 9) ...(9,0) (9, 1) ...(9, 9)
⇓

	(0,0)	(0,1)	(0,2)	(0,3)	(0,4)	(0,5)	(0,6)	(0,7)	(0,8)	(0,9)
↪	(1,0)	(1,1)	(1,2)	(1,3)	(1,4)	(1,5)	(1,6)	(1,7)	(1,8)	(1,9)
↪	(2,0)	(2,1)	(2,2)	(2,3)	(2,4)	(2,5)	(2,6)	(2,7)	(2,8)	(2,9)
↪	(3,0)	(3,1)	(3,2)	(3,3)	(3,4)	(3,5)	(3,6)	(3,7)	(3,8)	(3,9)
↪	(4,0)	(4,1)	(4,2)	(4,3)	(4,4)	(4,5)	(4,6)	(4,7)	(4,8)	(4,9)
↪	(5,0)	(5,1)	(5,2)	(5,3)	(5,4)	(5,5)	(5,6)	(5,7)	(5,8)	(5,9)
↪	(6,0)	(6,1)	(6,2)	(6,3)	(6,4)	(6,5)	(6,6)	(6,7)	(6,8)	(6,9)
↪	(7,0)	(7,1)	(7,2)	(7,3)	(7,4)	(7,5)	(7,6)	(7,7)	(7,8)	(7,9)
↪	(8,0)	(8,1)	(8,2)	(8,3)	(8,4)	(8,5)	(8,6)	(8,7)	(8,8)	(8,9)
↪	(9,0)	(9,1)	(9,2)	(9,3)	(9,4)	(9,5)	(9,6)	(9,7)	(9,8)	(9,9)

Figure 4.13 Within a submatrix, the (3,4) element is still one step from the (3,3) element, and ten steps from the (2,4) element.

As you know, `size1` and `size2` are simply the count of rows and columns. The `data` pointer is a single pointer to a stream of numbers. Since memory addresses are linear, the top of Figure 4.13 is closer to what is actually in memory: the first row of data, followed immediately afterward by the second row, then the third row, and so on, forming one long row of data. By adding line breaks, we humans can think of this one long row of data as actually being a grid, like the second half of Figure 4.13.

Stepping along the row means simply stepping along by `sizeof(double)` units, and stepping down a matrix column means stepping by `sizeof(double)* size2` steps from the current element. For example, to reach the (3,5) element of a ten by ten matrix, the processor must skip three rows and then skip five items, so it would jump `sizeof(double)*35` steps from the base element.

Modern computers are proactive about data gathering. When they read data from slower types of memory, they also check the neighbors as well. If the code is relatively predictable, the system can gather the next bit of data at the same time as

---

mand line. Between this and the compiler's optimization routines, the function call will reduce to the appropriate array operation.

it is crunching the current data element. The `gsl_matrix` structure works wonderfully with such a system, because steps are predictable and of fixed size, so the processor has a good chance of correctly guessing what data to put into its faster caches.

Now say that we have a 100×10 matrix, which would have the following information:

$$
\begin{array}{l}
\text{size1} = 100; \\
\text{size2} = 10; \\
\text{tda} = 10; \\
\text{data} = [\text{location of } (0,0)]; \\
\text{owner} = 1;
\end{array}
$$

With `tda` equal to `size2`, jumping down a column would require a jump of `sizeof(double)*tda`.[20]

If we wanted to pull out the 4×6 submatrix that begins at $(3, 2)$, then the resulting submatrix data would look like this:

$$
\begin{array}{l}
\text{size1} = 4; \\
\text{size2} = 6; \\
\text{tda} = 10; \\
\text{data} = [\text{location of } (3,2) \text{ in the original matrix}]; \\
\text{owner} = 0;
\end{array}
$$

We can use this matrix exactly as with the full matrix: For example, to get the third elements in the first row, we would step `sizeof(double) * 2` forward from the base element pointed to by `data`, and to get to the beginning of the next column, we would jump `sizeof(double)*tda`. Thus, the process of pulling a subset of the data merely required finding the first point and writing down arbitrary limits for `size1` and `size2`. No actual data was copied. This is how `gsl_matrix_row`, `APOP_ROW`, `APOP_COL`, `APOP_SUBMATRIX`, and other such routines work.

The `owner` variable now becomes important, because there could be multiple submatrices all pointing to the same data. Since the submatrix is not the owner of its data, the GSL will not allow it to free the data set to which it points.

The `gsl_vector` has a similar structure, including a starting point and a stride to indicate how far to jump to the next element. Thus, taking a row or column subset of a matrix also merely requires writing down the correct coordinates and lengths.

The GSL's structures are good for fast access, because the next element is always

---

[20]The abbreviation `tda` stands for *trailing dimension of array*. For a `gsl_vector`, the analogous element is the *stride*.

a fixed jump relative to the current element, whether you are traversing by rows or columns. They are exceptionally good for describing submatrices and subvectors, because doing so merely requires writing down new coordinates. They handle a $1 \times 10,000$ matrix just as easily as a $10,000 \times 1$ matrix or a 100×100 matrix.

They are not good for non-contiguous subsets like the first, second, and fifth columns of a data set, since the relative jumps from one column to the next are not identical. Similarly, they are not good for holding various types of data, where some jumps could be `sizeof(int)` and others could be `sizeof(double)`. Also, since space is always allocated for every element, there is no way to efficiently represent sparse matrices.

Systems that deal well with variable-sized jumps have the pros and cons reversed from the above. For example, one solution is to let the overall table be a list of column vectors, where each column has its own type. But traversing along a row could involve jumping all over memory, so common operations like finding the sum for each row becomes a significantly slower operation.

Other designers with different goals have used different means of representing a data matrix, and the `gsl_matrix` is by no means the best for all needs. But it does very well for the goal of allowing the hardware to process rows and columns of homogeneous data with maximal efficiency.

> ➤ The GSL's matrix and vector structures are very good for efficient computation, because each element has a fixed size and is a fixed distance from the neighboring elements.

> ➤ It is very easy to take contiguous subvectors or submatrices of these structures. Doing so requires copying only a few bits of metadata, but not the data itself.

> ➤ There is no simple way to take non-contiguous subsets of `gsl_-matrices` or `gsl_vectors`. You will either need to copy the data manually (i.e., using a `for` loop), or do the manipulations in the database before your data is in `gsl_matrix` form.

4.7 MODELS   Recall the one-sentence summary of statistical analysis from the first page of the introduction: estimate the parameters of a model using data. The Apophenia library provides functions and data structures at exactly this level of abstraction, in the form of the `apop_model` and `apop_data` structures and the functions that operate on them.

You have already met the `apop_data` structure, which lent a hand to operations on the matrix algebra layer of abstraction; the remainder of the chapter introduces the `apop_model` structure, which provides similar forms of strength through constraint: it encapsulates model information in a uniform manner, allows models to be used interchangeably in functions that can take any model as an input, and allows sensible defaults to be filled in as necessary.

A great deal of statistical work consists of converting or combining existing models to form new ones. That is, models can be filtered to produce models just as data can be filtered to provide new information. We can read estimation as filtering an un-parameterized model into a parameterized one. Bayesian updating (discussed more thoroughly on page 258) takes in a prior model, a likelihood function, and data, and outputs a new model—which can then be used as the input to another round of filtering when new data comes in.

Another example discussed below, is the imposition of a constraint: begin by estimating a general model, then generate a new model with a constraint imposed on some of the parameters, and re-estimate. The difference in log likelihoods of the constrained and unconstrained models can then be used for hypothesis testing.

The structure of the model struct   In the usage of this book, a model intermediates between data and parameters. From there, the model can go in three directions:

   i) $\mathbf{X} \Rightarrow \beta$: Given data, estimate parameters.

   ii) $\beta \Rightarrow \mathbf{X}$: Given parameters, generate artificial data (e.g., make random draws from the model, or find the expected value).

   iii) $(\mathbf{X}, \beta) \Rightarrow p$: Given both data and parameters, estimate their likelihood or probability.

To give a few examples, form (i) is the descriptive problem, such as estimating a covariance or OLS parameters. Monte Carlo methods use form (ii): producing a few million draws from the model given fixed parameters. Bayesian estimation is based on form (iii), describing a posterior probability given both data and parameters, as are the likelihoods in a maximum likelihood estimation.

For many common models, there are `apop_models` already written, including distributions like the Normal, Multivariate Normal, Gamma, Zipf, et cetera, and generalized linear models like OLS, WLS, probit, and logit. Because they are in a standardized form, they can be sent to model-handling functions, and be applied to data in sequence. For example, you can fit the data to a Gamma, a Lognormal, and an Exponential distribution and compare the outcomes (as in the exercise on page 257).

Every model can be estimated via a form such as

```
apop_model *est = apop_estimate(data, apop_normal);
```

Examples of this form appear throughout the book—have a look at the code later in this section, or on pages 133, 289, 352, or 361, for example.

Discussion of the other directions—making random draws of data given parameters and finding likelihoods given data and parameters—will be delayed until the chapters on Monte Carlo methods and maximum likelihood estimation, respectively.

Changing the defaults   A complete model includes both the model's functions and the environment in which those functions are evaluated (Gentleman & Ihaka, 2000). The `apop_model` thus includes both the outputs, the functions, and everything one would need to replicate one from the other. For the purposes of Apophenia's model estimations, an $\mathcal{N}(0,1)$ is a separate model from an $\mathcal{N}(1,2)$, and a maximum likelihood model whose optimization step is done via a conjugate gradient method is separate from an otherwise identical model estimated via a simplex algorithm.

But a generic `struct` indended to hold settings for all models faces the complication that different methods of estimation require different settings. The choice of conjugate gradient or simplex algorithm is meaningless for an instrumental variable regression, while a list of instrumental variables makes no sense to a maximum likelihood search.

Apophenia standard `apop_model` struct thus has an open space for attaching different groups of settings as needed. If the model's defaults need tweaking, then you can first add an MLE, OLS, histogram, or other settings group, and then change whatever details need changing within that group. Again, examples of the usage and syntax of this two-step processs abound, both in online documentation and throughout the book, such as on pages 153, 339, or 352.

Writing your own   It would be nice if we could specify a model in a single form and leave the computer to work out the best way to implement all three of the directions at the head of this section, but we are rather far from such computational nirvana. Recall the example of OLS from the first pages of Chapter 1. The first form of the model—find the value of β such that $(\mathbf{y} - \mathbf{X}\beta)^2$ is minimized—gave no hint that the correct form in the other direction would be $(\mathbf{X}'\mathbf{X})^{-1}\mathbf{X}'\mathbf{y}$. Other models, such as the probit model elaborated in Chapter 10, begin with similar $\mathbf{X}\beta$-type forms, but have no closed form solution.

Thus, writing down a model for a computable form requires writing down a procedure for one, two, or all three of the above directions, such as an `estimate` function, a `log_likelihood` and p (probability) function, and a `draw` function to make random draws. You can fill in those that you can solve in closed form, and can leave Apophenia to fill in computationally-intensive default procedures for the rest.

```
1   #include <apop.h>
2
3   apop_model new_OLS;
4
5   static apop_model *new_ols_estimate(apop_data *d, apop_model *params){
6       APOP_COL(d, 0, v);
7       apop_data *ydata = apop_data_alloc(d->matrix->size1, 0, 0);
8       gsl_vector_memcpy(ydata->vector, v);
9       gsl_vector_set_all(v, 1); //affine: first column is ones.
10      apop_data *xpx = apop_dot(d, d, 't', 0);
11      apop_data *inv = apop_matrix_to_data(apop_matrix_inverse(xpx->matrix));
12      apop_model *out = apop_model_copy(new_OLS);
13      out->data = d;
14      out->parameters = apop_dot(inv, apop_dot(d, ydata, 1), 0);
15      return out;
16  }
17
18  apop_model new_OLS = {.name ="A simple OLS implementation",
19                        .estimate = new_ols_estimate};
20
21  int main(){
22      apop_data *dataset = apop_text_to_data("data-regressme", 0, 1);
23      apop_model *est = apop_estimate(dataset, new_OLS);
24      apop_model_show(est);
25  }
```

Listing 4.14 A new implementation of the OLS model. Online source: `newols.c`.

For example, listing 4.14 shows a new implementation of the OLS model. The math behind OLS is covered in detail on page 274.

- In this case, only the `estimate` function is specified.
- The procedure itself is simply a matter of pulling out the first column of data and replacing it with ones, and calculating $(\mathbf{X'X})^{-1}\mathbf{X'y}$.
- Lines 12–14 allocate an `apop_model` and set its `parameter` element to the correct value. Line 13 keeps a pointer to the original data set, which is not used in this short program, but often comes in handy.
- The allocation of the output on line 12 needs the model, but we have not yet declared it. The solution to such circularity is to simply give a declaration of the

model on line 3.

- Given the function, line 18 initializes the model itself, a process which is rather simple thanks to designated initializers (see p 32).

- Although the `main` function is at the bottom of this file, the typical model deserves its own file. By using the `static` keyword in line five, the function name will only be known to this file, so you can name it what you wish without worrying about cluttering up the global name space. But with is a pointer to the function in the model object itself, a routine in another file could use the `apop_new_-OLS.estimate(...)` form to call this function (which is what `apop_estimate` on line 23 will do internally).

- Lines 1–3 provide the complete header an external file would need to use the new model, since the structure of the `apop_model` is provided in `apop.h`.

Compare the `new_OLS` model with the `apop_ols` model.

$\mathbb{Q}_{4.7}$

- Modify Listing 4.14 to declare an array of `apop_models`. Declare the first element to be `apop_ols` and the second to be `new_OLS`. [The reverse won't work, because `new_OLS` destroys the input data.]

- Write a `for` loop to fill a second array of pointers-to-models with the estimate from the two models.

- Calculate and display the difference between `estimate[0]->parameters->vector` and `estimate[1]->parameters->vector`.

✳ **AN EXAMPLE: NETWORK DATA** The default, for models such as `apop_ols` or `apop_probit`, is that each row of the data is assumed to be one observation, the first column of the data is the dependent variable, and the remaining columns are the independent variable.

For the models that are merely a distribution, the rule that one row equals one observation is not necessary, so the data matrix can have any form: $1 \times 10{,}000$, or $10{,}000 \times 1$, or 100×100. This provides maximum flexibility in how you produce the data.

But for data describing ranks (score of first place, second place, ...) things get more interesting, because such data often appears in multiple forms. For example, say that we have a classroom where every student wrote down the ID number his or her best friend, and we tallied this list of student numbers:
1 1 2 2 2 2 3 4 4 4 6 7 7 7.
First, we would need to count how often each student appeared:

id_no count
2   4
4   3
7   3
1   2
3   1
6   1.

In SQL:

```
select id_no, count(*) as ct
    from surveys
    group by id_no
    order by ct desc
```


If we were talking about city sizes (another favorite for rank-type analysis), we would list the size of the largest city, the second largest, et cetera. The labels are not relevant to the analysis; you would simply send the row of counts for most popular, second most popular, et cetera:
4 3 3 2 1 1.
Each row of the data set would be one classroom like the above, and the column number represents the ranking being tallied.

As mentioned above, you can add groups of settings to a model to tweak its behavior. In the case of the models commonly used for rank analysis, you can signal to the model that it will be getting rank-ordered data. For example:

```
apop_model *rank_version = apop_model_copy(apop_zipf);
Apop_settings_add_group (rank_version, apop_rank, NULL);
apop_model_show(apop_estimate(ranked_draws, rank_version));
```


Alternatively, some data sets are provided with one entry listing the rank for each observation. There would be four 1's, three 2's, three 3's, et cetera:
1 1 1 1 2 2 2 3 3 3 4 4 5 5.
In the city-size example, imagine drawing people uniformly at random from all cities, and then writing down whether each person drawn is from the largest city, the second largest, et cetera. Here, order of the written-down data does not matter. You can pass this data directly to the various estimation routines without adding a group of settings; e.g., apop_estimate(*ranked_draws*, apop_gamma).

The `nominee` column in the file `data-classroom` is exactly the sort of data on which one would run a network density analysis.

- Read the data into a database.

- Query the vector of ranks to an `apop_data` set, using a query like the one above.

$\mathbb{Q}_{4.8}$

- Transpose the matrix (*hint*: `gsl_matrix_transpose_memcpy`), because `apop_zipf`'s `estimate` function requires each classroom to be a row, with the nth-ranked in the nth column. This data set includes only one classroom, so you will have only one row of data.

- Call `apop_estimate(`*yourdata*`, apop_zipf)`; show the resulting estimate. How well does the Zipf model fit the data?

❋ **MLE MODELS**   To give some examples from a different style of model, here are some notes on writing models based on a maximum likelihood estimation.

- Write a likelihood function. Its header will look like this:

static double apop_new_log_likelihood(**gsl_vector** ∗beta, **apop_data** ∗d)

Here, `beta` holds the parameters to be maximized and `d` is the fixed parameters—the data. This function will return the value of the log likelihood function at the given parameters and data.
In some cases, it is more natural to express probabilities in log form, and sometimes in terms of a plain probability; use the one that works best, and most functions will calculate the other as needed.

- Declare the model itself:

apop_model new_model = { "The Me distribution", 2, 0, 0,
        .log_likelihood = new_log_likelihood };

  – If you are using a probability instead of a log likelihood, hook it into your model with `.p = `*new_p*.

  – The three numbers after the name are the size of the parameter structure, using the same format as `apop_data_alloc`: size of vector, rows in matrix, then columns in matrix. If any of these is `-1`, then the `-1` will be replaced with the number of columns in the input data set's matrix (i.e., *your_-*

data->matrix->size2).[21] This is what you would use for an OLS regression, for example, where there is one parameter per independent variable.

With this little, a call like apop_estimate(*your_data*, new_model) will work, because apop_estimate defaults to doing a maximum likelihood search if there is no explicitly-specified new_model.estimate function.

• For better estimations, write a gradient for the log likelihood function. If you do not provide a closed-form gradient function, then the system will fill in the blank by numerically estimating gradients, which is slower and has less precision. Calculating the closed-form gradient is usually not all that hard anyway, typically requiring just a few derivatives. See Listing 4.17 (or any existing apop_model) for an example showing the details of syntax.

SETTING CONSTRAINTS   A constraint could either be imposed because the author of the model declared an arbitrary cutoff ('we can't spend more than $1,000.') or because evaluating the likelihood function fails ($\ln(-1)$). Thus, the system needs to search near the border, without ever going past it, and it needs to be able to arbitrarily impose a constraint on an otherwise unconstrained function.

Apophenia's solution is to add a constraint function that gets checked before the actual function is evaluated. It does two things if the constraint is violated: it nudges the point to be evaluated into the valid area, and it imposes a penalty to be subtracted from the final likelihood, so the system will know it is not yet at an optimum. The unconstrained maximization routines will then have a continuous function to search but will never find an optimum beyond the parameter limits.[22]

To give a concrete example, Listing 4.15 adds to the apop_normal model a constraint function that will ensure that both parameters of a two-dimensional input are greater than given values.

Observe how the constraint function manages all of the requisite steps. First, it checks the constraints and quickly returns zero if none of them binds. Then, if they do bind, it sets the return vector to just inside the constrained region. Finally, it returns the distance (on the *Manhattan metric*) between the input point and the point returned.[23] The unconstrained evaluation system should repeatedly try points closer and closer to the zero-penalty point, and the penalty will continuously decline as we approach that point.

---

[21]if your model has a more exotic parameter count that needs to be determined at run-time, use the prep method of the apop_model to do the allocation.

[22]This is akin to the common penalty function methods of turning a constrained problem into an unconstrained one, as in Avriel (2003), but the formal technique as commonly explained involves a series of optimizations where the penalty approaches zero as the series progresses. It is hard to get a computer to find the limit of a sequence; the best you could expect would be a series of estimations with decreasing penalties; apop_estimate_restart can help with the process.

[23]If you need to calculate the distance to a point in your own constraint functions, see either apop_vector_distance or apop_vector_grid_distance.

```
1    #include <apop.h>
2
3    double linear_constraint(apop_data * d, apop_model *m){
4        double limit0 = 2.5,
5            limit1 = 0,
6            tolerance = 1e−3; // or try GSL_EPSILON_DOUBLE
7        double beta0 = apop_data_get(m−>parameters, 0, −1),
8            beta1 = apop_data_get(m−>parameters, 1, −1);
9        if (beta0 > limit0 && beta1 > limit1)
10            return 0;
11       //else create a valid return vector and return a penalty.
12       apop_data_set(m−>parameters, 0, −1,GSL_MAX(limit0 + tolerance, beta0));
13       apop_data_set(m−>parameters, 1, −1, GSL_MAX(limit1 + tolerance, beta1));
14       return GSL_MAX(limit0 + tolerance − beta0, 0)
15                       + GSL_MAX(limit1 + tolerance − beta1, 0);
16   }
17
18   int main(){
19       apop_model *constrained = apop_model_copy(apop_normal);
20       constrained−>estimate = NULL;
21       constrained−>constraint = linear_constraint;
22       apop_db_open("data−climate.db");
23       apop_data *dataset = apop_query_to_data("select pcp from precip");
24       apop_model *free = apop_estimate(dataset, apop_normal);
25       apop_model *constr = apop_estimate(dataset, *constrained);
26       apop_model_show(free);
27       apop_model_show(constr);
28       double test_stat = 2 * (free−>llikelihood − constr−>llikelihood);
29       printf("Reject the null (constraint has no effect) with %g%% confidence\n",
30              gsl_cdf_chisq_P(test_stat, 1)*100);
     }
```

Listing 4.15 An optimization, a constrained optimization, and a likelihood ratio test comparing the two. Online source: `normallr.c`.

In the real world, set linear constraints using the `apop_linear_constraint` function, which takes in a set of contrasts (as in the form of the F test) and does all the requisite work from there. For an example of its use, have a look at the budget constraint in Listing 4.17.

The `main` portion of the program does a likelihood ratio test comparing constrained and unconstrained versions of the Normal distribution.

- Lines 19–21 copy off the basic Normal model, and add the constraint function to the copy. The `estimate` routine for the Normal doesn't use any constraint, so it is invalid in the unconstrained case, so line 20 erases it.

- Lacking an explicit `estimate` routine in the model, line 25 resorts to maximum likelihood estimation (MLE). The MLE routine takes the model's constraint into account.

- Lines 28–29 are the hypothesis test. Basically, twice the difference in log likelihoods has a χ^2 distribution; page 350 covers the details.

AN EXAMPLE: UTILITY MAXIMIZATION   The process of maximizing a function subject to constraints is used extensively outside of statistical applications, such as for economic agents maximizing their welfare or physical systems maximizing their entropy. With little abuse of the optimization routines, you could use them to solve any model involving maximization subject to constraints. This section gives an extended example that numerically solves an Econ 101-style utility maximization problem.

The consumer's utility from a consumption pair (x_1, x_2) is $U = x_1^\alpha x_2^\beta$. Given prices P_1 and P_2 and B dollars in cash, she has a budget constraint that requires $P_1 x_1 + P_2 x_2 \leq B$. Her goal is to maximize utility subject to her budget constraint.

The data (i.e., the `apop_data` set of fixed elements) will have a one-element vector and a 2×2 matrix, structured like this: $\left[\begin{array}{c|cc} \text{budget} & \text{price}_1 & \alpha \\ & \text{price}_2 & \beta \end{array} \right]$.

Once the models are written down, the estimation is one function call, and calculating the marginal values is one more. Overall, the program is overkill for a problem that can be solved via two derivatives, but the same framework can be used for problems with no analytic solutions (such as for consumers with a stochastic utility function or dynamic optimizations with no graceful closed form).

Because the estimation finds the slopes at the optimum, it gives us comparative statics, answering questions about the change in the final decision given a marginal rise in price P_1 or P_2 (or both).

Listing 4.16 shows the model via numerical optimization, and because the model is so simple, Listing 4.17 shows the analytic version of the model.

- The `econ101_estimate` routine just sets some optimization settings and calls `apop_maximum_likelihood`.
- The budget constraint, in turn, is a shell for `apop_linear_constraint`. That function requires a constraint matrix, which will look much like the matrix of equations sent in to the F tests on page 310. In this case, the equations are

$$\left[\begin{array}{ccccc} -\text{budget} & < & -p_1\beta_1 & - & p_2\beta_2 \\ 0 & < & \beta_1 & & \\ 0 & < & & & \beta_2 \end{array} \right].$$

All inequalities are in less-than form, meaning that the first—the cost of goods is

```
#include <apop.h>
apop_model econ_101;

static apop_model * econ101_estimate(apop_data *choice, apop_model *p){
    Apop_settings_add_group(p, apop_mle, p);
    Apop_settings_add(p, apop_mle, tolerance, 1e−4);
    Apop_settings_add(p, apop_mle, step_size, 1e−2);
    return apop_maximum_likelihood(choice, *p);
}

static double budget(apop_data *beta, apop_model* m){
  double price0 = apop_data_get(m−>data, 0, 0),
           price1 = apop_data_get(m−>data, 1, 0),
           cash = apop_data_get(m−>data, 0, −1);
  apop_data *constraint = apop_data_alloc(3, 3, 2);
    apop_data_fill(constraint,
                        −cash, −price0, −price1,
                    0., 1., 0.,
                    0., 0., 1.);
    return apop_linear_constraint(m−>parameters−>vector, constraint, 0);
}

static double econ101_p(apop_data *d, apop_model *m){
  double alpha = apop_data_get(d, 0, 1),
           beta = apop_data_get(d, 1, 1),
           qty0 = apop_data_get(m−>parameters, 0, −1),
           qty1 = apop_data_get(m−>parameters, 1, −1);
    return pow(qty0, alpha) * pow(qty1, beta);
}

apop_model econ_101 = {"Max Cobb−Douglass subject to a budget constraint", 2, 0, 0,
    .estimate = econ101_estimate, .p = econ101_p, .constraint= budget};
```

Listing 4.16 An agent maximizes its utility. Online source: `econ101.c`.

less than the budget—had to be negated. However, the next two statemets, β_1 is positive and β_2 is positive, are natural and easy to express in these terms. Converting this system of inequalities into the familiar vector/matrix pair gives

$$\left[\begin{array}{c|cc} -\text{budget} & -p_1 & -p_2 \\ 0 & 1 & 0 \\ 0 & 0 & 1 \end{array} \right].$$

• The budget consraint as listed has a memory leak that you won't notice at this scale: it re-specifies the constraint every time. For larger projects, you can ensure that `constraint` is only allocated once by declaring it as `static` and initially setting it to `NULL`, then allocating it only if it is not `NULL`.

• The analytic version of the model in Listing 4.17 is a straightforward translation

```
#include <apop.h>
apop_model econ_101_analytic;
apop_model econ_101;

#define fget(r, c) apop_data_get(fixed_params, (r), (c))

static apop_model * econ101_analytic_est(apop_data * fixed_params, apop_model *pin){
    apop_model *est = apop_model_copy(econ_101_analytic);
    double budget = fget(0, −1), p1 = fget(0, 0), p2 = fget(1, 0),
                alpha = fget(0, 1), beta = fget(1, 1);
    double x2 = budget/(alpha/beta + 1)/p2,
            x1 = (budget − p2*x2)/p1;
    est−>data = fixed_params;
    est−>parameters = apop_data_alloc(2,0,0);
    apop_data_fill(est−>parameters, x1, x2);
    est−>llikelihood = log(econ_101.p(fixed_params, est));
    return est;
}

static void econ101_analytic_score(apop_data *fixed_params, gsl_vector *gradient,
        apop_model *m){
    double x1 = apop_data_get(m−>parameters, 0, −1);
    double x2 = apop_data_get(m−>parameters, 1, −1);
    double alpha = fget(0, 1), beta = fget(1, 1);
    gsl_vector_set(gradient, 0, alpha*pow(x1,alpha−1)*pow(x2,beta));
    gsl_vector_set(gradient, 1, beta*pow(x2,beta−1)*pow(x1,alpha));
}

apop_model econ_101_analytic = {"Analytically solve Cobb−Douglass maximization subject
        to a budget constraint",
        .vbase =2, .estimate=econ101_analytic_est, .score = econ101_analytic_score};
```

Listing 4.17 The analytic version. Online source: `econ101.analytic.c`.

of the solution to the constrained optimization. If you are familiar with Lagrange multipliers you should have little difficulty in verifying the equations expressed by the routines. The routines are in the natural slots for estimating parameters and estimating the vector of parameter derivatives.

- The term *score* is defined in Chapter 10, at which point you will notice that its use here is something of an abuse of notation, because the score is defined as the derivative of the log-utility function, while the function here is the derivative of the utility function.

- The preprocessor can sometimes provide quick conveniences; here it abbreviates the long function call to pull parameters to `fget`. Section 6.4 (page 211) gives the details of the preprocessor's use and many caveats.

- The process of wrapping library functions in standardized model routines, and oth-

```
#include <apop.h>
apop_model econ_101, econ_101_analytic;

void est_and_score(apop_model m, apop_data *params){
    gsl_vector *marginals = gsl_vector_alloc(2);
    apop_model *e = apop_estimate(params, m);
        apop_model_show(e);
        printf("\nThe marginal values:\n");
        apop_score(params, marginals, e);
        apop_vector_show(marginals);
        printf("\nThe maximized utility: %g\n", exp(e->llikelihood));
}

int main(){
    double param_array[] = {8.4, 1, 0.4,
                                    0, 3, 0.6}; //0 is just a dummy.
    apop_data *params = apop_line_to_data(param_array, 2,2,2);
        sprintf(apop_opts.output_delimiter, "\n");
        est_and_score(econ_101, params);
        est_and_score(econ_101_analytic, params);
}
```

Listing 4.18 Given the models, the `main` program is but a series of calls. Online source: `econ101.main.c`.

erwise putting everything in its place, pays off in the `main` function in Listing 4.18. Notably, the `est_and_score` function can run without knowing anything about model internals, and can thus run on both the closed-form and numeric-search versions of the model. It also displays the maximized utility, because—continuing the metaphor that likelihood=utility—the econ101 model's maximization returned the log utility in the `llikelihood` element of the output model.

- The `econ101.analytic` model calculates the parameters without calculating utility, but there is no need to write a separate calculation to fill it in—just call the utility calculation from the `econ101` model. Thanks to such re-calling of other models' functions, it is easy to produce variants of existing models.

- The only public parts of `econ101.c` and `econ101.analytic.c` are the models, so we don't have to bother with a header file, and can instead simply declare the models themselves at the top of `econ101.main.c`.

- The model files will be compiled separately, and all linked together, using a Makefile as per Appendix A, with an `OBJECTS` line listing all three `.o` files.

Q4.9   Use one of the models to produce a plot of marginal change in x_0 and x_1 as α expands.

> ➤ The `apop_model` aggregates methods of estimating parameters from data, drawing data given parameters, and estimating likelihoods given both.

> ➤ Most `apop_models` take in a data structure with an observation on each row and a variable on each column. If the model includes a dependent variable, it should be the first column.

> ➤ Given a prepackaged model, you can estimate the parameters of the model by putting your data into an appropriate `apop_data` structure, and then using `apop_estimate(data, your_model)`. This will produce an `apop_model` that you can interrogate or display on screen.

> ➤ If closed-form calculations for any of the model elements are available, then by all means write them in to the model. But the various functions that take `apop_models` as input make their best effort to fill in the missing methods. For example, the score function is not mandatory to use gradient-based optimization methods.

GRAPHICS

Graphs are friendly.

—Tukey (1977, p 157)

Graphics is one of the places where the computing world has not yet agreed on a standard, and so instead there are a dozen standards, including JPG, PNG, PDF, GIF, and many other TLAs. You may find yourself in front of a computer that readily handles everything, including quick displays to the screen, or you may find yourself logging in remotely to a command-line at the university's servers, that support only SVG graphics. Some journals insist on all graphics being in EPS format, and others require JPGs and GIFs.

The solution to the graphics portability problem is to use a handful of external programs, easily available via your package manager, that take in plain text and output an image in any of the panoply of graphics formats. The text-to-graphics programs here are as open and freely available as gcc, so you can be confident that your code will be portable to new computers.[1]

But this chapter is not just about a few plotting programs: it is about how you can control any text-driven program from within C. If you prefer to create graphics or do other portions of your analytic pipeline using a separate package (like one of the stats packages listed in the introduction), then you can use the techniques here to do so.

---

[1]There is some politics about how this is not strictly true: the maintainers of Gnuplot will not allow you to modify their code and then distribute the modified package independently (i.e., to *fork* the code base). The project is entirely unrelated to the GNU project, and the name is simply a compromise between the names that the two main authors preferred: nplot and llamaplot.

Appendix B takes a different approach to converting data to an executable script, via command-line tools to modify text. If your data is not going through any computations or transformations, you may be better off using a shell script as per Appendix B instead of writing a C program as per this chapter.

The plot-producing program with which this chapter will primarily be concerned is Gnuplot. Its language basically has only two verbs—`set` and `plot`—plus a few variants (`unset`, `replot`, et cetera). To give you a feel for the language, here is a typical Gnuplot script; you can see that it consists of a string of `set` commands to eliminate the legend, set the title, et cetera, and a final `plot` command to do the actual work.

```
unset key
set title "US national debt"
set term postscript color
set out 'debt.eps'
plot 'data−debt' with lines
```

In the code supplement, you will find the `plots` file, which provides many of the plots in this chapter in cut-and-paste-able form. You will also find the `data-debt` file plotted in this example. But first, here are a few words on the various means of sending these plot samples to Gnuplot.

PRELIMINARIES As with SQLite or mySQL, there is a command-line interpreter for Gnuplot's commands (`gnuplot`), so you can interactively try different `set` commands to see what options look best. After you have finished shopping for the best settings, you can put them into a script that will always produce the perfect plot, or better still, write a C program to autogenerate a script that will always produce the perfect plot.

There are various ways by which you can get Gnuplot to read commands from a script (e.g., a file named *plotme*).

- From the shell command line, run `gnuplot -persist <`*plotme*. This runs the script and exits, but the plot persists on the screen. Without the `-persist` option, the plot will disappear after a split second. If you are writing to a file rather than looking at the plot on screen, then the `-persist` option is unnecessary.
- Run `gnuplot` with no options, and from its prompt, type `load` *'plotme'*. This leaves you at the Gnuplot prompt to experiment with settings.
- The hybrid: `gnuplot` *plotme* `-` from the command line. This executes the instructions in *plotme*, but leaves you at the Gnuplot prompt to play with different settings.

set term ...	Meaning	Output goal
x11	Window system; most POSIX OSes	display on screen
windows	Window system; Windows OSes	display on screen
aqua	Window system; Mac OS X	display on screen
png	Portable network graphics	browser
gif	Graphics interchange format	browser, word processor
svg	scalable vector graphics	browser, word processor
ps	Postscript or encapsulated postscript	PDF
latex	LATEX graphics sub-language	LATEX docs

Table 5.1 Some of the terminal types that Gnuplot supports.

If you are at a Gnuplot prompt, you can exit via either the `exit` command or <ctrl-d>. On many systems, you can also interact with the plot, spinning 3-D plots with the mouse or zooming in to selected subregions of 2-D plots.

Check your Gnuplot installation. Write a one-line text file named `plotme` whose text reads: `plot sin(x)`. Execute the script using one of the above methods. Once that works, try the national debt example above.
If your system is unable to display plots to the screen, read on for alternative output formats.

`set term` **AND** `set out`   Gnuplot defaults to putting plots on the screen, which is useful for looking at data, but is not necessarily useful for communicating with peers. Still worse, some systems are not even capable of screen output. There are many potential solutions to the problem.

The `set terminal` command dictates the language with which to write the output. Table 5.1 presents the more common options, and Gnuplot's help system describes the deatils regarding each of them; e.g., `help set term latex` or `help set term postscript`. The default is typically the on-screen format appropriate for your system.[2]

The `set out` command provides a file name to write to. For example, if a Gnuplot file has

```
set term postscript color
set out 'a_plot.eps'
```

---

[2]The popular JPG format is not listed because its compression is designed to work well with photographic images, and makes lines and text look fuzzy. Use it for plots and graphs only as a last resort.

at its head, then the script will not display to the screen and will instead write the
designated file, in Postscript format.

Comments

Gnuplot follows the commenting standards of many
scripting languages: everything on a line after a # is
ignored. For example, if a script begins with
```
#set terminal postscript color
#set out 'printme.eps'
```
then these lines are ignored, and Gnuplot will display
plots on the screen as usual. If you later decide to print
the plot, you can delete the #s and the script will write
to `printme.eps`.

You can rest assured that no matter where you are, there is some
way to view graphics, but it may
take some experimentation to find
out how. For example, if you are
dialing in to a remote server, you
may be able to copy the graphics to
a `public_html` directory, make the
file publicly readable (chmod 644
plot.png) and then view the plot
from a web browser. Or, you could
produce Postscript output, and then run `ps2pdf` to produe a PDF file, which you
can open via your familiar PDF viewer.

Now that you know how to run a Gnuplot script and view its output, we can move
on to what to put in the script.

5.1 plot   The `plot` command will set the basic shape of the plot.

To plot a basic scatterplot of the first two columns of data in a file, such as the
`data-debt` file in the online code supplement, simply use `plot 'data-debt'`.

You will often dump more columns of data to your datafile than are necessary.
For example, the first portion of the `data-debt` file includes three columns: the
year, the debt, and the deficit. To plot only the first and third columns of data,
use `plot 'datafile' using 1:3`. This produces a scatterplot with X values
from column one and Y values from column 3. Notice that Gnuplot uses index
numbering instead of offset numbering: the first column is one, not zero.

Say that you just want to see a single data series, maybe column three; then `plot
'datafile' using 3`. With only one column given, the plot will assume that the
X values are the ordinal series $1, 2, 3, \ldots$ and your data are the Y values.

replot   You will often want multiple data sets on the same plot, and Gnuplot does
this easily using the `replot` command. Just change every use of `plot` after
the first to `replot`. Page 170 presents a few more notes on using this function.

```
set xrange [−4:6]
plot sin(x)
replot cos(x)
replot log(x) + 2*x − 0.5*x**2
```

- Gnuplot always understands the variable x to refer to the first axis and y to the second. If you are doing a parametric plot (see the example in Figure 11.4, page 360), then you will be using the variables t, u, and v.
- Gnuplot knows all of the functions in the standard C math library. For example, the above sequence will produce a set of pleasing curves. Notice that x**2 is common math-package notation for x^2.

splot   The plot command prints flat, 2-D plots. To plot 3-D surfaces, use splot. All of the above applies directly, but with three dimensions. If your data set has three columns, then you can plot it with splot 'datafile'. If your data set has more than three columns, then specify the three you want with a form like splot 'datafile' using 1:5:4.

- There is also the crosstab-like case where the data's row and column dimensions represent the X and Y axes directly: the $(1,1)$st element in the data file is the height at the Southwest corner of the plot, and the (n,n)th element is the height at the Northeast corner. To plot such data, use splot 'datafile' matrix.
- Surface plotting goes hand-in-hand with the pm3d (palette-mapped 3-D) option, that produces a pleasing color-gradient surface. For example, here is a simple example that produces the sort of plot used in advertisements for math programs. Again, if you run this from the Gnuplot prompt and a system that supports it, you should be able to use the mouse to spin the plot.

```
set pm3d
splot sin(x) * cos(y) with pm3d
```

Here is a more extended script, used to produce Figure 5.2. [This example appears in a slightly-modified form in agentgrid.gnuplot in the code supplement. The simulation that produced it is available upon request.]

```
1   set term postscript color;
2   set out 'plot.eps';
3   set pm3d; #for the contour map, use set pm3d map;
4   unset colorbox
5   set xlabel 'percent acting'; set ylabel 'value of emulation (n)';
```

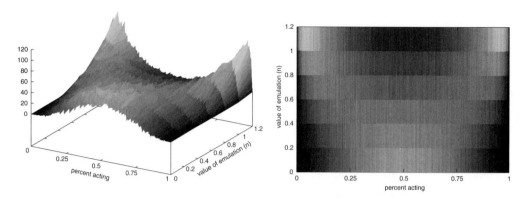

Figure 5.2 Two views of the same density plot from a few thousand simulations of a group of agents. Agents act iff $t_i + nk_i > C$, where t_i is a Normally distributed private preference, k_i is the percentage of other people acting, n is the agent's preference for emulating others, and C is a Uniformly-distributed cutoff. The vertical axis shows the density of simulations with the given percent acting and value of n. When the value of emulation is low, outcomes are unimodally distributed, but outcomes become bimodally distributed as emulation becomes more valuable.

```
6   set palette gray;
7   set xtics ('0' 0,'0.25' 250,'0.5' 500,'0.75' 750, '1' 999);
8   set ytics ('0' 0, '0.2' 1, '0.4' 2, '0.6' 3, '0.8' 4, '1' 5, '1.2' 6)
9   splot 'datafile' matrix with pm3d
```

- Lines one and two send the output to a Postscript file instead of the screen. Given line six, the `color` modifier is optional in this case.

- Line three tells Gnuplot to prepare for a palette-mapped surface, and is necessary before the `splot` command on line nine.

- Line four deletes the legend, which in the pm3d case is known as the colorbox.

- Line five sets the labels, and demonstrates that Gnuplot commands may put each either on a separate line or separated with a semicolon. As you can see from this script, ending a line with a semicolon is optional.

- Line six sets the color scheme to something appropriate for a book printed in black and white. There are many palettes to be had; the default, omitting the `set palette` line entirely, works for most purposes.

- Lines seven and eight fix the axis labels, because Gnuplot defaults to using the index of the column or row as the label. The format requires a text label, followed by the index at which the label will be placed; see below for further notes.

- Changing line three to `set pm3d map` produces an overhead view of the same surface, sometimes known as a *contour plot*, as in the second plot of Figure 5.2.

The short version   The sample lines above were all relatively brief, but you can put a huge amount of information onto one line. For example,

```
set style data bars
set style function lines
set linetype 3
set xrange [−10:10]
set yrange [0:2]
plot 'data' using 1:3 title 'data'
replot sin(x) title 'sine'

# can be rewritten as:
plot 'data' using 1:3 [−10:10][0:2] with bars title 'data', sin(x) with lines linetype 3 title 'sine'

# or as
plot 'data' using 1:3 [−10:10][0:2] w bars title 'data', sin(x) w l lt 3 title 'sine'
```

All of these settings will be discussed below. The purpose of this example is to show that style information can be put above the plot command, or it can be mixed in on the line defining the `plot`. The `replot` command is also technically optional, because you can add additional steps on one `plot` line using a comma. Finally, once you have everything on one line, you can abbreviate almost anything, such as replacing `with lines` with `w l`. This is yet another minimalism-versus-clarity tradeoff, and you are encouraged to stick with the clear version at first.

5.2   ✳ SOME COMMON SETTINGS   At this point, you can produce a basic plot of data or a function (or both at once). But you may have in mind a different look from Gnuplot's default, which means you will need to put a few `set` commands before the final `plot`.

This section catalogs the most common settings. For more information on the settings here and many more, see the very comprehensive Gnuplot documentation, which can be accessed from inside the Gnuplot command-line program via `help`, optionally followed by any of the headers below (e.g., `help set style`, `help set pointtype`).

Finally, if you are interactively experimenting with settings—which you are encouraged to do while reading this section—bear in mind that you may have to give a `replot` command (one word with no options) before the settings take effect.

`set style`   The basic style of the plot may be a simple line or points, a bar plot, boxes with error bars, or many other possibilities. Gnuplot keeps track of two types of style: that for function plotting (`set style function`) and for data plotting (`set style data`). For example, to plot something that looks like a bar chart, try `set style data boxes`, followed on the next line with `plot` *yourdata*. As above, this is equivalent to the slightly shorter form `plot` *yourdata* `with boxes`, but it is often useful to separate the style-setting from the plot content.

Other favorite data styles include `lines`, `dots`, `impulses` (lines from the x-axis to the data level), `steps` (a continuous line that takes no diagonals), `linespoints` (a line with the actual data point marked), and `errorbars` (to be discussed below).

If you are plotting a function, like `plot sin(x)`, then use `set style function lines` (or `dots` or `impulses`, et cetera). Because there are separate styles for functions and data, you can easily plot data overlaid by a function in a different style.

`set pointtype, set linetype`   You can set the width and colors of your lines, and whether your points will display as balls, triangles, boxes, stars, et cetera.[3]

The `pointtype` and `linetype` commands, among a handful of other commands, may differ from on-screen to Postscript to PNG to other formats, depending upon what is easy in the different formats. You can see what each terminal can do via the `test` command. E.g.:

```
set terminal postscript
set out 'testpage.ps'
test
```

Among other things, the test page displays a numbered catalog of points and lines available for the given terminal.

`set title, set xlabel, set ylabel`   These simple commands label the X and Y axes and the plot itself. If the plot is going to be a figure in a paper with a paper-side caption, then the title may be optional, but there is rarely an excuse for omitting axis labels. Sample usage:

---

[3] As of this writing, Gnuplot's default for the first two plotted lines is to use `linetype` 1=red and `linetype` 2=green. Seven percent of males and 0.4% of females are red–green colorblind and therefore won't be able to distinguish one line from the other. Try, e.g., `plot sin(x); replot cos(x) linetype 3`, to bypass `linetype` 2=green, thus producing a red/blue plot.

```
  ⎡ set xlabel 'Time, days'
  ⎢ set ylabel 'Observed density, picograms/liter'
  ⎢ set title 'Density over time'
  ⎣
```

set key   Gnuplot puts a legend on your plot by default. Most of the time, it is rea-
           sonably intelligent about it, but sometimes the legend gets in the way. Your
first option in this case is to just turn off the key entirely, via unset key.

The more moderate option is to move the key, using some combination of left,
right, or outside to set its horizontal position and top, bottom, or below to
set its vertical; ask help set key for details on the precise meaning of these
positions (or just experiment).

- The key also sometimes benefits from a border, via set key box.
- For surface plots with pm3d, the key is a thermometer displaying the range of colors
  and their values. To turn this off, use unset colorbox. See help set colorbox
  on moving the box or changing its orientation.

set xrange, set yrange   Gnuplot generally does a good job of selecting a default
                          range for the plot, but you can manually override this
using set range[min:max].

- Sometimes, you will want to leave one end of the range to be set by Gnuplot, but
  fix the other end, in which case you can use a * to indicate the automatically-set
  bound. For example, set yrange [*:10] fixes the top of the plot at ten, but lets
  the lower end of the plot fall where it may.
- You may want the axes to go backward. Say that your data represents rankings, so
  1 is best; then set yrange [*:1] reverse will put first place at the top of the
  plot, and autoset the bottom of the plot to just past the largest ranking in the data
  set.

set xtics AND set ytics   Gnuplot has reasonably sensible defaults for how the
                           axes will be labeled, but you may also set the tick marks
directly. To do so, provide a full list in parens for the text and position of every last
label, such as on lines seven and eight of the code on page 161.

Producing this by hand is annoying, but as a first indication of producing Gnuplot
code from C, here is a simple routine to write a set ytics line:

```
static void deal_with_y_tics(FILE *f, double min, double max, double step){
    int j = 0;
        fprintf(f, "set ytics (");
        for (double i=n_min; i< n_max; i+=n_step){
            fprintf(f, "'%g' %i", i, j++);
            if (i+n_step <n_max−1)
                fprintf(f, ", ");
        }
        fprintf(f, ")\n");
}
```

ASSORTED   Here are a few more settings that you may find handy.

unset border	*#Delete the border of the plot.*
unset grid	*#Make the plot even more minimalist.*
set size square	*#Set all axes to have equal length on screen or paper.*
set format y "%.3g"	*#You can use printf strings to format axis labels.*
set format y ""	*#Or just turn off printing on the Y axis entirely.*
set zero 1e−20	*#Set limit at which a point is rounded to zero (default: 1e−8).*

5.3   FROM ARRAYS TO PLOTS

The scripts above gave a file name from which to read the data (plot 'data-debt'). Alternatively, plot '-' tells Gnuplot to plot data to be placed immediately after the plot command. With the '-' trick, the process of turning a matrix into a basic plot is trivial. In fact, the principle is so simple that there are several ways of implementing it.

Write to a file   Let data be an apop_data set whose first and fifth columns we would like to plot against each other. Then we need to create a file, put a plot '-' command in the first line (perhaps preceded by a series of static set commands), and then fill the remainder with the data to be plotted. Below is the basic code to create a Gnuplot file. Since virtually anything you do with Gnuplot will be a variant of this code, it will be dissected in detail.

```
1   FILE *f = fopen("plot_me", "w");
2   if (!f) exit(0);
3   fprintf(f, "set key off; set ylabel 'picograms/liter'\n set xrange [−10:10]\n");
4   fprintf(f, "plot '−' using 1:5 title 'columns one and five'\n");
5   fclose(f);
6   apop_matrix_print(data−>matrix, "plot_me");
```

- The first argument to `fopen` is the file to be written to, and the second option should be either `"a"` for append or `"w"` for write anew; in this case, we want to start with a clean file so we use `"w"`.

- The `fopen` function can easily fail, for reasons including a mistyped directory name, no permissions, or a full disk. Thus, you are encouraged to check that `fopen` worked after every use. If the file handle is `NULL`, then something went wrong and you will need to investigate.

- As you can see from lines three and four, the syntax for `fprintf` is similar to that of the rest of the `printf` family: the first argument indicates to what you are writing, and the rest is a standard `printf` line, which may include the usual insertions via `%g`, `%i`, and so on.

- You can separate Gnuplot commands with a semicolon or newline, and can put them in one `fprintf` statement or many, as is convenient. However, you will need to end each line of data (and the `plot` line itself) with a newline.

- Since Gnuplot lets us select columns via the `using 1:5` clause, there is no need to pare down the data set in memory. Notice again that the first column in Gnuplot is 1, not 0.

- The `title` clause shows that Gnuplot accepts both "double quotes" and 'single quotes' around text such as file names or labels. Single quotes are nothing special to C, so this makes it much easier to enter such text.

- Line five closes the file, so there is no confusion when line six writes the data to the file. It prints the matrix instead of the full `apop_data` structure so that no names or labels are written. Since `apop_matrix_print` defaults to appending, the matrix appears after the `plot` header that lines two and three wrote to the file. You would need to set `apop_opts.output_append=0` to overwrite.

- At the end of this, `plot_me` will be executable by Gnuplot, using the forms like `gnuplot -persist < plot_me`, as above.

Instant gratification   The above method involved writing your commands and data to a file and then running Gnuplot, but you may want to produce plots as your program runs. This is often useful for simulations, to give you a hint that all is OK while the program runs, and to impress your friends and funders. This is easy to do using a pipe, so named because of UNIX's running data-as-water metaphor; see Appendix B for a full exposition.

The command `popen` does two things: it runs the specified program, and it produces a data pipe that sends a stream of data produced by your program to the now-running child program. Any commands you write to the pipe are read by the child as if someone had typed those commands into the program directly.

Listing 5.3 presents a sample function to open and write to a pipe.

```
1    #include <apop.h>
2
3    void plot_matrix_now(gsl_matrix *data){
4       static FILE *gp = NULL;
5         if (!gp)
6             gp = popen("gnuplot −persist", "w");
7         if (!gp){
8             printf("Couldn't open Gnuplot.\n");
9             return;
10        }
11        fprintf(gp,"reset; plot '−' \n");
12        apop_opts.output_type = 'p';
13        apop_opts.output_pipe = gp;
14        apop_matrix_print(data, NULL);
15        fflush(gp);
16   }
17
18   int main(){
19        apop_db_open("data−climate.db");
20        plot_matrix_now(apop_query_to_matrix("select (year*12+month)/12., temp from temp"));
21   }
```

Listing 5.3 A function to open a pipe to Gnuplot and plot a vector. Online source: `pipeplot.c`.

- The `popen` function takes in the location of the Gnuplot executable, and a `w` to indicate that you will be writing to Gnuplot rather than reading from it. Most systems will accept the simple program name, `gnuplot`, and will search the program path for its location (see Appendix A on paths). If gnuplot is not on the path, then you will need to give an explicit location like `/usr/local/bin/gnuplot`. In this case, you can find where Gnuplot lives on your machine using the command `which gnuplot`. The `popen` function then returns a `FILE*`, here assigned to gp.

- Since gp was declared to be a static variable, and `popen` is called only when gp==NULL, it will persist through multiple calls of this function, and you can repeatedly call the function to produce new plots in the same window.
  If gp is `NULL` after the call to `popen`, then something went wrong. This is worth checking for every time a pipe is created.

- But if the pipe was created properly, then the function continues with the now-familiar process of writing `plot '-'` and a matrix to a file. The `reset` command to Gnuplot (line 11) ensures that next time you call the function, the new plot will not have any strange interactions with the last plot.

- Lines 12 and 13 set the output type to `'p'` and the output pipe to gp; `apop_-matrix_print` uses these global variables to know that it should write to that pipe instead of a file or `stdout`.

- Notice the resemblance between the form here and the form used to write to a file above. A FILE pointer can point to either a program expecting input or a file, and the program writes to either in exactly the same manner. They are even both closed using fclose. The only difference is that a file opens via fopen and a program opens with popen. Thus, although the option is named apop_opts.output_-pipe, it could just as easily point to a file; e.g., FILE *f=fopen(*plotme*, "w"); apop_opts.output_pipe = f;.

- One final detail: piped output is often kept in a *buffer*, a space in memory that the system promises to eventually write to the file or the other end of the pipe. This improves performance, but you want your plot now, not when the system deems the buffer worth writing. The function on line 15, fflush, tells the system to send all elements of the gp buffer down the pipeline. The function also works when you are expecting standard output to the screen, by the way, via fflush(stdout).[4]

The main drawback to producing real-time plots is that they can take over your computer, as another plot pops up and grabs focus every half second, and can significantly slow down the program. Thus, you may want to settle for occasional redisplays, such as every fifty periods of your simulation, via a form like

```
for (int period =0; period< 1000; period++){
    gsl_vector *output = run_simulation();
    if (!(period % 50))
        plot_vector_now(output);
}
```

Q5.2

Now would be a good time to plot some data series.

- Query a two-column table from data-tattoo.db giving the birth year in the first column and the mean number of tattoos for respondents in that birth year in the second.

- Dump the data to a Gnuplottable file using the above fprintf techniques.

- Plot the file, then add set commands that you fprintf to file to produce a nicer-looking plot. E.g., try boxes and impulses.

- Modify your program to display the plot immediately using pipes. How could you have written the program initially to minimize the effort required to switch between writing to a file and a pipe?

- How does the graph look when you restrict your query to include only those with a nonzero number of tattoos?

---

[4]Alternatively, fflush(NULL) will flush all buffers at once. For programs like the ones in this book, where there are only a handful of streams open, it doesn't hurt to just use fflush(NULL) in all cases.

※*Self-executing files*   This chapter is about the many possible paths from a data
set to a script that can be executed by an external program;
here's one more.

Those experienced with POSIX systems know that a script can be made executable
by itself. The first line of the script must begin with the special marker #! followed by the interpreter that will read the file, and the script must be given execute
permissions. Listing 5.4 shows a program that produces a self-executing Gnuplot
script to plot $\sin(x)$. You could even use `system("./plotme")` to have the script
execute at the end of this program.

```
#include <apop.h>
#include <sys/stat.h> //chmod

int main(){
    char filename[] = "plot_me";
    FILE *f = fopen(filename, "w");
    fprintf(f, "#!/usr/bin/gnuplot −persist\n\
                plot sin(x)");
    fclose(f);
    chmod(filename, 0755);
}
```

Listing 5.4 Write to a file that becomes a self-executing Gnuplot script. Run the script using
`./plot_me` from the command line. Online source: `selfexecute.c`.

※ `replot` **REVISITED**   `replot` and the `plot ’-’` mechanism are incompatible, because Gnuplot needs to re-read the data upon replotting, but
can not back up in the stream to reread it. Instead, when replotting data sets, write
the data to a separate file and then refer to that file in the file to be read by Gnuplot.
To clarify, here is a C snippet that writes to a file and then asks Gnuplot to read the
file:

```
apop_data_print(data, "datafile");
FILE *f = fopen("gnuplot", "w");
fprintf(f, "plot ’datafile’ using 1 title ’data column 1’;\n \
            replot ’datafile’ using 5 title ’data column 5’;\n");
fclose(f);
```

Also, when outputting to a paper device, replotting tends to make a mess. Set the
output terminal just before the final replot:

```
plot ’datafile’ using 1
replot ’datafile’ using 2
replot ’datafile’ using 3
set term postscript color
set out ’four_lines.eps’
replot ’datafile’ using 4
```

$\sum$

> ➤ A Gnuplot command file basically consists of a group of `set` commands to specify how the plot will look, and then a single `plot` (2-D version) or `splot` (3-D version) command.

> ➤ You can run external text-driven programs from within C using `fopen` and `popen`.

> ➤ Using the `plot '-'` form, you can put the data to be plotted in the command file immediately after the command. You can use this and `apop_data_print` to produce plot files of your data.

5.4 A SAMPLING OF SPECIAL PLOTS

For a system that basically only has a `set` and a `plot` command, Gnuplot is surprisingly versatile. Here are some specialized visualizations that go well beyond the basic 2-D plot.

LATTICES   Perhaps plotting two pairs of columns at a time is not sufficient—you want bulk, displaying every variable plotted against every other. For this, use the `apop_plot_lattice` function.

```
#include "eigenbox.h"

int main(){
    apop_plot_lattice(query_data(), "out");
}
```

Listing 5.5 The code to produce Figure 5.6. Link with the code in Listing 8.2, p 267. Online source: `lattice.c`.

Listing 5.5 produces a file that (via `gnuplot -persist <out`) produces the plot in Figure 5.6. Yes, it is one line of code, but you will need to link it with the data-querying code in Listing 8.2, p 267. Each variable is plotted against every other; e.g., the upper-middle plot shows males per 100 females versus state population, and the middle-left plot shows a mirror image (along the diagonal line) of the same plot.

What is a lattice plot good for? Some call it *getting a lay of the land*, while others call it *data snooping*. Given ten perfectly random variables, there is a good chance that at least one pair of lattice plots will look to you as if it demonstrates a nice correlation. A formal regression on the chosen pair of variables will likely verify your initial visual impression. For more on this conflict, see page 316.

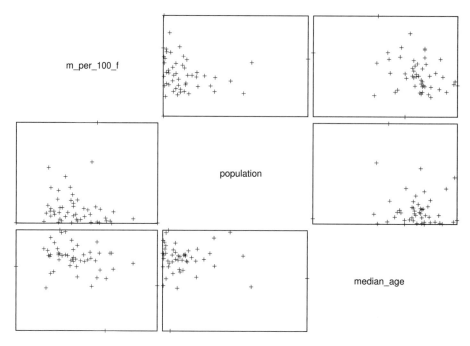

Figure 5.6 The Census data, as queried on page 267.

ERROR BARS   The typical error bar has three parts: a center, a top limit, and a bottom limit. Gnuplot supports this type of data directly, via `set style data errorbars`. You can provide the necessary information in a variety of formats; the most common are (x, y center, y top, y bottom) and (x, y center, y range), where the range is typically a standard deviation. Listing 5.7, which produces Figure 5.8, takes the second approach, querying out a month, the mean temperature for the month, and the standard deviation of temperature for the month. Plotting the data shows both the typical annual cycle of temperatures and the regular fluctuation of variances of temperature.

HISTOGRAMS   A histogram is a set of X- and Y-values like any other, so plotting it requires no special tools. However, Gnuplot will not take a list of data and form a histogram for you—you have to do this on the C-side and then send the final histogram to Gnuplot. Fortunately, `apop_plot_histogram` does the binning for you. Have a look at Listing 11.2, page 359, for an example of turning a list of data items into a histogram (shown in Figure 11.3).

```
#include <apop.h>

int main(){
    apop_db_open("data−climate.db");
    apop_data *d = apop_query_to_data("select \
            (yearmonth/100. − round(yearmonth/100.))∗100 as month, \
            avg(tmp), stddev(tmp) \
            from precip group by month");
    printf("set xrange[0:13]; plot '−' with errorbars\n");
    apop_matrix_show(d−>matrix);
}
```

Listing 5.7 Query out the month, average, and variance, and plot the data using errorbars. Prints
to stdout, so pipe the output through Gnuplot: ./errorbars | gnuplot -persist.
Online source: errorbars.c.

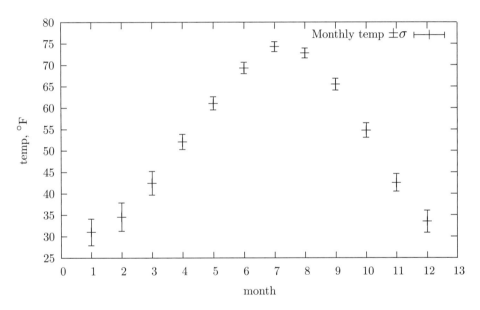

Figure 5.8 Monthly temperature, $\pm\sigma$.

The *leading digit* of a number is simply its most significant digit: the leading digit of 3247.8 is 3, and the leading digit of 0.098 is 9. *Benford's law* (Benford, 1938) states that the digit $d \in \{0, 1, \ldots, 9\}$ will be a leading digit with frequency

$$F_d \propto \ln\left((d+1)/d\right). \tag{5.4.1}$$

Q5.3 Check this against a data set of your choosing, such as the population column in the World Bank data, or the Total_area column from the US Census data. The formal test is the exercise on page 322; since this is the chapter on plotting, just produce a histogram of the chosen data set and verify that it slopes sharply downward. (*Hint*: if $d = 3.2e7$, then $10^{(\text{int})\log10(d)} = 1e7$.)

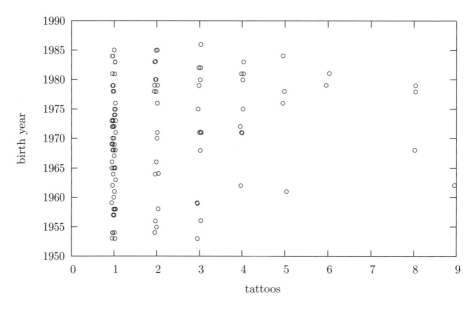

Figure 5.9 Each point represents a person.

LOG PLOTS   You can plot your data on a log scale by either transforming it before it gets to Gnuplot or by transforming the plot.

Log plots work best on a log-base-10 scale, rather than the typical natural logarithm, because readers can immediately convert a 2 on the scale to 1e2, a −4 to 1e−4, et cetera. From C, you can use the `log10(x)` function to calculate $\log_{10}x$, and if your data is in a `gsl_vector`, you can use `apop_vector_log10` to transform the entire vector at once.

In Gnuplot, simply `set logscale y` to use a log-scaled Y axis, `set logscale x` to use a log-scaled X axis, or `set logscale xy` for both.

Q5.4   Redo the `data-debt` plot from the beginning of this chapter using a log scale.

PRUNING AND JITTERING   Plotting the entire data set may be detrimental for a few reasons. One is the range problem: there is always that one data point at $Y = 1e20$ throws off the whole darn plot. If you are using an interactive on-screen plot, you can select a smaller region, but it would be better to just not plot that point to begin with.

The second reason for pruning is that the data set may be too large for a single page. The black blob you get from plotting ten million data points on a single piece of paper is not very informative. In this case, you want to use only a random subset of the data.[5]

Both of these are problems of selecting data, and so they are easy to handle via SQL. A simple `select * from plotme where value < 1e7` will eliminate values greater than a million, and page 84 showed how to select a random subset of the data.

In Gnuplot, you can add the `every` keyword to a plot, such as `plot 'data' every 5` to plot every fifth data point. This is quick and easy, but take care that there are no every-five patterns in the data.

Now consider graphing the number of tattoos a person has against her year of birth. Because both of these are discrete values, we can expect that many people will share the same year of birth and the same tattoo count, meaning that the plot of those people would be exactly one point. A point at (1965, 1 tattoo) could represent one person or fifty.

```
1    #include <apop.h>
2    gsl_rng *r;
3
4    void jitter(double *in){
5        *in += (gsl_rng_uniform(r) − 0.5)/10;
6    }
7
8    int main(){
9        apop_db_open("data−tattoo.db");
10       gsl_matrix *m = apop_query_to_matrix("select \
11                   tattoos.'ct tattoos ever had' ct, \
12                   tattoos.'year of birth'+1900 yr \
13                   from tattoos \
14                   where yr < 1997 and ct+0.0 < 10");
15       r = apop_rng_alloc(0);
16       apop_matrix_apply_all(m, jitter);
17       printf(set key off; set xlabel 'tattoos'; \n\
18               set ylabel 'birth year'; \n\
19               plot '−' pointtype 6\n");
20       apop_matrix_show(m);
21   }
```

Listing 5.10 By adding a bit of noise to each data point, the plot reveals more data. Online source: `jitter.c`.

---

[5]Another strategy for getting less ink on the page is to change the point type from the default cross to a dot. For the typical terminal, do this with `plot ... pointtype 0`.

One solution is to add a small amount of noise to every observation, so that two points will fall exactly on top of each other with probability near zero. Figure 5.9 shows such a plot. Without jittering, there would be exactly one point on the one-tattoo column for every year from about 1955 to 1985; with jittering, it is evident that there are generally more people with one tattoo born from 1965–1975 than in earlier or later years, and that more than half of the sample with two tattoos was born after 1975.

Further, the process of jittering is rather simple. Listing 5.10 shows the code used to produce the plot. Lines 10–14 are a simple query; lines 17–20 produce a Gnuplot header and file that get printed to `stdout` (so run the program as `./jitter | gnuplot`). In between these blocks, line 16 applies the `jitter` function to every element of the matrix. That function is on lines 4–6, and simply adds a random number $\in [-0.05, 0.05]$ to its input. Comment out line 16 to see the plot without jitter.

Q5.5   Modify the code in Listing 5.10 to do the jittering of the data points in the SQL query instead of in the `gsl_matrix`.

➤ Apophenia will construct a grid of plots, plotting every variable against every other, via `apop_plot_lattice`.

➤ With Gnuplot's `set style errorbars` command, you can plot a range or a one-σ spread for each data point.

➤ You need to aggregate data into histograms outside of Gnuplot, but once that is done, plotting them is as easy as with any other data set.

➤ If the data set has an exceptional range, you can take the log of the data in C via `log10`, plot the log in Gnuplot via `set logscale y`, or trim the data in SQL via, e.g., `select cols from table where var<1e20`.

➤ If data falls on a grid (e.g., integer-valued rows and columns), then you can add *jitter* to the plot to reveal the density at each point.

5.5 ANIMATION Perhaps three dimensions is not quite enough, and you need one more. Gnuplot easily supports animation: just stack matrices one after the next and call `plot` in between. However, many media (such as paper) do not yet support animation, meaning that your output will generally be dependent on your display method. The GIF format provides animation and is supported by all major web browsers, so you can also put your movies online.[6]

There are two details that will help you with plotting multiple data sets. First, Gnuplot reads an e alone on a line to indicate the end of a data set. Second, Gnuplot allows you to define constants via a simple equals sign; e.g., the command p = 0.6 creates the variable p and sets it to 0.6. Third, Gnuplot has a pause *p* command that will wait p seconds before drawing the next plot.

Tying it all together, we want a Gnuplottable file that looks something like this:

```
set pm3d;
  p = 1;
splot '−' with pm3d
{data[0] here}
e
  pause p;
splot '−' with pm3d
{data[1] here}
e
  pause p;
splot '−' with pm3d
{data[2] here}
e
```

You can run the resulting output file through the Gnuplot command line as usual. If a one-second pause is too long or too short, you only need to change the single value of p at the head of the file to change the delay throughout.

You can write the above plots to a paper-oriented output format, in which case each plot will be on a separate page. You could then page through them with a screen viewer, or perhaps print them into a flipbook. When writing a file, set p=0 in the above, since there is no use delaying between outputs.

In addition to the example below, Listings 7.12 (page 253) and 9.1 (page 298) also demonstrate the production of animations.

---

[6]For GIFs, you will need to request animation in the `set term` line, e.g., `set term gif animate delay 100`. The number at the end is the pause between frames in hundredths of a second.

 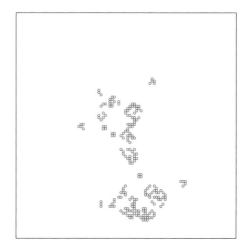

Figure 5.11 The Game of Life: at left is the original colony; at right is the colony after 150 periods. The vaguely V-shaped figures (such as the groups farthest left, right, and top) are known as *gliders*, because they travel one step forward every four periods.

An example: the game of life   There is a tradition of agent-based modeling built around plotting agents on a grid, pioneered by Epstein & Axtell (1996). A simple predecessor is Conway's Game of Life, a *cellular automaton* discussed at length in Gardner (1983). The game is played on a grid, where each point on the grid can host a single blob. These blobs can not move, and are somewhat delicate: if the blob has only zero or one neighbors, it dies of loneliness, and if it has four or more neighbors, it dies of overcrowding. If an empty cell is surrounded by exactly three blobs, then a new blob is born on that cell.

These simple rules produce complex and interesting patterns. At left in Figure 5.11 is the so-called *r pentomino*, a simple configuration of five blobs. At right is the outcome after 150 periods of life. Listing 5.12 presents the code to run the game. Because the program prints Gnuplot commands to `stdout`, run it using `./life | gnuplot`.

- The game uses two grids: the completed grid from the last period, named `active` in the code, and the incomplete grid that will soon represent the state of life in the next period, named `inactive`.
- Both grids are initialized to zero (lines 27–28), and then lines 29–31 define the r pentomino.
- Line 32 is the Gnuplot header, and line 34 tells Gnuplot that data points are coming.
- The main work each period is preparing the inactive grid, which is what the `calc_-grid` function does. It sets everything to zero, and then the two loops (`i`-indexed for rows and `j`-indexed for columns) checks every point in the grid except the borders.
- The `area_pop` function calculates the population in a 3×3 space; line 16 needs

```
1    #include <apop.h>
2    int area_pop(gsl_matrix *a, int row, int col){
3       int i, j, out = 0;
4          for (i=row−1; i<= row+1; i++)
5              for (j=col−1; j<= col+1; j++)
6                  out += gsl_matrix_get(a, i, j);
7          return out;
8    }
9
10   void calc_grid(gsl_matrix* active, gsl_matrix* inactive, int size){
11      int i, j, s, live;
12         gsl_matrix_set_all(inactive, 0);
13         for(i=1; i< size−1; i++)
14             for(j=1; j< size−1; j++){
15                 live = gsl_matrix_get(active, i, j);
16                 s = area_pop(active, i, j) − live;
17                 if ((live && (s == 2 || s == 3))
18                     || (!live && s == 3)){
19                         gsl_matrix_set(inactive, i, j, 1);
20                         printf("%i %i\n", i, j);
21                 }
22             }
23   }
24
25   int main(){
26      int i, gridsize=100, periods = 550;
27      gsl_matrix *t, *active = gsl_matrix_calloc(gridsize,gridsize);
28      gsl_matrix *inactive = gsl_matrix_calloc(gridsize,gridsize);
29         gsl_matrix_set(active, 50, 50, 1); gsl_matrix_set(active, 49, 51, 1);
30         gsl_matrix_set(active, 49, 50, 1); gsl_matrix_set(active, 51, 50, 1);
31         gsl_matrix_set(active, 50, 49, 1);
32         printf("set xrange [1:%i]\n set yrange [1:%i]\n", gridsize, gridsize);
33         for (i=0; i < periods; i++){
34             printf("plot '−' with points pointtype 6\n");
35             calc_grid(active, inactive, gridsize);
36             t = inactive;
37             inactive = active;
38             active = t;
39             printf("e\n pause .02\n");
40         }
41   }
```

Listing 5.12 Conway's game of life. Online source: `life.c`.

to subtract the population (if any) at the central point to get the population in the point's neighborhood.

- Notice that the loops in both `area_pop` and `calc_grid` never consider the edges of the grids. That means that `area_pop` does not need to concern itself with edge conditions like if (i!=0)....

- The rules of the Game of Life are summarized in the if statement in lines 17–18. There will be a blob at this point if either there is currently a blob and it has two or three neighbors, or there is no living blob there but there are three neighbors.
- If the if statement finds that there will be a blob in this space next period, it prints a point to Gnuplot, and marks it in the currently inactive grid.
- Lines 36–38 is the classic swap of the active and inactive grids, using a temp location to help make the exchange. Since we are only shunting the addresses of data, the operation takes zero time.

$\mathbb{Q}_{5.6}$ | Other rules for life or death also produce interesting results. For example, rewrite the rules so that a living blob stays alive only if there are 2, 3, 4, or 5 neighbors, and an empty space has a birth only if there is one neighbor. Or try staying-alive rules of 2 and 6 with birth rules of 1 and 3.

5.6 ON PRODUCING GOOD PLOTS   Cleveland & McGill (1985) offer some suggestions for producing plots for the purpose of perceiving the patterns among the static. Their experiments were aimed at how well people could compare data presented in various formats, and arrived at an ordering of graphical elements from those that were most likely to allow accurate perception to layouts that inhibited accurate perception:

- *i*) Position along a common scale (e.g., the height of the means in the temperature plot on page 173)
- *ii*) Position on identical but nonaligned scales (such as comparing points on two separate graphs)
- *iii*) Length (e.g., the height of the error bars in the temperature plot on page 173)
- *iv*) Angle
- *v*) Slope (when not too close to vertical or horizontal)
- *vi*) Area
- *vii*) Volume, Density, Color saturation (e.g., a continuous scale from light blue to dark blue)
- *viii*) Color hue (e.g., a continuous scale from red to blue)

Data should be presented using techniques as high up on the scale as possible. Pie charts (representing data via angle and area) are a bad idea because there are many better ways to present the same data. Gnuplot does provide a means of setting data-dependent color, but given that color is at the bottom of Cleveland and

McGill's list, it should be used as a last resort. They did not run experiments with animation, but our eyes have cells exclusively dedicated to sensing motion, so it seems sensible that if movement were included on the list, it would rank highly.

As for angles and slopes, consider these plots of the US national debt since 1995. The top plot has a y axis starting at $0, while the second has a y axis starting at $4.5 trillion. Is the rate of change from 1995–1996 larger or smaller than the rate of change from 2005–2006? Was growth constant or decelerating from 1995–2000? It is difficult to answer these questions from the top graph, because everything is somewhat flat—the change in angles is too small to be perceived, and we are bad at discerning slopes. Differences in slope on the second scale are more visible. The lesson is that plots show their patterns most clearly when the axes are set such that the slope is around 45°.

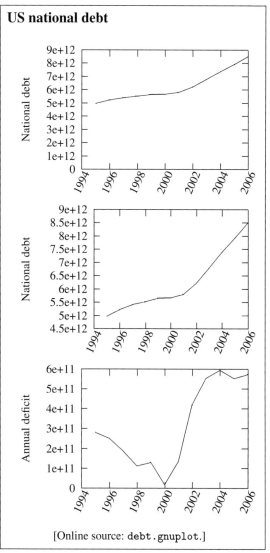

[Online source: `debt.gnuplot`.]

The bottom plot is the US national deficit. This is the amount the government spends above its income, and is thus the rate of change of the debt. The rate of change in the first two graphs (a slope in those graphs) is the height of this plot for each year, and position along a common scale is number one on Cleveland and McGill's list. The answers to the above questions are now obvious: the rate of change for the debt slowed until 2000 and then quickly rose to about double 1995 rates.

Darrell Huff, in the classic *How to Lie With Statistics* (Huff & Geis, 1954, Chapter 5), has a different goal and so makes different recommendations. He points out that the top two graphs tell a different narrative. The second graph tells a story that the national debt is rapidly increasing, because the height of the point at 2006 is about eight times the height at 1995. The top graph shows that the debt rose, but not at a fast-multiplying rate. Huff concludes that the full story is best told when the zero of the y axis always in the picture, even if this means blank space on the page—advice that directly contradicts the advice above. So the choice for any given plot depends on the context and intent, although some rules—like avoiding pie charts and continuous color scales—are valid for almost all situations.

5.7 ✳ GRAPHS—NODES AND FLOWCHARTS In common conversation, we typically mean the word *graph* to be a plot of data like every diagram to this point in the chapter. The mathematician's definition of *graph*, however, is a set of nodes connected by edges, as in Figure 5.13. Gnuplot can only plot; if you have network data that you would like to visualize, Graphviz is the package to use. Like all of the tools in this book, Graphviz installs itself gracefully via your package manager.

The package includes various executables, the most notable of which are dot and neato. Both take the same input files, but dot produces a flowchart where there is a definite beginning and end, while neato produces more amorphous plots that aim only to group nodes via the links connecting them.

The programs are similar to Gnuplot in that they take in a plain text description of the nodes and edges, and produce an output file in any of a plethora of graphics formats. The syntax for the input files is entirely different from Gnuplot's, but the concept is familiar: there are elements to describe settings interspersed with data elements.

For example, flip back to page 3 and have a look at Figure 1.3. Produce the first graph in the figure using the following input to dot:

```
digraph {
    rankdir = LR;
    node [shape=box];
    "Data" -> "Estimation" -> "Parameters";
}
```

The lines with = in them set parameters, stating that the graph should read left-to-right instead of the top-to-bottom default, and that the nodes should be boxes instead of the default ellipses. The line with the ->s defines how the nodes should link, and already looks like a text version of Figure 1.3. The command line

```
dot -Tpng <graphdata.dot > output.png
```

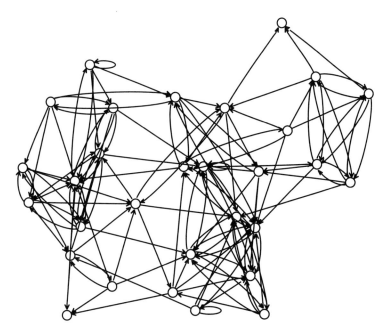

Figure 5.13 The social network of a Junior High classroom.

produces a graph much like that in Figure 1.3, in the PNG format.[7]

At this point, you have all the tools you need to autogenerate a graph. For example, say that you have an $n \times n$ grid where a one in position (i, j) indicates a link between agents i and j and a zero indicates no link. Then a simple `for` loop would convert this data into a `neato`-plottable file, with a series of rows with a form like, e.g., `node32 -> node12`.

```
void produce_network_graph(apop_data *link_data, char *outfile){
    FILE *g = fopen(outfile, "w");
    fprintf(g, "digraph{\n");
    for (int i=0; i< link_data−>matrix−>size1; i++)
        for (int j=i+1; j< link_data−>matrix−>size2; j++)
            if (apop_data_get(link_data, i,j))
                fprintf(g, "node%i −> node%i;\n", i,j);
    fprintf(g, "}\n");
    fclose(g);
}
```

In the code supplement, you will find a file named `data-classroom`, which lists the survey results from a Junior High classroom in LA, in which students listed

---

[7]The second half of Figure 1.3 was produced using exactly the same graph, plus the `psfrag` package to replace text like *Data* with the OLS-specific math shown in the figure.

their five best friends. The *ego* column is the student writing down the name of his or her best friend, and the *nominee* column gives the number of the best friend. Figure 5.13 graphs the classroom, using `neato` and the following options for the graph:

```
digraph{
    node [label="",shape=circle,height=0.12,width=0.12];
    edge [arrowhead=open,arrowsize=.4];
    ...
    }
```

A few patterns are immediately evident in the graph: at the right is a group of four students who form a complete clique, but do not seem very interested in the rest of the class. The student above the clique was absent the day of the survey and thus is nominated as a friend but has no nominations him or herself; similarly for one student at the lower left. Most of the graph is made from two large clumps of students who are closely linked, at the left and in the center, probably representing the boys and the girls. There are two students who nominated themselves as best friends (one at top right and one at the bottom), and those two students are not very popular.

Write a program to replicate Figure 5.13.

- Read the `data-classroom` file into an `apop_data` set, using either `apop_text_to_db` and a query, or `apop_text_to_data`.

- Open your output file, and copy in the header from above.

- Write a single `for` loop to write one line to the output file for each line in the data set. Base your printing function on the one from page 183.

- Close the file.

Running the program will produce a `neato`-formatted file.

- From the command line, run `neato -Tps <my_output > out.eps`. The output graph should look just like Figure 5.13.

- In the `dot` manual (`man dot` from the command line), you will see that there are many variant programs to produce different types of graph. What does the classroom look like via `dot`, `circo`, and `twopi`?

The exercise on page 414 will show you another way to produce the Graphviz-readable output file.

Internal use   If you have been sticking to the philosophy of coding via small, simple
                  functions that call each other, your code will look like a set of elements
linked via function calls—exactly the sort of network for which Graphviz was
written. If you feel that your code files are getting a bit too complex, you can use
Graphviz to get the big picture.

For example, say that your database is growing involved, with queries that merge
tables into new tables, other queries to split the tables back into still more tables, et
cetera. For each query that creates a table, it is easy to write down a line (or lines)
like `base_tab -> child_tab`. Then, `dot` can sort all those individual links into
a relatively coherent flow from raw data to final output.[8]

You could also graph the calling relationships among the functions in your C
code—but before you start manually scanning your code, you should know that
there is a program to do this for you. Ask your package manager for `doxygen`,
which generates documentation via specially-formatted comments in the source
file. If configured correctly, it will use Graphviz to include call graphs in the documentation.

The online code supplement includes a few more examples of Graphviz at work,
including the code used to create Figures 1.1, 6.5, and 6.7.

➤ The Graphviz package produces graphs from a list of nodes and
   edges. Such lists are easy to autogenerate from C.

➤ You can also use Graphviz to keep track of relationships among functions in your code or tables in your database.

5.8   ⁂ PRINTING AND LATEX   This book focuses on tools to write replicable,
                                   portable analyses, where every step is described
in a handful of human-legible text files that are sent to programs that behave in the
same manner on almost any computer. The TEX document preparation system (and
the set of macros built on top of it, LATEX) extend the pipeline to the final writeup.
For example, you can easily write a script to run an analysis and then regenerate
the final document using updated tables and plots.[9]

---

[8]SQL does not have the sort of metadata other systems have for describing a table's contents in detail (e.g.,
the `apop_data` structure's `title` element). But you can set up a metadata table, with a column for the table
name, its description, and the tables that generated that table. Such a table is reasonably easy to maintain, because
you need only add an `insert into metadata ...` query above any query that generates a table. ℚ: Write a
function to take in such a table and output a flowchart demonstrating the flow of data through the database tables.

[9]To answer some questions you are probably wondering: yes, this book is a LATEX document. Most of the plots
were produced via `set term latex` in Gnuplot, to minimize complications with sending Postscript to the press.
Save for the pointer-and-box diagrams in the C chapter, Student's hand-drawn menagerie, and the snowflake at

A complete tutorial on LaTeX would be an entire book—which has already been written dozens of times. But this chapter's discussion of the pipeline from raw data to output graphs is incomplete without mention of a few unpleasant details regarding plots in LaTeX.

You have two options when putting a plot in a TeXed paper: native LaTeX and Postscript.

Native format   Just as you can set the output device to a screen or a Postscript printer, you can also send it to a file written using LaTeX's graphics sub-language. One the plus side, the fonts will be identical to those in your document, and the resolution is that of TeX itself (100 times finer than the wavelength of visible light). On the minus side, some features, such as color, are currently not available.

Producing a plot in LaTeX format requires setting the same `term`/`out` settings as with any other type of output: `set term latex; set out 'plot.tex'`.[10]

Just as you can dump one C file into another via `#include`, you can include the Gnuplot output via the `\input` command:

```
\documentclass{article}
\usepackage{latexsym}
\begin{document}
...
\begin{figure}
\input outfile.tex
\caption{This figure was autogenerated by a C program.}
\end{figure}
...
\end{document}
```

Another common complaint: the Y-axis label isn't rotated properly. The solution provides a good example of how you can insert arbitrary LaTeX code into your Gnuplot code. First, in the Gnuplot file, you can set the label to any set of instructions that LaTeX can understand. Let λ be an arbitrary label; then the following command will write the label and tell LaTeX to rotate it appropriately:

---

the head of every chapter, I made a point of producing the entire book using only the tools it discusses.

The snowflake was generated by covering a triangle with a uniform-by-area distribution of dots, each with a randomly selected color and size, and then rotating the triangle to form the figure. Therefore, any patterns you see beyond the six-sided rotational symmetry are purely apophenia.

And in my experience helping others build data-to-publication pipelines, the detail discussed in this section about rotating the Y-axis label really *is* a common complaint.

[10]By the way, for a single plot, the `set out` command is optional, since you could also use a pipe: `gnuplot < plotme > outfile.tex`.

```
set ylabel '\rotatebox{90}{Your $\lambda$ here.}'
```

Two final notes to complete the example: `\rotatebox` is in the `graphicx` package, so it needs to be called in the document preamble:

```
\usepackage{latexsym, graphicx}
```

Second, many dvi viewers do not support rotation, so if you are viewing via TEX's native dvi format, the rotation won't appear. Use either `pdflatex` or `dvips` to view the output as it will print.

The Postscript route   Which brings us to the second option for including a graphic: Postscript.

Use the *graphicx* package to incorporate the plot. E.g.:

```
\documentclass{article}
\usepackage{graphicx}
\begin{document}
...
\begin{figure}
\rotatebox{90}{\scalebox{.35}{\includegraphics{outfile.eps}}}}
\caption{This figure was autogenerated by a C program.}
\end{figure}
...
\end{document}
```

Notice that you will frequently need to rotate the plot $90°$ and scale the figure down to a reasonable size.

The first option for generating PDFs is to use `epstopdf`. First, convert all of your eps files to pdf files on the command line. In `bash`, try

```
for i in *.eps; do
    epstopdf $i;
done
```

Then, in your LATEX header, add

```
\usepackage[pdftex]{epsfig}
```

The benefit to this method is that you can now run `pdflatex my_document` without incident; the drawback is that you now have two versions of every figure cluttering up your directory, and must regenerate the PDF version of the graphic every time you regenerate the Postscript version.

The alternative is to go through Postscript in generating the document:

```
latex a_report
dvips < a_report.dvi > a_report.ps
ps2pdf a_report.ps
```

Either method is a lot of typing, but there is a way to automate the process: the `make` program, which is discussed in detail in Appendix A. Listing 5.14 is a sample makefile for producing a PDF document via the Postscript route. As with your C programs, once the makefile is in place, you can generate final PDF documents by just typing `make` at the command line. The creative reader could readily combine this makefile with the sample C makefile from page 387 to regenerate the final report every time the analysis or data are updated. (*Hint*: add a `gen_all` target that depends on the other targets. That target's actions may be blank.)

```
DOCNAME = a_report

pdf: $(DOCNAME).pdf

$(DOCNAME).dvi: $(DOCNAME).tex
        latex $(DOCNAME); latex $(DOCNAME)

$(DOCNAME).ps: $(DOCNAME).dvi
        dvips −f < $(DOCNAME).dvi > $(DOCNAME).ps

$(DOCNAME).pdf: $(DOCNAME).ps
        ps2pdf $(DOCNAME).ps $(DOCNAME).pdf

clean:
        rm −f $(DOCNAME).blg $(DOCNAME).log $(DOCNAME).ps
```

Listing 5.14 A makefile for producing PDFs from LaTeX documents. Online source: `Makefile.tex`.

➤ Once a plot looks good on screen, you can send it to an output file using the `set term` and `set out` commands.

➤ For printing, you will probably want to use `set term postscript`. For online presentation, use `set term png` or `set term gif`. For inserting into a LaTeX document, you can either use Postscript or `set term latex`.

6

✳ MORE CODING TOOLS

If you have a good handle on Chapter 2, then you already have what you need to write some very advanced programs. But C is a world unto itself, with hundreds of utilities to facilitate better coding and many features for the programmer who wishes to delve further.

This chapter covers some additional programming topics, and some details of C and its environment. As with earlier chapters, the syntax here is C-specific, but it is the norm for programming languages to have the sort of features and structures discussed here, so much of this chapter will be useful regardless of language.

The statistician reader can likely get by with just a skim over this chapter (with a focus on Section 6.1), but readers working on simulations or agent-based models will almost certainly need to use the structures and techniques described here.

The chapter roughly divides into three parts. After Section 6.1 covers functions that operate on other functions, Section 6.2 will use such functions to build structures that can hold millions of items, as one would find in an agent-based model. Section 6.3 shows the many manners in which your programs can take in parameters from the outside world, including parameter files, enivornment variables, and the command line. Sections 6.4 and 6.5 cover additional resources that make life in front of a computer easier, including both syntactic tricks in C and additional programs useful to programmers.

6.1   FUNCTION POINTERS   A data point d is stored somewhere in memory, so we can refer to its address, &d. Similarly, a function f is stored somewhere in memory, so we can refer to its address as well.

What is a pointer to a function good for? It lets us write Functions that will accept any function pointer and then use the pointed-to function. [Functions calling functions is already confusing enough, so I will capitalize Function to indicate a parent function that takes a lower-case function as an input.] For example, a Function could search for the largest value of an input function over a given range, or a bootstrap Function could take in a statistic-calculating function and a data set and then return the variance of the statistic.

TYPES   Before we can start writing Functions to act on functions, we need to take the type of input function into consideration. If a function expects ints, then the compiler needs to know this, so it can block attempts to send the function a string or array.

The syntax for declaring a function pointer is based on the syntax for declaring a function. Say that we want to write a Function that will take in an array of doubles plus a function, and will apply the function to every element of the array, returning an array of ints. Then the input function has to have a form like

$$\text{int double\_to\_int (\textbf{double} x);} \tag{A}$$

Recall that a pointer declaration is just like a declaration for the pointed-to type but with another star, like int *i; the same goes for declaring function pointers, but there are also extra parens. Here is a type for a function pointer that takes in a double and returns a pointer to int; you can see it is identical to line A, but for the addition of a star and parens:

$$\text{int (*double\_to\_int) (\textbf{double} x)} \tag{B}$$

By the way, if the function returned an int* instead of a plain int, the declaration would be:

$$\text{int *(*double\_to\_int) (\textbf{double} *x)}$$

The type declarations do nothing by themselves, just as the word int does nothing by itself. But now that you know how to define a function type, you can put the declaration of the function into your header line. A Function that applies a function to an array of doubles would have a header like this:

$$\text{int* apply (\textbf{double} *v, int (*instance\_of\_function) (\textbf{double} x));} \tag{C}$$

Putting typedef *to work*   Are you confused yet? Each component basically makes sense, but together it is cluttered and confusing. There is a way out: typedef. By putting that word before line B—

> **typedef int** (∗double_to_int) (**double** x);

—we have created an new type named double_to_int that we can use like any other type. Now, line C simplifies to

> **int**∗ apply (**double** ∗v, **double_to_int** instance_of_function);

```
1    #include <apop.h>
2
3    typedef double (*dfn) (double);
4
5    double sample_function (double in){
6        return log(in)+ sin(in);
7    }
8
9    void plot_a_fn(double min, double max, dfn plotme){
10       double val;
11       FILE *f = popen("gnuplot −persist", "w");
12       if (!f)
13           printf("Couldn't find Gnuplot.\n");
14       fprintf(f, "set key off\n plot '−' with lines\n");
15       for (double i=min; i<max; i+= (max−min)/100.0){
16           val = plotme(i);
17           fprintf(f, "%g\t%g\n", i, val);
18       }
19       fprintf(f, "e\n");
20    }
21
22   int main(){
23       plot_a_fn(0, 15, sample_function);
24   }
```

Listing 6.1 A demonstration of a Function that takes in any function $\mathbb{R} \to \mathbb{R}$ and plots it. Online source: plotafunction.c.

Listing 6.1 shows a program to plot any function of the form $\mathbb{R} \to \mathbb{R}$, using the Gnuplot program described in Chapter 5. With a typedef in place, the syntax is easy. You don't need extra stars or ampersands in either the declaration of the Function-of-a-function or in the call to that Function, and you can call the pointed-to function like any other.

- Line 3: To make life easier, the `dfn` type is declared at the top of the file.
- Line 5: The header for the `sample_function` matches the format of the `dfn` function type (i.e., `double in, double out`).
- Line 9: The `plot_a_fn` Function specifies that it takes in a function of type `dfn`.
- Line 16: Using the passed-in function is as simple as using any other function: this line gives no indication that `plotme` is in any way special.
- Line 23: Finally, in `main`, you can see how `plot_a_fn` is called. The `sample_function` is passed in with just its name.

For another example, have a look at `jackiteration.c` on page 132.

Q6.1

Turn `plotafunction.c` into a library function callable by other programs.

- Comment out the `main` function.

- Write a header `plotafunction.h` with the necesary type and function definitions.

- Write a test program that `#includes plotafunction.h` and plots the `calc_taxes` function from `taxes.c` (p 118).

- Modify the makefile to produce the final program by creating and linking both *your_code.o* and `plotafunction.o`.

Q6.2

Define a type `dfn` as in line three of Listing 6.1. Then write a Function with header `void apply(dfn fn, double *array, int array_len)` that takes as arguments a function, an array, and the length of the array, and changes each element `array[i]` to `fn(array[i])`. [Apophenia provides comparable functions; see page 117 for details.]
Test your Function by creating an array of the natural numbers $1, 2, 3, \ldots 20$ and transforming it to a list of squares.

$\sum$

➤ You can pass functions as function arguments, just as you would pass arrays or numbers.

➤ The syntax for declaring a function pointer is just like the syntax for declaring a function, but the name is in parens and is preceded by a star.   ⫸

$\sum$

⮛

> ➤ Defining a new type to describe the function helps immensely. This requires putting `typedef` in front of the function pointer declaration in the last summary point.

> ➤ Once you have a `typedef` in place, you can declare Functions that take functions, use the passed-in functions, and call the parent Function as you would expect. Given the `typedef`, you need neither stars nor ampersands for these operations.

6.2 DATA STRUCTURES

Say that you have a few million observations to store on your computer. You want to find any given item quickly, add or delete elements easily, and not worry too much about a complicated organization system. There are several options for balancing these goals, and choosing among them is not trivial. This section will consider three: the array, the linked list, and the binary tree.

They will be implemented here via Glib, a library of general-use functions that every C programmer seems to re-implement. It includes a few features for string handling and other such conveniences, and modules to handle the data structures described here.[1] The extended example below provides documentation-by-example of initializing, adding to, removing from, and finding elements within the various structures, but your package manager will be happy to install the complete documentation, as well as Glib itself.

AN EXAMPLE

This game consists of a series of meetings between pairs of birds, who compete over r *utils* of resource.[2] If two doves meet, they split the resource evenly between them. If a dove and a hawk meet, the dove backs down and the hawk gets the resource. If two hawks meet, then the hawks fight, destroying c utils in resources before finally splitting what is left. Table 6.2 shows a payoff table summarizing the outcomes. For each pairing, the row player gets the first payoff, and the column player gets the second.

---

[1]Glib also provides a common data structure known as a *hash table*, which is another technique for easy data retrieval. It converts a piece of data, such as a string, into a number that can then be used to jump to the string's data in a table very quickly. Binary trees tend to work better in the context of agent-based modeling, so I have omitted hashes from this chapter. See Kernighan & Pike (1999), Chapter 3, for an extended example of hash tables that produce nonsense text. It was intended as an amusement (compare with the Exquisite Corpse-type game played by Pierce (1980, p 262)), but is now commonly used to produce spam email. Q: Try implementing Kernighan and Pike's nonsense generator using Glib's hash tables, string hash functions, and list structures.

[2]The util is the unit of measurement for the quantity of utility an agent gets from an action.

	dove	hawk
dove	$(\frac{r}{2}, \frac{r}{2})$	$(0, r)$
hawk	$(r, 0)$	$(\frac{r-c}{2}, \frac{r-c}{2})$

Table 6.2 The payoff matrix for the hawk/dove game. If $c < r$, then this is a prisoner's dilemma.

With $c < r$, the game is commonly known as a *prisoner's dilemma*, due to a rather contrived story about two separated prisoners who must choose between providing evidence about the other prisoner and remaining silent. Its key feature is that being a dove (cooperating) always makes the agent worse off than being a hawk (not cooperating, which the literature calls defection). The only equilibrium to the P.D. game is when nobody cooperates, destroying resources every period, but the societal optimum is when everyone cooperates, producing r utils of utility total every time.

On top of this we can add an evolutionary twist: say that a bird that is very success-ful will spawn chicks. In any one interaction, a bird gets an equal or better payoff as a hawk than as a dove, so it seems that over time, the hawks would approach 100% of the population. In the simulation below, a bird's odds of reproducing are proportional to the percentage of total flock wealth the bird holds, and its odds of dying are inversely proportional to the same.

To simulate the game, we will need a flock of birds. Have a look at the header file `birds/birds.h` in the online code supplement. It begins by describing the basic structure that the rest of the functions depend upon, describing a single bird:

```
typedef struct {
    char type;
    int wealth;
    int id;
} bird;
```

The header then lists two types of function. The first are functions for each relevant action in the simulation: startup, births, deaths, and actual plays of the hawk/dove game. The second group are functions for flock management, such as counting the flock or iterating over every member of the flock.

The first set of functions are implemented in Listing 6.3. Each function, taken individually, should make sense: `play_hd_game` takes in two birds and modifies their payoff according to the game rules above; `bird_plays` takes in a single bird, finds an opponent, and then has them play against each other; et cetera.

```c
#include "birds.h"
#include <time.h>

gsl_rng *r;
int periods = 400;
int initial_pop = 1000;
int id_count = 0;

void play_hd_game(bird *row, bird *col){
    double resource = 2,
              cost = 2.01;
    if (row->type == 'd' && col->type == 'h')
          col->wealth += resource;
    else if (row->type == 'h' && col->type == 'd')
          row->wealth += resource;
    else if (row->type == 'd' && col->type == 'd'){
          col->wealth += resource/2;
          row->wealth += resource/2;
    } else { // hawk v hawk
          col->wealth += (resource-cost)/2;
          row->wealth += (resource-cost)/2;
    } }

void bird_plays(void *in, void *dummy_param){
    bird *other;
      while(!(other = find_opponent(gsl_rng_uniform_int(r,id_count))) && (in != other))
          ;//do nothing.
      play_hd_game(in, other); }

bird *new_chick(bird *parent){
    bird *out = malloc(sizeof(bird));
      if (parent)
          out->type = parent->type;
      else{
          if (gsl_rng_uniform(r) > 0.5)
                out->type = 'd';
          else
                out->type = 'h';
      }
      out->wealth = 5* gsl_rng_uniform(r);
      out->id = id_count;
      id_count ++;
      return out; }

void birth_or_death(void *in, void *t){
    bird *b = in; //cast void to bird;
    int *total_wealth = t;
      if (b->wealth*20./ *total_wealth >= gsl_rng_uniform(r))
          add_to_flock(new_chick(b));
      if (b->wealth*800./ *total_wealth <= gsl_rng_uniform(r))
          free_bird(b); }

void startup(int initial_flock_size){
      flock_init();
      r = apop_rng_alloc(time(NULL));
      printf("Period\tHawks\tDoves\n");
      for(int i=0; i< initial_flock_size; i++)
          add_to_flock(new_chick(NULL)); }
```

```
int main(){
    startup(initial_pop);
    for (int i=0; i< periods; i++){
        flock_plays();
        count(i);
    } }
```

Listing 6.3 The birds. Online source: `birds/birds.c`

Now for the flock management routines, which will be implemented three times: as an array, as a list, and as a binary tree.

ARRAYS An array is as simple as data representation can get: just write each element right after the other. The matrices and vectors throughout this book keep their data in arrays of this type.

The system can retrieve an item from an array faster than from any other data structure, since the process consists of simply going to a fixed location and reading the data there. On the other hand, adding and deleting elements from an array is difficult: the simulation has to call `realloc` every time the list expands. If you are lucky, `realloc` will not move the array from its current location, but will simply find that there is more free space for the array to grow. If you are not lucky, then the array will have to be moved in its entirety to a new, more spacious home.

An array can not have a hole in the middle, so elements can not be deleted by freeing the memory. There are a few solutions, none of which are very pleasant. The last element of the list could be moved in to the space, requiring a copy, a shrinking of the array, and a loss of order in the elements. If order is important, every element could be shifted down a notch, so if item 50 is deleted, item 51 is put in slot 50, item 52 is put in slot 51, et cetera. Thus, every death could mean a call to `memmove` to execute thousands or millions of copy operations.

Listing 6.4 marks dead birds by setting their id to -1. This means that as the program runs, more and more memory is used by dead elements, and the rest of the system must check for the marker at every use.

As for finding an element, `add_to_flock` takes pains to ensure that the id and array index will always match one-to-one, so finding a bird given its id number is trivial. As with the code above, the code consists of a large number of short functions, meaning that it is reasonably easy to understand, read, and write each function by itself.

```
#include "birds.h"

bird *flock;
int size_of_flock, hawks, doves;

void flock_plays(){
    for (int i=0; i< size_of_flock; i++)
        if (flock[i].id >= 0)
            bird_plays(&(flock[i]), NULL); }

void add_to_flock(bird* b){
    size_of_flock = b->id;
    flock = realloc(flock, sizeof(bird)*(size_of_flock+1));
    memcpy(&(flock[b->id]), b, sizeof(bird));
    free(b); }

void free_bird(bird* b){ b->id = -1; }

bird * find_opponent(int n){
    if (flock[n].id >= 0)
        return &(flock[n]);
    else return NULL; }

int flock_size(){ return size_of_flock; }

int flock_wealth(){
  int i, total =0;
    for (i=0; i< size_of_flock; i++)
        if (flock[i].id >= 0)
            total += flock[i].wealth;
    return total; }

double count(int period){
  int i, tw = flock_wealth();
    hawks = doves = 0;
    for(i=0; i< size_of_flock; i++)
        if (flock[i].id>=0)
            birth_or_death(&(flock[i]), &tw);
    for(i=0; i< size_of_flock; i++)
        if (flock[i].id>=0){
            if (flock[i].type == 'h')
                hawks ++;
            else doves ++;
        }
    printf("%i\t%i\t%i\n", period, hawks, doves);
    return (doves+0.0)/hawks;}

void flock_init(){
  flock = NULL;
  size_of_flock = 0; }
```

Listing 6.4 The birds, array version. The fatal flaw is that birds are copied in, but never eliminated. Dead birds will eventually pile up. Online source: `birds/arrayflock.c`

Q6.3

Because the `play_hd_game` function sets $r == 2$ and $c == 2.01$, hawks lose 0.005 utils when they fight, so it is marginally better to be a dove when meeting a hawk, and the game is not quite a prisoner's dilemma. After running the simulation a few times to get a feel for the equilibrium number of birds, change c to 2.0 and see how the equilibrium proportion of doves changes. [For your convenience, a sample makefile is included in the `birds` directory of the code supplement.]

Finally, rename `main` to `one_run` and wrap it in a function that takes in a value of c and returns the proportion of doves at the end of the simulation. Send the function to the `plot_a_function` function from earlier in the chapter to produce a plot.

Q6.4

Rewrite `arrayflock.c` to delete birds instead of just mark them as dead. Use `memmove` to close holes in the array, then renumber birds so their `id` matches the array index. Keep a counter of current allocated size (which may be greater than the number of birds) so you can `realloc` the array only when necessary.

LINKED LISTS A *linked list* is a set of `structs` connected by pointers. The first `struct` includes a `next` pointer that points to the next element, whose `next` pointer points to the next element, et cetera; see Figure 6.5.

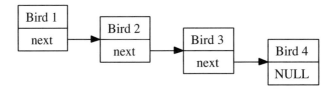

Figure 6.5 The archetypal linked list. Online source: `list.dot`.

The linked list is a favorite for agent-based simulation, because birth and death is easy to handle. To add an element to a linked list, just create a new node and replace the `NULL` pointer at the end of the list with a pointer to the new node. Deleting a node is also simple: to delete bird 2, simply reroute the `next` pointer from bird $1 \rightarrow 2$ so that it points from bird $1 \rightarrow 3$, and then free the memory holding bird 2.

But the real failing of the linked list is the trouble of finding an arbitrary element. In an array, finding the ten thousandth element is easy: `flock[9999]`. You can see in the code of Listing 6.6 that glib provides a `g_list_nth_data` function to return the nth element of the list, which makes it look simple, but the only way that that function can find the ten thousandth member of the `flock` is to start at the head of the list and take 9,999 `->next` steps. In fact, if you compile and run this program, you will see that it runs much more slowly than the array and tree versions.

```
#include "birds.h"
#include <glib.h>

GList *flock;
int hawks, doves;

void flock_plays(){ g_list_foreach(flock, bird_plays, NULL); }

void add_to_flock(bird* b){ flock = g_list_prepend(flock, b); }

void free_bird(bird* b){
    flock = g_list_remove(flock, b);
    free(b); }

bird * find_opponent(int n){ return g_list_nth_data(flock, n); }

void wealth_foreach(void *in, void *total){
    *((int*)total) += ((bird*)in)->wealth; }

int flock_wealth(){
  int total = 0;
    g_list_foreach(flock, wealth_foreach, &total);
    return total; }

int flock_size(){ return g_list_length(flock); }

void bird_count(void *in, void *v){
  bird *b = in;
    if (b->type == 'h')
        hawks ++;
    else doves ++;
}

double count(int period){
  int total_wealth =flock_wealth();
    hawks = doves = 0;
    g_list_foreach(flock, birth_or_death, &total_wealth);
    g_list_foreach(flock, bird_count, NULL);
    printf("%i\t%i\t%i\n", period, hawks, doves);
    return (doves+0.0)/hawks;}

void flock_init(){ flock = NULL; }
```

Listing 6.6 The birds, linked list version. The fatal flaw is that finding a given bird requires traversing the entire list every time. Online source: `birds/listflock.c`

- The `g_list_foreach` function implements exactly the sort of apply-function-to-list setup implemented in Section 6.1. It takes in a list and a function, and internally applies the function to each element.

- The folks who wrote the Glib library could not have known anything about the `bird` structure, so how could they write a linked list that would hold it? The solution is `void` pointers—that is, a pointer with no type associated, which could therefore point to a location holding data of any type whatsoever. For example, `bird_count` takes in two void pointers, the first being the element held in the list, and the second being any sort of user-specified data (which in this case is just ignored).

- The first step in using a void pointer is casting it to the correct type. For example, the first line in `bird_count`—`bird *b = in;`—points b to the same address as `in`, but since b has a type associated, it can be used as normal.

- As for adding and removing, the Glib implementation of the list takes in a pointer to a `GList` and a pointer to the data to be added, and returns a new pointer to a `GList`. The input and output pointers could be identical, but since this is not guaranteed, use the form here to reassign the list to a new value for every add/delete. For example, the flock starts in `flock_init` as NULL, and is given its first non-NULL value on the first call to `add_to_flock`.

BINARY TREES The *binary tree* takes the linked list a step further by giving each node two outgoing pointers instead of one. As per Figure 6.7, think of these pointers as the left pointer and the right pointer. The branching allows for a tree structure. The directions to an element are now less than trivial—to get to `bird5`, start at the head (`bird1`), then go left, then go right. But with eight data points in a linked list, you would need up to seven steps to get to any element, and on average 3.5 steps. In a tree, the longest walk is three steps, and the average is 1.625 steps. Generally, the linked list will require on the order of n steps to find an item, and a b-tree will require on the order of $\ln(n)$ steps (Knuth, 1997, pp 400–401).[3]

The tree arrangement needs some sort of order to the elements, so the system knows whether to go left or right at each step. In this case, the `id` for each bird provides a natural means of ordering. For text data, `strcmp` would provide a similar ordering. More generally, there must be a *key value* given to each element, and the tree structure must have a function for comparing keys.

Eariler, you saw an interesting implementation of a set of binary trees: a database. Since databases require fast access to every element, it is natural that they would internally structure data in a binary tree, and this is exactly how SQLite and mySQL operate internally: each new index is its own tree.

This adds some complication, because you now need to associate with each tree a function for comparing keys. In the code below, `g_tree_new` initializes a tree using the `compare_birds` function.

---

[3]For those who follow the reference, notice that Knuth presents the equation for the sum of path lengths, which he calls the internal path length. He finds that it is of order $n \ln(n) + \mathcal{O}(n)$ for complete binary trees; the average path length is thus $\ln(n) + \mathcal{O}(1)$.

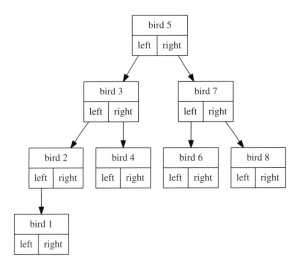

Figure 6.7 The archetypal binary tree. Online source: `btree.dot`.

What if two birds have the same id? Then there is no way to order them uniquely, and therefore there is no way to reliably store and retrieve them. Thus, the key for each element must be unique.[4]

The added complication of a tree solves many of the problems above. As with the list, inserting and deleting elements does not require major `reallocing`, although there is often minor internal reshuffling to keep the branches of the tree at about even length. With the key and short chains, finding an element is much faster.

The `const` keyword

The `const` modifiers in the header for `compare_-keys` indicate that the data to which the pointers point will not be changed over the course of the function. As you can see by the fact that it has not appeared until page 201, the `const` keyword is mostly optional, though it is good form and provides one more check that your functions don't do what you hadn't intended. However, when conforming with function specifications elsewhere, like GLib's function header for key-comparison functions, you may need to use it. If you then get an error like *'subfunction' discards qualifiers from pointer target type*, then you will need to rewrite the subfunction so that it too takes `const` inputs (and does not modify them).

```
#include "birds.h"
#include <glib.h>

GTree *flock = NULL;
int hawks, doves;

static gint compare_keys(const void *L, const void *R){
    const int *Lb = L;
    const int *Rb = R;
        return *Lb − *Rb;
}
```

---

[4]There exist tree implementations that do not require unique keys, but it is a requirement for GLib. Similarly, some databases are very strict about requiring that each table have a field representing a key, and some are not.

```
static gboolean tree_bird_plays(void *key, void *in, void *v){
    bird_plays(in, NULL);
    return 0;
}

void flock_plays(){ g_tree_foreach(flock, tree_bird_plays, NULL); }

void add_to_flock(bird* b){ g_tree_insert(flock, &(b->id), b); }

bird * find_opponent(int n){return g_tree_lookup(flock, &n);}

int flock_size(){ return g_tree_nnodes(flock); }

static gboolean wealth_foreach(void *key, void *in, void *t){
    int *total = t;
    *total += ((bird*)in)->wealth;
    return 0; }

int flock_wealth(){
    int total = 0;
    g_tree_foreach(flock, wealth_foreach, &total);
    return total; }

static gboolean tree_bird_count(void *key, void *in, void *v){
    if (((bird *)in)->type == 'h')
        hawks ++;
    else doves ++;
    return 0; }

GList *dying_birds;

void free_bird(bird* b){dying_birds = g_list_prepend(dying_birds, b);}

static gboolean tree_birth_or_death(void *key, void *in, void *t){
    birth_or_death(in, t);
    return 0; }

static void cull_foreach(void *b, void *v){
    bird* a_bird = b;
    g_tree_remove(flock, &(a_bird->id));
    free(a_bird); }

double count(int period){
    int total_wealth =flock_wealth();
    hawks = doves = 0;
    dying_birds = NULL;
    g_tree_foreach(flock, tree_birth_or_death, &total_wealth);
    g_list_foreach(dying_birds, cull_foreach, NULL);
    g_list_free(dying_birds);
    g_tree_foreach(flock, tree_bird_count, NULL);
    printf("%i\t%i\t%i\n", period, hawks, doves);
    return (doves+0.0)/hawks;}

void flock_init(){ flock = g_tree_new(compare_keys); }
```

Listing 6.8 The birds, binary tree version. The fatal flaw is the complication in maintaining the key for every bird. Online source: `birds/treeflock.c`

• Culling the flock is especially difficult because a tree can internally re-sort when an element is added/deleted, so it is impossible to delete elements while traversing a tree. In the implementation of Listing 6.8, the `free_bird` function actually freed the bird; here it just adds dying birds to a `GList`, and then another post-traversal step goest through the `GList` and cull marked birds from the tree.

> ➤ There are various means of organizing large data sets, such as collections of agents in an agent-based model.

> ➤ Arrays are simply sequential blocks of structs. Pros: easy to implement; you can get to the $10,000$th element in one step. Cons: no easy way to add, delete, or reorganize elements.

> ➤ A linked list is a sequence of structs, where each includes a pointer to the next element in the list. Pro: adding/deleting/resorting elements is trivial. Con: Getting to the $10,000$th element takes 9,999 steps.

> ➤ A binary tree is like a linked list, but each struct has a left and right successor. Pros: adding and deleting is only marginally more difficult than with a linked list; getting to the $10,000$th element takes at most 13 steps. Con: Each element must be accessed via a unique key, adding complication.

6.3 PARAMETERS   Your simulations and analyses will require tweaking. You will want to try more agents, or you may want your program to load a data set from a text file to a database for one run and then use the data in the database for later runs.

This section will cover a cavalcade of means of setting parameters and specifications, in increasing order of ease of use and difficulty in implementation.

The first option—a default of sorts—is to set variables at the top of your `.c` file or a header file. This is trivial to implement, but you will need to recompile every time you change parameters.

Interactive   The second option is to interactively get parameters from the user, via `scanf` and `fgets`. Listing 6.9 shows a program that asks data of the user and then returns manipulated data. The `scanf` function basically works like `printf` in reverse, reading text with the given format into pointers to variables. Unfortunately, the system tends to be rather fragile in the real world, as a stray comma or period can entirely throw off the format string. The `fgets` function will read an entire line into a string, but has its own quirks. In short, the interactive input features are good for some quick interrogations or a bit of fun, but are not to be heavily relied upon.

```
#include <stdio.h>
#include <string.h> //strlen

int main(){
    float indata;
    char s[100];
        printf("Give me a number: ");
        scanf("%g", &indata);
        printf("Your number squared: %g\n", indata*indata);
        printf("OK, now give me a string (max length, 100):\n");
        fgets(s, 99, stdin); //eat a newline.
        fgets(s, 99, stdin);
        printf("Here it is backward:\n");
        for (int i=strlen(s)-2; i>=0; i--)
            printf("%c", s[i]);
        printf("\n");
}
```

Listing 6.9 Reading inputs from the command line. Online source: `getstring.c`.

Environment variables   These are variables passed from the shell (aka the command prompt) to the program. They are relatively easy to set, but are generally used for variables that are infrequently changing, like the username. Environment variables are discussed at length in Appendix A.

Parameter files   There are many libraries that read parameter files; consistent with the rest of this chapter, Listing 6.10 shows a file in Glib's *key file* format, which will be read by the program in Listing 6.11. The configuration file can be in a human language like English, you can modify it as much as you want without recompiling the code itself, it provides a permanent record of parameters for each run, and you can quickly switch among sets of variables.

• The payoff for Listing 6.11 is on line 22: printing the name of a distribution, a parameter, and the mean of that distribution given that parameter. The program to that point finds these three items.

• Line seven indicates which section of Listing 6.10 the following code will read. By commenting out line seven and uncommenting line eight, the code would read the Exponential section. Below, you will see that setting the `config` variable on the command line is not difficult.

• Line 10 reads the entire `glib.config` file into the `keys` structure. If something

```
#gkeys.c reads this file

[chi squared configuration]
distribution name = Chi squared
parameter = 3

[exponential configuration]
distribution name = Exponential
parameter = 2.2
```

Listing 6.10 A configuration in the style of Glib's key files. Online source: `glib.config`.

```
1    #include <glib.h>
2    #include <apop.h>
3
4    int main(){
5       GKeyFile *keys = g_key_file_new();
6       GError *e = NULL;
7       char *config = "chi squared configuration";
8    // char *config = "exponential configuration";
9       double (*distribution)(double, double);
10        if (!g_key_file_load_from_file(keys, "glib.config", 0, &e))
11            fprintf(stderr, e->message);
12        double param = g_key_file_get_double(keys, config, "parameter", &e);
13        if (e) fprintf(stderr, e->message);
14        char* name = g_key_file_get_string(keys, config, "distribution name", &e);
15        if (e) fprintf(stderr, e->message);
16
17        if (!strcmp(name, "Chi squared"))
18            distribution = gsl_cdf_chisq_Pinv;
19        else if (!strcmp(name, "Exponential"))
20            distribution = gsl_cdf_exponential_Pinv;
21
22        printf("Mean of a %s distribution with parameter %g: %g\n", name,
23                param, distribution(0.5, param));
24    }
```

Listing 6.11 A program that reads Listing 6.10. Online source: `gkeys.c`.

goes wrong, then line 11 prints the error message stored in e. Properly, the program should exit at this point; for the sake of brevity the `return 0` lines have been omitted.

- Now that `keys` holds all the values in the config file, lines 12 and 14 can get individual values. The two `g_key_file_get...` functions take in a filled key structure, a section name, a variable name, and a place to put errors. They return the requested value (or an error).

- Unfortunately, there is no way to specify functions in a text file but by name, so lines 17–20 set the function pointer `distribution` according to the name from the config file.

Q_{6.5}

Rewrite the code in Listing 6.11 to set parameters via database, rather than via the command line.

- Write a text file with three columns: configuration, parameters, and data.

- Read in the file using `apop_text_to_db` at the beginning of `main`.

- Write a function with header `double get_param(char *config, char *p)` that queries the database for the parameter named `p` in the configuration group `config` and returns its value. Then modify the program to get the distribution and its parameter using the `get_param` function.

Command line   Reading parameters from the command line can take the most effort to implement among the parameter-setting options here, but it is the most dynamic, allowing you to change parameters every time you run the program. You can even write batch files in Perl, Python, or a shell-type language to run the program with different parameter variants (and thus keep a record of those variants).

The `main` function takes inputs and produces an output like any other. The output is an integer `returned` at the end of `main`, which is typically zero for success or a positive integer indicating a type of failure. The inputs are always an integer, giving the number of command-line elements, and a `char**`—an array of strings—listing the command-line elements themselves. Like any function specification, the types are non-negotiable but the internal name you choose to give these arguments is arbitrary. However, the universal custom is to name them `argc` (argument count) and `argv` (argument values).[5] This is an ingrained custom, and you can expect to see those names everywhere.[6]

Listing 6.12 shows the rudimentary use of `argc` and `argv`. Here is a sample usage from my command line:

---

[5]They are in alphabetical order in the parameters to `main`, which provides an easy way to remember that the count comes first.

[6]Perl uses `argv`, Python uses `sys.argv`, and Ruby uses `ARGV`. All three structures automatically track array lengths, so none of these languages uses an `argc` variable.

```
#include <stdio.h>

int main(int argc, char **argv){
    for (int i=0; i< argc; i++)
        printf("Command line argument %i: %s\n", i, argv[i]);
}
```

Listing 6.12 This program will simply print out the command line arguments given to it. Online source: `argv.c`.

```
>>> /home/klemens/argv one 2 −−three fo\ ur "cinco −− \"five\""
command line argument 0: /home/klemens/argv
command line argument 1: one
command line argument 2: 2
command line argument 3: −−three
command line argument 4: fo ur
command line argument 5: cinco −− "five"
```

- Argument zero (`argv[0]`) is always the name of the command itself. Some creative programs run differently if they are referred to by different names, but the norm is to just skip over `argv[0]`.

- After that, the elements of `argv` are the command line broken at the spaces, and could be dashes, numbers, or any other sort of text.

- As you can see from the parsing of `fo\ ur`, a space preceded by a backslash is taken to be a character like any other, rather than an argument separator.

- Argument five shows that everything between a pair of quotation marks is a single argument, and a backslash once again turns the quotation mark into plain text.

For some purposes, this is all you will need to set program options from your command line. For example you could have one program to run three different actions with a `main` like the following:

```
int main(int argc, int **argv){
    if (argc == 1){
        printf("I need a command line argument.\n")
        return 1;
    }
    if (!strcmp(argv[1], "read"))
        read_data();
    else if (!strcmp(argv[1], "analysis_1"))
        run_analysis_1();
    else if (!strcmp(argv[1], "analysis_2"))
        run_analysis_2();
}
```

$\mathbb{Q}_{6.6}$ The `callbyval.c` program (Listing 2.5, page 38) calculated the factorial of a number which was hard-coded into `main`. Rewrite it to take the number from the command line, so `factorial 15` will find 15!. (*Hint*: the `atoi` function converts a text string to the integer it represents; for example, `atoi("15") == 15`.)

getopt   For more complex situations, use `getopt`, which parses command lines for *switches* of the form -x. . . . It is part of the standard C library, so it is fully portable.[7]

Listing 6.13 shows a program that will display a series of exponents. As explained by the message on lines 9–14, you can set the minimum of the series via -m, the maximum via -M, and the increment via -i. Specify the base of the exponents after the switches. Sample usage (which also demonstrates that spaces between the switch and the data are optional):

```
>>> ./getopt −m 3 −M4 −i 0.3 2
2^3: 8
2^3.3: 9.84916
2^3.6: 12.1257
2^3.9: 14.9285
```

There are three steps to the process:

- `#include <unistd.h>`.

- Specify a set of letters indicating valid single-letter switches in a crunched-together string like line 15 of Listing 6.13. If the switch takes in additional info (here, every switch but -h), indicate this with a colon after the letter.

- Write a `while` loop to call `getopt` (line 27), and then act based upon the value of the char that `getopt` returned.

- `argv` is text, but you will often want to specify numbers. The functions `atoi`, `atol`, and `atof` convert ASCII text to an `int`, `long int`, or `double`, respectively.[8]

---

[7]The GNU version of the standard C library provides a sublibrary named `argp` that provides many more functions and does more automatically, but is correspondingly more complex and less portable. Glib also has a subsection devoted to command-line arguments, which also provides many more features so long as the library is installed.

[8]Getopt readily handles negative numbers that are arguments to a flag (-m -3), but a negative number after the options will look like just another flag, e.g., ./getopt -m -3 -M 4 -2 looks as if there is a flag named 2. The special flag -- indicates that `getopt` should stop parsing flags, so ./getopt -m -3 -M 4 -- -2 will work. This is also useful elsewhere, such as handling files that begin with a dash; e.g., given a file named *-a_mistake*, you can delete it with `rm -- -a_mistake`.

```
1   #include <stdio.h> //printf
2   #include <unistd.h> //getopt
3   #include <stdlib.h> //atof
4   #include <math.h> //powf
5
6   double min = 0., max = 10.;
7   double incr = 1., base = 2.;
8
9   void show_powers(){
10      for (double i=min; i<=max; i+= incr)
11          printf("%g^%g: %g\n", base, i, powf(base, i));
12  }
13
14  int main(int argc, char ** argv){
15      char c, opts[]= "M:m:i:h";
16      char help[]= "A program to take powers of a function. Usage:\n"
17                  "\t\tgetopt [options] [a number]\n"
18                  "-h\t This help\n"
19                  "-m\t The minimum exponent at which to start.\n"
20                  "-M\t The maximum exponent at which to finish.\n"
21                  "-i\t Increment by this.\n";
22
23      if (argc==1) {
24          printf(help);
25          return 1;
26      }
27      while ( (c=getopt(argc, argv, opts)) != -1)
28          if (c=='h'){
29              printf(help);
30              return 0;
31          } else if (c=='m'){
32              min = atof(optarg);
33          } else if (c=='M'){
34              max = atof(optarg);
35          } else if (c=='i'){
36              incr = atof(optarg);
37          }
38      if (optind < argc)
39          base = atof(argv[optind]);
40      show_powers();
41  }
```

Listing 6.13 Command-line parsing with getopt. Online source: getopt.c.

- optarg also sets the variable optind to indicate the position in argv that it last visited. Thus, line 38 was able to check whether there are any non-switch arguments remaining, and line 39 could parse the remaining argument (if any) without getopt's help.

- The program provides human assistance. If the user gives the -h switch or leaves off all switches entirely, then the program prints a help message and exits. Every variable that the user could forget to set via the command line has a default value.

 $\mathbb{Q}_{6.7}$ Listing 6.1 (page 191) is hard-coded to plot a range from $x = 0$ to $x = 15$. Modify it to use getopt to get a minimum and maximum from the user, with zero and fifteen as defaults. Provide help if the user uses the -h flag.

Σ

➤ There are many ways to change a program's settings without having to recompile the program.

➤ Environment variables, covered in Appendix A, are a lightweight means of setting variables in the shell that the program can use.

➤ There are many libraries for parsing parameter files, or you could use SQL to pull settings from a database.

➤ The main function takes in arguments listed on the command line, and some C functions (like getopt) will help you parse those arguments into program settings.

6.4   ❋ SYNTACTIC SUGAR   Returning to C syntax, there are several ways to do almost everything in Chapter 2. For example, you could rewrite the three lines

```
b = (i > j);
a += b;
i++;
```

as the single expression a+=b=i++>j;. The seventh element of the array k can be called k[6], *(k+6), or—for the truly perverse—6[k]. That is, this book overlooks a great deal of C syntax, which is sometimes useful and even graceful, but confuses as easily as it clarifies.

This section goes over a few more details of C syntax which are also not strictly necessary, but have decent odds of being occasionally useful. In fact, they are demonstrated a handful of times in the remainder of the book. If your interest is piqued and you would like to learn more about how C works and about the many alternatives that not mentioned here, see the authoritative and surprisingly readable reference for the language, Kernighan & Ritchie (1988).

The obfuscatory if   There is another way to write an `if` statement:

```
if (a < b)
        first_val;
else
        second_val;

/* is equivalent to */

a < b ? first_val : second_val;
```

Both have all three components: first the condition, then the 'what to do if the condition is true' part, and then the 'what to do if the condition is false' part. However, the first is more-or-less legible to anybody who knows basic English, and the second takes the reader a second to parse every time he or she sees it. On the other hand, the second version is much more compact.

The condensed form is primarily useful because you can put it on the right side of an assignment. For example, in the `new_chick` function of Listing 6.3 (p 195), you saw the following snippet:

```
if (gsl_rng_uniform(r) > 0.5)
    out->type = 'd';
else
    out->type = 'h';
```

Using the obfuscatory if, these four lines can be reduced to one:

```
out->type = gsl_rng_uniform(r) > 0.5 ? 'd' : 'h';
```

Macros   As well as `#include`-ing files, the preprocessor can also do text substitution, where it simply replaces one set of symbols with another.

Text substitution can do a few things C can't do entirely by itself. For example, you may have encountered the detail of C that all global variables must have constant size (due to how the compiler sets them up).[9] Thus, if you attempt to compile the following program:

---

[9]Global and static variables are initialized before the program calls `main`, meaning that they have to be allocated without evaluating any non-constant expressions elsewhere in the code. Local variables are allocated as needed during runtime, so they can be based on evaluated expressions as usual.

```
int array_size = 10;
int a[array_size];

int main(){ }
```

you will get an error like `variable-size type declared outside of any function`.

The easy alternative is to simply leave the declaration at the top of the file but move the initialization into `main`, but you can also fix the problem with `#define`. The following program will compile properly, because the preprocessor will substitute the number 10 for `ARRAY_SIZE` before the compiler touches the code:

```
#define ARRAY_SIZE 10
int a[ARRAY_SIZE];

int main(){ }
```

Do not use an equals sign or a semicolon with `#defines`.

You can also `#define` function-type text substitutions. For example, here is the code for the `GSL_MIN` macro from the `<gsl/gsl_math.h>` header file:

```
#define GSL_MIN(a,b) ((a) < (b) ? (a) : (b))
```

It would expand every instance in the code of `GSL_MIN(a,b)` to the if-then-else expression in parens. `GSL_MAX` is similarly defined.

This is a *macro*, which is evaluated differently from a function. A function evaluates its arguments and then calls the function, so `f(2+3)` is guaranteed to evaluate exactly as `f(5)` does. A macro works by substituting one block of text for another, without regard to what that text means. If the macro is

```
#define twice(x) 2*x
```

then `twice(2+3)` expands to `2*2+3`, which is not equal to `twice(5) = 2*5`.

We thus arrive at the first rule of macro writing, which is that everything in the macro definition should be in parentheses, to prevent unforseen interactions between the text to be inserted and the rest of the macro.

Repeated evaluation is another common problem to look out for. For example, GSL_MIN(a++, b) expands to ((a++) < (b) ? (a++) : (b)), meaning that a may be incremented twice, not once as it would with a function. Again, the first solution is to not use macros except as a last resort, and the second is to make sure calls to macros are as simple as possible.

The one thing that a macro can do better than a function is take a type as an argument, because the preprocessor just shunts text around without regard to whether that text represents a type, a variable, or whatever else. For example, recall the form for reallocating a pointer to an array of doubles:

var_array = realloc(var_array, new_length * **sizeof**(**double**))

This can be rewritten with a macro to create a simpler form:

#define REALLOC(ptr, length, type) ptr = realloc((ptr), (length) * **sizeof**(type))
//which is used like this:
REALLOC(var_array, new_length, **double**);

It gives you one more moving part that could break (and which now needs to be #included with every file), but may make the code more readable. This macro also gives yet another demonstration of the importance of parens: without parens, a call like REALLOC(ptr, 8 + 1, double) would allocate $8 + \texttt{sizeof(double)}$ bytes of memory instead of $9 \cdot \texttt{sizeof(double)}$ bytes.

- If you need to debug a macro, the -E flag to gcc will run only the preprocessor, so you can see what expands to what. You probably want to run gcc -E *onefile.c* | less.
- The custom is to put macro names in capitals. You can rely on this in code you see from others, and are encouraged to stick to this standard when writing your own, as a reminder that macros are relatively fragile and tricky. Apophenia's macros can be written using either all-caps or, if that looks too much like yelling to you, using only an initial capital.

➤ Short if statements can be summarized to one line via the condition ? true_value : false_value form.

➤ You can use the preprocessor to #define constants and short functions.

6.5 MORE TOOLS Since C is so widely used, there is an ecosystem of tools built around helping you easily write good code. Beyond the debugger, here are a few more programs that will make your life as a programmer easier.

MEMORY DEBUGGER The setup is this: you make a mistake in memory handling early in the program, but it is not fatal, so the program continues along using bad data. Later on in the program, you do something innocuous with your bad data and get a segfault. This is a pain to trace using gdb, so there are packages designed to handle just this problem.

If you are using the GNU standard library (which you probably are if you are using gcc), then you can use the shell command

> export MALLOC_CHECK_=2

to set the `MALLOC_CHECK_` enviornment variable; see Appendix A for more on environment variables. When it is not set or is set to zero, the library uses the usual `malloc`. When it is set to one, the library uses a version of `malloc` that checks for common errors like double-freeing and off-by-one errors, and reports them on `stderr`. When the variable is set to two, the system halts on the first error, which is exactly what you want when running via gdb.

Another common (and entirely portable) alternative is Electric Fence, a library available via your package manager. It also provides a different version of `malloc` that crashes on any mis-allocations and mis-reads. To use it, you would simply recompile the program using the efence library, by either adding `-lefence` to the compilation command or the `LINKFLAGS` line in your makefile (see Appendix A).

REVISION CONTROL The idea behind the revision control system (RCS) is that your project lives in a sort of database known as a repository. When you want to work, you check out a copy of the project, and when you are done making changes, you check the project back in to the repository and can delete the copy. The repository makes a note of every change you made, so you can check out a copy of your program as it looked three weeks ago as easily as you could check out a current copy.

This has pleasant psychological benefits. Don't worry about experimenting with your code: it is just a copy, and if you break it you can always check out a fresh copy from the repository. Also, nothing matches the confidence you get from making major changes to the code and finding that the results precisely match the results from last month.

Finally, revision control packages facilitate collaboration with coauthors. If your changes are sufficiently far apart (e.g., you are working on one function and your coauthor on another), then the RCS will merge all changes to a single working copy. If it is unable to work out how to do so, then it will give you a clearly demarcated list of changes for you to accept or reject.

This method also works for any other text files you have in your life, such as papers written in LaTeX, HTML, or any other text-based format. For example, this book is under revision control.

There is no universal standard revision control software, but the Subversion package is readily available via your package manager. For usage, see Subversion's own detailed manual describing set-up and operation from the command line, or ask your search engine for the various GUIs written around Subversion.

THE PROFILER   If you feel that your program is running too slowly, then the first step in fixing it is measurement. The *profiler* times how long every function takes to execute, so you know upon which functions to focus your clean-up efforts.

First, you need to add a flag to the compilation to include profiler symbols, -pg. Then, execute your program, which will produce a file named gmon.out in the directory, with the machine-readable timings that the profiler will use.[10] Unlike the debugger's -g option, the -pg option may slow down the program significantly as it writes to gmon.out, so use -g always and -pg only when necessary.

Finally, call gprof ./my_executable to produce a human-readable table from gmon.out.[11] See the manual (man gprof) for further details about reading the output.

As with the debugger, once the profiler points out where the most time is being taken by your program, what you need to do to alleviate the bottleneck often becomes very obvious.

If you are just trying to get your programs to run, optimizing for speed may seem far from your mind. But it can nonetheless be an interesting exercise to run a modestly complex program through the profiler because, like the debugger's backtrace, its output provides another useful view of how functions call each other.

---

[10]If the program is too fast for the profiler, then rename main to internal_main and write a new main function with a for loop to call internal_main ten thousand times.

[11]gprof outputs to stdout; use the usual shell tricks to manipulate the output, such as piping output through a pager—gprof ./my_executable | less—or dumping it to a text file—gprof ./my_executable > outfile—that you can view in your text editor.

Optimization   The gcc compiler can do a number of things to your code to make
            it run faster. For example, it may change the order in which lines of
code are executed, or if you assign x = y + z, it may replace every instance of x
with y + z. To turn on optimization, use the -O3 flag when compiling with gcc.
[That's an 'O' as in optimization, not a zero. There are also -O1 and -O2, but as
long as you are optimizing, why not go all out?]

The problem with optimization, however, is that it makes debugging difficult. The
program jumps around, making stepping through an odd trip, and if every instance
of x has been replaced with something else, then you can not check its value. It
also sometimes happens that you did not do your memory allocation duties quite
right, and things went OK without optimization, but suddenly the program crashes
when you have optimization on. A memory debugger may provide some clues, but
you may just have to re-scour your code to find the problem. Thus, the -O3 flag
is a final step, to be used only after you are reasonably confident that your code is
debugged.

$Q_{6.8}$ | Add the -pg switch to the makefile in the birds directory and check the timing of the three different versions. It may help to comment out the `printf`
function and run the simulation for more periods. How does the -O3 flag
change the timings?

II

Statistics

DISTRIBUTIONS FOR DESCRIPTION

This chapter covers some methods of describing a data set, via a number of strategies of increasing complexity. The first approach, in Section 7.1, consists of simply looking at summary statistics for a series of observations about a single variable, like its mean and variance. It imposes no structure on the data of any sort. The next level of structure is to assume that the data is drawn from a distribution; instead of finding the mean or variance, we would instead use the data to estimate the parameters that describe the distribution. The simple statistics and distributions in this chapter are already sufficient to describe rather complex models of the real world, because we can chain together multiple distributions to form a larger model. Chapter 8 will take a slightly different approach to modeling a multidimensional data set, by projecting it onto a subspace of few dimensions.

7.1 MOMENTS   The first step in analyzing a data set is always to get a quick lay of the land: where do the variables generally lie? How far do they wander? As variable A goes up, does variable B follow?

※ **ESTIMATOR VOCABULARY**   A *statistic* is the output of a function that takes in data, typically of the form $f : \mathbb{R}^n \to \mathbb{R}$. That is, a statistic takes in data and summarizes it to a single dimension. Common statistics include the mean of $\mathbf{x}$, its variance, $\max(\mathbf{x})$, the covariance of $\mathbf{x}$ and $\mathbf{y}$, or the regression parameter β_2 from a regression of $\mathbf{X}$ on $\mathbf{y}$ (which could be written in the form $\beta_2(\mathbf{X}, \mathbf{y})$).

220

CHAPTER 7

Thus, many of the means of describing a data set, such as writing down its mean, could be described as the generation of statistics. One goal of writing down a statistic is dimension reduction: simply summarizing the data via a few human-comprehensible summary statistics, such as the data set's mean and variance.

Another goal, which is more often the case, is to use the statistics of the data, $\mathbf{X}$, to estimate the same statistic of the population. Let $\mathcal{P}$ signify the population. When the US Census Bureau said in an August 2006 press release[1] that 46.6 million people in the United States have no health insurance, they meant that the count of people in the Current Population Survey that did not have health insurance (a sample statistic) indicated that the count of people in the United States without health insurance (a population statistic) was 46.6 million. Is the estimate of the statistic based on the sample data, $\hat{\beta} \equiv f(\mathbf{X})$, a valid estimate of the population statistic, $\beta \equiv f(\mathcal{P})$? There are several ways to make this a precise question. For example, as $\mathbf{X}$ grows larger, does $\hat{\beta} \to \beta$? Do there exist estimates of β that have smaller variance than $\hat{\beta}$? After discussing some desirable qualities in an estimator, we will begin with some simple statistics.

⁎ *Evaluating an estimator*   From a given population, one could take many possible samples, say $\mathbf{X}_1, \mathbf{X}_2, \ldots$, which means that there could be many possible calculated statistics, $\hat{\beta}_1 = f(\mathbf{X}_1)$, $\hat{\beta}_2 = f(\mathbf{X}_2)$, ...

There are many means of describing whether the collection of statistics $\hat{\beta}_i$ (for $i = 1, i = 2, \ldots$) is a precise and accurate estimate of the true value of β. You will see in the sections to follow that intuitive means of estimating a population statistic sometimes work on all of these scales at once, and sometimes fail.

- Unbiasedness: The expected value of $\hat{\beta}$ (discussed in great detail below) equals the true population value: $E(\hat{\beta}_i) = \beta$.
- Variance: The variance is the expected value of the squared distance to the expected value: $E\left((\hat{\beta}_i - E(\hat{\beta}))^2\right)$. The variance is also discussed in detail below.
- Mean squared error: MSE of $\hat{\beta} \equiv E(\hat{\beta} - \beta)^2$. Below we will see that the MSE equals the variance plus the square of bias. So if $\hat{\beta}$ is an unbiased estimator of β, meaning that $E(\hat{\beta}) = \beta$, then the MSE is equivalent to the variance, but as the bias increases, the difference between MSE and variance grows.
- Efficiency: An estimator $\hat{\beta}$ is efficient if, for any other estimator $\tilde{\beta}$, $MSE(\hat{\beta}) \leq MSE(\tilde{\beta})$. If $\hat{\beta}$ and $\tilde{\beta}$ are unbiased estimators of the same β, then this reduces to $\text{var}(\hat{\beta}) \leq \text{var}(\tilde{\beta})$, so some authors describe an efficient estimator as the unbiased

---

[1]US Census Bureau, "Income Climbs, Poverty Stabilizes, Uninsured Rate Increases," Press release #CB06-136, 29 August 2006, `http://www.census.gov/Press-Release/www/releases/archives/income_wealth/007419.html`.

estimator with minimum variance among all unbiased estimators.

We test this using inequality 10.1.7 (page 333), that the variance must be greater than or equal to the Cramér–Rao lower bound. If $\text{var}(\hat{\beta})$ equals the CRLB, we know we have a minimum variance.

- BLUE: $\hat{\beta}$ is the Best Linear Unbiased Estimator if $\text{var}(\hat{\beta}) \leq \text{var}(\tilde{\beta})$ for all linear unbiased estimators $\tilde{\beta}$, and $\hat{\beta}$ is itself a linear function and unbiased.

- Aymptotic unbiasedness: Define $\hat{\beta}_n = f(x_1, \ldots, x_n)$. For example, $\hat{\mu}_1 = x_1$, $\hat{\mu}_2 = (x_1 + x_2)/2$, $\hat{\mu}_3 = (x_1 + x_2 + x_3)/3$, …. Then $\lim_{n \to \infty} E(\hat{\beta}_n) = \beta$. Clearly, unbiasedness implies asymptotic unbiasedness.

- Consistency: $\text{plim}(\hat{\beta}_n) = \beta$, i.e., for a fixed small ϵ, $\lim_{n \to \infty} P(|\hat{\beta}_n - \beta| > \epsilon) = 0$. Equivalently, $\lim_{n \to \infty} P((\hat{\beta}_n - \beta)^2 > \epsilon^2) = 0$.
One can verify consistency using *Chebychev's inequality*; see, e.g., Casella & Berger (1990, p 184).
In a sense, consistency is the asymptotic analog to the low MSE condition. If MSE goes to zero, then consistency follows (but not necessarily vice versa).
However, a biased estimator or a high-variance estimator may have a few things going for it, but an inconsistent estimator is just a waste. You could get yourself two near-infinite samples and find that $\hat{\beta}$ is different for each of them—and then what are you supposed to pick?

- Asymptotic efficiency: $\text{var}(\hat{\beta}) \to$ the Cramér–Rao lower bound. This makes sense only if $\hat{\beta}$'s asymptotic distribution has a finite mean and variance and $\hat{\beta}$ is consistent.

EXPECTED VALUE   Say that any given value of x has probability $p(x)$. Then if $f(x)$ is an arbitrary function,

$$E(f(x)) = \int_{\forall x} f(x)p(x)dx.$$

The $p(x)dx$ part of the integral is what statisticians call a *measure* and everyone else calls a weighting. If $p(x)$ is constant for all x, then every value of x gets equal weighting, as does every value of $f(x)$. If $p(x_1)$ is twice as large as $p(x_2)$, meaning that we are twice as likely to observe x_1, then $f(x_1)$ gets double weighting in the integral.

If we have a vector of data points, $\mathbf{x}$, consisting of n elements x_i, then we take each single observation to be equally likely: $p(x_i) = \frac{1}{n}$, $\forall i$. The expected value for a sample then becomes the familiar calculation

$$E(\mathbf{x}) = \frac{\sum_i x_i}{n},$$

and (given no further information about the population) is the Best Unbiased Estimator of the true mean μ.[2]

---

[2]The term *expected value* implies that the mean is what we humans actually expect will occur. But if I have

VARIANCE AND ITS DISSECTIONS   The *variance* for discrete data is the familiar formula of the mean of the squared distance to the average. Let $\overline{\mathbf{x}}$ indicate the mean of the data vector $\mathbf{x}$; then the best unbiased estimate of the variance of the sample is

$$\text{var}(\mathbf{x}) = \frac{1}{n} \sum_i (x_i - \overline{\mathbf{x}})^2 . \tag{7.1.1}$$

Degrees of freedom

Rather than calculating the variance of a sample, say that we seek the variance of a population, and have only a sample from which to estimate the variance. The best unbiased estimate of the variance of the population is

$$\widehat{\text{var}}(\mathbf{x}) = \frac{1}{n-1} \sum_i (x_i - \overline{\mathbf{x}})^2 . \tag{7.1.2}$$

We can think of the sum being divided by $n - 1$ instead of n (as in Equation 7.1.1) because there are only $n - 1$ random elements in the sum: given the mean $\overline{\mathbf{x}}$ and $n - 1$ elements, the nth element is deterministically solved. That is, there are only $n - 1$ *degrees of freedom*. An online appendix to this book provides a more rigorous proof that Equation 7.1.2 is an unbiased estimator of the population variance.

As $n \to \infty$, $1/n \approx 1/(n - 1)$, so both the estimate of variance based on $1/n$ and on $1/(n - 1)$ are asymptotically unbiased estimators of the population variance.

The number of degrees of freedom (*df*) will appear in other contexts throughout the book. The *df* indicates the number of dimensions in which the data could possibly vary. With no additional information, this is just the number of independently drawn variables, but there may be more restrictions on the data. Imagine three variables, which would normally have three dimensions, with the added restriction that $x_1 + 2x_2 = x_3$. Then this defines a plane (which happens to be orthogonal to the vector $(1, 2, -1)$ and goes through the origin). That is, by adding the restriction, the data points have been reduced to a two-dimensional surface. For the sample variance, the restriction is that the mean of the sample is $\hat{\mu}$.

The square root of the variance is called the *standard deviation*. It is useful because the Normal distribution is usually described in terms of the standard deviation (σ) rather than the variance (σ^2). Outside of the context of the Normal, the variance is far more common.

The variance is useful in its own right as a familiar measure of dispersion. But it can also be decomposed various ways, depending on the situation, to provide still more information, such as how much of the variance is due to bias, or how much variation is explained by a linear regression. Since information is extracted from the decomposition of variance time and time again throughout classical statistics, it is worth going over these various dissections.

Recall from basic algebra that the form $(x + y)^2$ expands to $x^2 + y^2 + 2xy$. In some special cases the unsightly $2xy$ term can be eliminated or merged with another term, leaving the pleasing result that $(x + y)^2 = x^2 + y^2$.

---

a one in a million chance of winning a two million dollar lottery, there are no states of the world where I am exactly two dollars wealthier. Further, research pioneered by Kahneman and Tversky (e.g., Kahneman *et al.* (1982)) found that humans tend to focus on other features of a probability distribution. They will consider events with small probability to either have zero probability or a more manageable value (e.g., reading $p = 1e{-}7$ as $p = 1e{-}3$). Or, they may assume the most likely state of the world occurs with certainty. Bear in mind that human readers of your papers may be interested in many definitions of expectation beyond the mean.

Throughout the discussion below $\bar{x} = E[x]$; that is $\bar{x}$ is constant for a given data set. The expectation of a constant is the constant itself, so $E[\bar{x}]$ is simply $\bar{x}$; and $E[y^2\bar{x}]$ would expand to $\frac{1}{n}\sum_{i=1}^{n}[y_i^2\bar{x}] = \bar{x} \cdot \frac{1}{n}\sum_{i=1}^{n}y_i^2 = \bar{x} \cdot E[y^2]$.

The first breakdown of variance is the equation as above:

$$
\begin{aligned}
\mathrm{var}(x) &= E\left[(x - \bar{x})^2\right] \\
&= E\left[x^2 - 2x\bar{x} + \bar{x}^2\right] \\
&= E[x^2] - E[2x\bar{x}] + E[\bar{x}^2] \\
&= E[x^2] - 2E[x]^2 + E[x]^2 \\
&= E[x^2] - E[x]^2.
\end{aligned}
$$

Read this as: $\mathrm{var}(x)$ is the expectation of the squared values minus the square of the expected value. This form simplifies many transformations and basic results of the type that frequently appear in probability homework questions.

$\mathbb{Q}_{7.1}$ | Write a function to display $\mathrm{var}(x)$, $E[x^2]$, $E[x]^2$, and $E[x^2] - E[x]^2$ for any input data, then use it to verify that the first and last expressions are equal for a few columns of data selected from any source on hand.

Mean squared error   The next breakdown appears with the mean squared error. Say that we have a biased estimate of the mean, $\tilde{x}$; if you had the true mean $\bar{x}$, then you could define the bias as $(\tilde{x} - \bar{x})$. It turns out that the MSE is a simple function of the true variance and the bias. The value can be derived by inserting $-\bar{x} + \bar{x} = 0$ and expanding the square:

$$
\begin{aligned}
MSE &\equiv E\left[(x - \tilde{x})^2\right] \\
&= E\left[((x - \bar{x}) + (\bar{x} - \tilde{x}))^2\right] \\
&= E\left[(x - \bar{x})\right]^2 + 2E\left[(x - \bar{x})(\bar{x} - \tilde{x})\right] + E\left[(\bar{x} - \tilde{x})^2\right] \\
&= \mathrm{var}(x) + 2 \cdot \mathrm{bias}(\tilde{x})E(x - \bar{x}) + \mathrm{bias}(\tilde{x})^2 \\
&= \mathrm{var}(x) + \mathrm{bias}(\tilde{x})^2
\end{aligned}
$$

In this case the middle term drops out because $E(x - \bar{x}) = 0$, and the MSE breaks down to simply being the variance of x plus the bias squared.

Within group/among group variance   The next breakdown of variance, common in ANOVA estimations (where ANOVA is short for analysis of variance), arises when the data is divided into a set of groups. Then the total variance over the entire data set could be expressed as *among group variance* and *within group variance*. Above, $\mathbf{x}$ consisted of a homogeneous sequence of elements x_i, $i = \{1, \ldots, n\}$, but now break it down into subgroups x_{ij}, where j indicates the group and i indicates the elements within the group. There is thus a mean $\bar{\mathbf{x}}_j$ for each group j, which is the simple mean for the n_j elements in that group. The unsubscripted $\bar{\mathbf{x}}$ continues to represent the mean of the entire sample. With that notation in hand, a similar breakdown to those given above can be applied to the groups:

$$\mathrm{var}(\mathbf{x}) = E\left[(\mathbf{x} - \bar{\mathbf{x}})^2\right]$$

$$= \frac{1}{n}\sum_j\left[\sum_{i=1}^{n_j}(x_{ij} - \bar{\mathbf{x}}_j + \bar{\mathbf{x}}_j - \bar{\mathbf{x}})^2\right]$$

$$= \frac{1}{n}\sum_j\left[\sum_{i=1}^{n_j}(x_{ij} - \bar{\mathbf{x}}_j)^2 + \sum_{i=1}^{n_j}(\bar{\mathbf{x}}_j - \bar{\mathbf{x}})^2 + \sum_{i=1}^{n_j}(x_{ij} - \bar{\mathbf{x}}_j)(\bar{\mathbf{x}}_j - \bar{\mathbf{x}})\right]$$

$$(7.1.3)$$

$$= \frac{1}{n}\sum_j\left[\sum_{i=1}^{n_j}(x_{ij} - \bar{\mathbf{x}}_j)^2 + \sum_{i=1}^{n_j}(\bar{\mathbf{x}}_j - \bar{\mathbf{x}})^2\right] \qquad (7.1.4)$$

$$= \frac{1}{n}\sum_j\left[n_j\,\mathrm{var}(\mathbf{x}_j)\right] + \frac{1}{n}\sum_j\left[n_j(\bar{\mathbf{x}}_j - \bar{\mathbf{x}})^2\right]$$

The transition from Equation 7.1.3 to 7.1.4 works because $(\bar{\mathbf{x}}_j - \bar{\mathbf{x}})$ is constant for a given j, and $\sum_{i=1}^{n_j}(x_{ij} - \bar{\mathbf{x}}_j) = \bar{\mathbf{x}}_j - \bar{\mathbf{x}}_j = 0$. Once again, the unsightly middle term cancels out, and we are left with an easily interpretable final equation. In this case, the first element is the weighted mean of within-group variances, and the second is the weighted among-group variance, where each group is taken to be one unit at $\bar{\mathbf{x}}_j$, and then the variance is taken over this set of group means.

The `data-metro.db` set gives average weekday passenger boardings at every station in the Washington Metro subway system, from the founding of the system in November 1977 to 2007.[3] The system has five lines (Blue, Green, Orange, Red, Yellow), and Listing 7.1 breaks down the variance of ridership on the Washington Metro into within-line and among-line variances.

- Line 20 is the query to join the `riders` and `lines` tables. The parens mean that it

---

[3] As a Washington-relevant detail, all post-'77 measurements were made in May, outside the peak tourist season.

```
1    #include <apop.h>
2
3    void variance_breakdown(char *table, char *data, char *grouping){
4        apop_data* aggregates = apop_query_to_mixed_data("mmw",
5                    "select var_pop(%s) var, avg(%s) avg, count(*) from %s group by %s",
6                    data, data, table, grouping);
7        APOP_COL_T(aggregates, "var", vars);
8        APOP_COL_T(aggregates, "avg", means);
9        double total= apop_query_to_float("select var_pop(%s) from %s", data, table);
10       double mean_of_vars = apop_vector_weighted_mean(vars, aggregates->weights);
11       double var_of_means = apop_vector_weighted_var(means, aggregates->weights);
12       printf("total variance: %g\n", total);
13       printf("within group variance: %g\n", mean_of_vars);
14       printf("among group variance: %g\n", var_of_means);
15       printf("sum within+among: %g\n", mean_of_vars + var_of_means);
16   }
17
18   int main(){
19       apop_db_open("data−metro.db");
20       char joinedtab[] = "(select riders/100 as riders, line from riders, lines \
21                   where lines.station =riders.station)";
22       variance_breakdown(joinedtab, "riders", "line");
23   }
```

Listing 7.1 Decomposing variance between among-group and within-group. Online source: `amongwithin.c`.

can comfortably be inserted into a `from` clause, as in the query on line four.[4]

• The query on line 9 pulls the total variance—the total sum of squares—and the query on lines 4–6 gets the within-group variances and means.

• Lines 10 and 11 take the weighted mean of the variances and the weighted variance of the means.

• Lines 14–17 print the data to screen, showing that these two sub-calculations do indeed add up to the total variance.

 Q7.2 | Rewrite the program to use the `data-wb.db` data set (including the `classes` table) to break down the variance in GDP per capita into within-class and among-class variance.

Within-group and among-group variance is interesting by itself. To give one example, Glaeser *et al.* (1996, Equation 8) break down the variance in crime into

---

[4]Using a subquery like this may force the SQL interpreter to re-generate the subtable for every query, which is clearly inefficient. Therefore, when using functions like `apop_anova` in the wild, first run a `create table` ... query to join the data, perhaps index the table, and then send that table to the `apop_anova` function.

within-city and among-city variances, and find that among-city variance is orders of magnitude larger than within-city variance.

Returning to the Metro data, we could group data by year, and look for within- and among-group variation in that form, or we could group data by line and ask about within- and among-group variation there.

```
#include <apop.h>

int main(){
    apop_db_open("data−metro.db");
    char joinedtab[] = "(select year, riders, line \
                        from riders, lines \
                        where riders.station = lines.station)";
    apop_data_show(apop_anova(joinedtab, "riders", "line", "year"));
}
```

Listing 7.2 Produce a two-way ANOVA table breaking variance in per-station passenger boardings into by-year effects, by-line effects, an interaction term, and the residual. Online source: `metroanova.c`.

Listing 7.2 produces an *ANOVA table*, which is a spreadsheet-like table giving the within-group and among-group variances. The form of the table dates back to the mid-1900s—ANOVA is basically the most complex thing that one can do without a matrix-inverting computer, and the tabular form facilitates doing the calculation with paper, pencil, and a desk calculator. But it still conveys meaning even for those of us who have entirely forgotten how to use a pencil.

The first three rows of the output present the between-group sum of squares. That is, if we were to aggregate all the data points for a given group into the mean, how much variation would remain? With *grouping1* and *grouping2*, there are three ways to group the data: group by *grouping1* [(Green line), (Red line), . . .], group by *grouping2* [(1977), (1978), . . . , (2007)], and the *interaction*: group by *grouping1, grouping2* [(Green Line, 1977), (Red Line, 1977), . . . (Green Line, 2007), (Red Line, 2007)]. Using algebra much like that done above, we can break down the total sum of squares into (weighted sum of squares, *grouping1*) + (weighted sum of squares, *grouping2*) + (weighted sum of squares, *grouping1, grouping2*) + (weighted sum of squares, residual).

We can compare the weighted grouped sums to the residual sum, which is listed as the F statistic in the ANOVA table. As will be discussed in the chapter on testing (see page 309), an F statistic over about two is taken to indicate that the grouping explains more variation than would be explained via a comparable random grouping of the data. The output of this example shows that the grouping by year is very significant, as is the more refined interaction grouping by line and year, but the grouping by line is not significant, meaning that later studies may be justified in not focusing on how the subway stations are broken down into lines.

$Q_{7.3}$ | Most stations are on multiple lines, so a station like Metro Center is included in the Blue, Orange, and Red groups. In fact, the Yellow line has only two stations that it doesn't share with other lines. [You can easily find an online map of the Washington Metro to verify this.] This probably causes us to underestimate the importance of the per-line grouping. How would you design a grouping that puts all stations in only one group? It may help in implementation to produce an intermediate table that presents your desired grouping. Does the ANOVA using your new grouping table show more significance to the line grouping?

By changing the second group in the code listing from `"year"` to `NULL`, we would get a one-way ANOVA, which breaks down the total sum of squares into just (weighted sum of squares, *grouping1*) + (weighted sum of squares, residual). The residual sum of squares is therefore larger, the *df* of the residual is also larger, and in this case the overall change in the F statistic is not great.

※ *Regression variance*   Next, consider the OLS model, which will be detailed in Section 8.2.1. In this case, we will break down the observed value to the estimated value plus the error: $\mathbf{y} = \mathbf{y}_{\mathrm{est}} + \boldsymbol{\epsilon}$.

$$\mathrm{var}(\mathbf{y}) = E\left[(\mathbf{y}_{\mathrm{est}} + \boldsymbol{\epsilon} - \bar{\mathbf{y}})^2\right]$$
$$= E\left[(\mathbf{y}_{\mathrm{est}} - \bar{\mathbf{y}})^2\right] + E[\boldsymbol{\epsilon}^2] + 2E\left[(\mathbf{y}_{\mathrm{est}} - \bar{\mathbf{y}})\boldsymbol{\epsilon}\right]$$

It will be shown below that $\mathbf{y}_{\mathrm{est}}\boldsymbol{\epsilon} = \beta_{\mathrm{OLS}}\mathbf{X}\boldsymbol{\epsilon} = 0$ (because $\mathbf{X}\boldsymbol{\epsilon} = 0$), and $\bar{\boldsymbol{\epsilon}} = 0$, so $E[\bar{\mathbf{y}}\boldsymbol{\epsilon}] = \bar{\mathbf{y}}E[\boldsymbol{\epsilon}] = 0$. Thus, the $2E[\ldots]$ term is once again zero, and we are left with

$$\mathrm{var}(\mathbf{y}) = E\left[(\mathbf{y}_{\mathrm{est}} - \bar{\mathbf{y}})^2\right] + E[\boldsymbol{\epsilon}^2] \tag{7.1.5}$$

Make the following definitions:

$$SST \equiv \text{total sum of squares}$$
$$= E\left[(\mathbf{y} - \bar{\mathbf{y}})^2\right]$$
$$= \text{var}(\mathbf{y})$$
$$SSR \equiv \text{Regression sum of squares}$$
$$= E\left[(\mathbf{y}_{\text{est}} - \bar{\mathbf{y}})^2\right]$$
$$SSE \equiv \text{Sum of squared errors}$$
$$= E[\epsilon^2]$$

Then the expansion of $\text{var}(\mathbf{y})$ in Equation 7.1.5 could be written as

$$SST = SSR + SSE.$$

This is a popular breakdown of the variance, because it is relatively easy to calculate and has a reasonable interpretation: total variance is variance explained by the regression plus the unexplained, error variance. As such, these elements will appear on page 311 with regard to the F test, and are used for the common *coefficient of determination*, which is an indicator of how well a regression fits the data. It is defined as:

$$R^2 \equiv \frac{SSR}{SST}$$
$$= 1 - \frac{SSE}{SST}.$$

You will notice that the terminology about the sum of squared components and the use of the F test matches that used in the ANOVA breakdowns, which is not just a coincidence: in both cases, there is a portion of the data's variation explained by the model (grouping or regression), and a portion that is unexplained by the model (residual). In both cases, we can use this breakdown to gauge whether the model explains more variation than would be explained by a random grouping. The exact details of the F test will be delayed until the chapter on hypothesis testing.

COVARIANCE   The population *covariance* is $\sigma^2_{\mathbf{xy}} = \frac{1}{n}\sum_i(x_i - \bar{x})(y_i - \bar{y})$, which is equivalent to $E[\mathbf{xy}] - E[\mathbf{x}]E[\mathbf{y}]$. [Q: Re-apply the first variance expansion above to prove this.] The variance is a special case where $\mathbf{x} = \mathbf{y}$.

As with the variance, the unbiased estimate of the sample covariance is $s^2_{xy} = \sigma^2_{xy} \cdot \frac{n}{n-1}$, i.e., $\sum_i(x - \bar{x})(y - \bar{y})$ divided by $n - 1$ instead of n.

Given a vector of variables $\mathbf{x}_1, \mathbf{x}_2, \ldots \mathbf{x}_n$, we typically want the covariance of every combination. This can neatly be expressed as a matrix

$$\begin{bmatrix} \sigma_1^2 & \sigma_{12}^2 & \cdots & \sigma_{1n}^2 \\ \sigma_{21}^2 & \sigma_2^2 & \cdots & \sigma_{2n}^2 \\ \vdots & & \ddots & \\ \sigma_{n1}^2 & \sigma_{n2}^2 & \cdots & \sigma_n^2 \end{bmatrix},$$

where the diagonal elements are the variances (i.e., the covariance of x_i with itself for all i), and the off-diagonal terms are symmetric in the sense that $\sigma_{ij}^2 = \sigma_{ji}^2$ for all i and j.

Correlation and Cauchy–Schwarz: The *correlation coefficient* (sometimes called the *Pearson correlation coefficient*) is

$$\rho_{\mathbf{xy}} \equiv \frac{\sigma_{\mathbf{xy}}}{\sigma_{\mathbf{x}}\sigma_{\mathbf{y}}}.$$

By itself, the statistic is useful for looking at the relations among columns of data, and can be summarized into a matrix like the covariance matrix above. The correlation matrix is also symmetric, and has ones all along the diagonal, since any variable is perfectly correlated with itself.

The *Cauchy–Schwarz inequality*, $0 \le \rho^2 \le 1$, puts bounds on the correlation coefficient. That is, ρ is in the range $[-1, 1]$, where $\rho = -1$ indicates that one variable always moves in the opposite direction of the other, and $\rho = 1$ indicates that they move in perfect sync.[5]

The matrix of correlations is another popular favorite for getting basic descriptive information about a data set; produce it via `apop_data_correlation`. The correlation matrix will be the basis of the Cramér–Rao lower bound on page 333.

MORE MOMENTS Given a continuous probability distribution from which the data was taken, you could write out the expectation in the variance equation as an integral,

$$\mathrm{var}(f(x)) = E\left(\left(f(x) - \overline{f(x)} \right)^2 \right)$$

$$= \int_{\forall x} \left(f(x) - \overline{f(x)} \right)^2 p(x)dx.$$

---

[5] It would be a digression to prove the Cauchy–Schwarz inequality here, but see Hölder's inequality in any probability text, such as Feller (1966, volume II, p 155).

Similarly for higher powers as well:

$$\text{skew}\,(f(x)) \equiv \int_{\forall x} \left(f(x) - \overline{f(x)}\right)^3 p(x)dx$$

$$\text{kurtosis}\,(f(x)) \equiv \int_{\forall x} \left(f(x) - \overline{f(x)}\right)^4 p(x)dx.$$

These three integrals are *central moments* of $f(x)$. They are central because we subtracted the mean from the function before taking the second, third, or fourth power.[6]

Transformed moments

Let $\mathcal{S}$ and $\mathcal{K}$ be the third and fourth central moments as given here. Some use a *standardized moment* for kurtosis, which may equal $\mathcal{K}_1' = \mathcal{K}/(\sigma^2)^2$, $\mathcal{K}_2' = \mathcal{K}/(\sigma^2)^2 - 3$, or whatever else the author felt would be convenient. Similarly, some call $\mathcal{S}' = \mathcal{S}/(\sigma^2)^{3/2}$ the skew. These adjustments are intended to ease comparisons to the standard Normal and to accommodate differences in scale.

The only way to know what a given source means when it refers to skew and kurtosis is to look up the definitions. The GSL uses $\mathcal{K}_2'$ (because engineers are probably comparing their data to a standard Normal); Apophenia uses $\mathcal{K}$ (because the corrections can add complication in situations outside the Normal distribution, and is easy to make when needed).

What information can we get from the higher moments? Section 9.1 will discuss the powerful Central Limit Theorem, which says that if a variable represents the mean of a set of independent and identical draws, then it will have an $\mathcal{N}(\mu, \sigma)$ distribution, where μ and σ are unknowns that can be estimated from the data. These two parameters completely define the distribution: the skew of a Normal is always zero, and the kurtosis is always $3\sigma^4$. If the kurtosis is larger, then this often means that the assumption of independent draws is false—the observations are interconnected. One often sees this among social networks, stock markets, or other systems where independent agents observe and imitate each other.

Positive skew indicates that a distribution is upward leaning, and a negative skew indicates a downward lean. Kurtosis is typically put in plain English as *fat tails*: how much density is in the tails of the distribution? For example, the kurtosis of an $\mathcal{N}(0, 1)$ is three, while the kurtosis of a Student's t distribution with n degrees of freedom is greater than three, and decreases as $n \to \infty$, converging to three (see page 365 for a full analysis). An un-normalized kurtosis $> 3\sigma^4$ is known as *leptokurtic* and $< 3\sigma^4$ is known as *platykurtic*; see Figure 7.3 for a mnemonic.

The caveats about unbiased estimates of the sample versus population variance (see box, page 222) also hold for skew and kurtosis: calculating the mean as done in the definitions above leads to a biased estimate of the population skew or kurtosis, but there are simple corrections that can produce an unbiased estimate. An online

---

[6]The central first moment is always zero; the non-central second, third, ..., moments are difficult to interpret and basically ignored. Since there is no ambiguity, some authors refer to the useful moments as *the nth moment*, $n \in \{1, 2, 3, 4\}$, and leave it as understood when the moment is central or non-central.

* In case any of my readers may be unfamiliar with the term "kurtosis" we may define mesokurtic as "having β_2 equal to 3," while platykurtic curves have $\beta_2 < 3$ and leptokurtic > 3. The
important property which follows from this is that platykurtic curves have shorter "tails" than the

normal curve of error and leptokurtic longer "tails." I myself bear in mind the meaning of the words
by the above *memoria technica*, where the first figure represents platypus, and the second kangaroos,
noted for "lepping," though, perhaps, with equal reason they should be hares!

Figure 7.3 Leptokurtic, mesokurtic and platykurtic, illustration by Gosset in *Biometrika* (Student,
1927, p 160). In the notation of the time, $\beta_2 \equiv$ kurtosis/(variance squared).

appendix to this book offers a few more facts about central moments, and derives
the correction factors.

But in all cases, the population vs sample detail is relevant only for small n. Efron
& Tibshirani (1993, p 43) state that estimating variance via n is "just as good"
as estimating it via $n - 1$, so there is highly-esteemed precedent for ignoring this
detail. For the higher moments, the sample and population estimates converge even
more quickly.

Coding it   Given a vector, Apophenia provides functions to calculate most of the
above, e.g.:

```
apop_data *set = gather_your_data_here();
apop_data *corr = apop_data_correlation(set);
APOP_COL(set, 0, v1);
APOP_COL(set, 1, v2);
double mean1 = apop_vector_mean(v1);
double var1 = apop_vector_var(v1);
double skew1 = apop_vector_skew(v1);
double kurt1 = apop_vector_kurtosis(v1);
double cov = apop_vector_cov(v1, v2);
double cor = apop_vector_correlation(v1, v2);
apop_data_show(apop_data_summarize(set));
```

The last item in the code, `apop_matrix_summarize`, produces a table of some
summary statistics for every column of the data set.

Your data may be aggregated so that one line of data represents multiple observations. For example, sampling efficiency can be improved by sampling subpopulations differently depending upon their expected variance (Särndal *et al.*, 1992). For this and other reasons, data from statistical agencies often includes weightings.

This is not the place to go into details about statistically sound means of weighting data, but if you have a separate vector with weights, you can use `apop_vector_weighted_mean`, `apop_vector_weighted_var`, et cetera, to use those weights. Or, if your `apop_data` set's `weights` vector is filled, `apop_data_summarize` will make use of it.

Q7.4

- Write a query that pulls the number of males per 100 females and the population density from the Census data (`data-census.db`). The query will
    - join together the `geography` and `demos` tables by county number, and
    - exclude states and the national total, and
    - return a two-column table.
- Write a function `void summarize_paired_data(char *q)` that takes in a query that produces a two-column table, and outputs some of the above summary information about both columns, including the mean, variance, and correlation coefficients.
- Write a `main()` that sends your query to the above function, and run the program. Is population density positively or negatively correlated to males per female?
- Write another query that pulls the ratio of (median income for full-time workers, female)/(median income for full-time workers, male) and the population density.
- Add a line to `main()` to send that query to your summarizing function as well. How is the new pair of variables correlated?

Quantiles   The mean and variance can be misleading for skewed data. The first option for describing a data set whose distribution is likely ab-Normal is to plot a histogram of the data, as per page 172.

A numeric option is to print the quartiles, quintiles, or deciles of the data. For quartiles, sort the data, and then display the values of the data points 0%, 25%, 50%, 75%, and 100% of the way through the set. The 0% value is the minimum of the data set, the 100% value is the maximum, and the 50% value is probably the median (see below). For quintiles, print the values of data points 0%, 20%, 40%, 60%, 80%, and 100% of the way through the set, and for deciles, print the values every ten percent.

Sorting your data is simple. If you have an `apop_data` set and a `gsl_vector`, then

```
apop_data_sort(my_data, 2, 'd');
gsl_vector_sort(my_vector);
```

would sort `my_data` in place so that column 2 is in 'd'escending order, and sort the vector in place to ascending order, so `gsl_vector_get(my_vector, 0)` is the minimum of the data, `gsl_vector_get(my_vector, my_vector->size)` is the maximum, and `gsl_vector_get(my_vector, my_vector->size/2)` is about the median.

Alternatively, the function `apop_vector_percentiles` takes in a `gsl_vector` and returns the percentiles—the value of the data point 0%, 1%, ..., 99%, 100% of the way through the data. It takes in two arguments: the data vector, and a character describing the rounding method—`'u'` for rounding up, `'d'` for rounding down, and `'a'` for averaging. Since the number of elements in a data set is rarely divisible by a hundred and one, the position of most percentile points likely falls between two data points. For example, if the data set has 107 points, then the tenth data point is 9.47% through the data set, and the eleventh data point is 10.38% through the set, so which is the tenth percentile? If you specify `'u'`, then it is the eleventh data point; if you specify `'d'` then it is the tenth data point, and if you specify `'a'`, then it is the simple average of the two.

The standard definition of the median is that it is the middle value of the data point if the data set has an odd number of elements, and it is the average of the two data points closest to the middle if the data set has an even number of elements. Thus, here is a function to find the median of a data set. It finds the percentiles using the averaging rule for interpolation, marks down the 50th percentile, then cleans up and returns that value.

```
double find_median(gsl_vector *v){
    double *pctiles = apop_vector_percentiles(v, 'a');
    double out = pctiles[50];
    free(pctiles);
    return out;
}
```

Q₇.₅

Write a function with header double show_quantiles(gsl_vector *v, char rounding_method, int divisions) that passes v and rounding_method to apop_vector_percentiles, and then prints a table of selected quantiles to the screen. For example, if divisions==4, print quartiles, if divisions==5, print quintiles, if divisions==10, print deciles, et cetera.

On page 88 you tabulated GDP per capita for the countries of the world. Use your function to print the deciles for country incomes from data-wb.

Q₇.₆

The *trimean* is $\frac{1}{4}$ the sum of the first quartile, third quartile, and two times the median (Tukey, 1977, p 46). It uses more information about the distribution than the median alone, but is still robust to extreme values (unlike the mean).

Write a function that takes in a vector of data points, applies apop_vector_percentiles internally, and returns the trimean. How does the trimean of GDP per capita compare to the mean and median, and why?

See also page 319, which compares percentiles of the data to percentiles of an assumed distribution to test whether the data were drawn from the distribution.

➤ The most basic means of describing data is via its moments. The basic moments should be produced and skimmed for any data set; in simple cases, there is no need to go further.

➤ The variance can often be decomposed into smaller parts, thus revealing more information about how a data set's variation arose.

➤ The mean and variance are well known, but there is also information in higher moments—the skew and kurtosis.

➤ It is also important to know how variables interrelate, which can be summarized using the correlation matrix.

➤ There is a one-line function to produce each of these pieces of information. Notably, apop_data_summarize produces a summary of each column of a data set.

➤ You can get a more detailed numerical description of a data set's distribution using quartiles or quintiles; to do so, use apop_vector_percentiles.

7.2 SAMPLE DISTRIBUTIONS   Here are some distributions that an observed variable may take on. They are not just here so you can memorize them before the next statistics test. Each has a story attached, which is directly useful for modeling. For example, if you think that a variable is the outcome of n independent, binary events, then the variable should be modeled as a Binomial distribution, and once you estimate the parameter to the distribution, you will have a full model of that variable, that you can even test if so inclined. Table 7.4 presents a table of what story each distribution is telling. After the catalog of models, I will give a few examples of such modeling.

The distribution	The story
Bernoulli	A single success/failure draw, fixed p.
Binomial	What are the odds of getting x successes from n Bernoulli draws with fixed p?
Hypergeometric	What are the odds of getting x successes from n Bernoulli draws, where p is initially fixed, but drawing is without replacement?
Normal/Gaussian	Binomial as $n \to \infty$; if $\mu_j \equiv \sum_{i=1}^{n} x_{ij}/n$, then $\mu_j \sim$ Normal.
Lognormal	If $\mu_j \equiv \prod_{i=1}^{n} x_{ij}$, then $\mu_j \sim$ Lognormal.
Multinomial	n draws from m possibilities with probabilities p_1, ..., p_m, $\sum_{i=1}^{m} p_i = 1$.
Multivariate Normal	Multinomial as $n \to \infty$.
Negative binomial	How many Bernoulli draws until n successes?
Poisson	Given λ events per period, how many events in t periods?
Gamma	The 'Negative Poisson': given a Poisson setup, how long until n events?
Exponential	A proportion λ of the remaining stock leaves each period; how much is left at time t?
Beta	A versatile way to describe the odds that p takes on any value $\in (0, 1)$.
Uniform	No information but the upper and lower bounds.

Table 7.4 Every probability distribution tells a story.

Common distributions of statistical parameters (as opposed to natural populations) are discussed in Section 9.2.

Bernoulli and Poisson events   The core of the system is an event, which sometimes happens and sometimes does not. Some peo-
ple have a disease, some do not; some days it rains, some days it does not. Events
add up to more-or-less continuous quantities: some cities see a 22% chance of rain
on a given day, and some see a 23.2% chance; some populations have high rates of
disease and some do not.

From there, there are variants: instead of asking how many successes we will see
in n trials (the Binomial and Poisson distributions), we can ask how many trials we
can expect to make to get n successes (the Negative binomial and Gamma distributions). We can look at sampling without replacement (Hypergeometric). We can
look at the case where the number of trials goes to infinity (Normal) or aggregate
trials and take their product (Lognormal). In short, a surprisingly wide range of
situations can be described from the simple concept of binary events aggregated in
various ways.

The snowflake problem   For a controlled study (typical of physical sciences), the
claim of a fixed probability (p or λ) is often plausible, but
for most social science experiments, where each observation is an individual or a
very different geographic region, the assumption that all observations are identical
is often unacceptable—every snowflake is unique.

It may seem like the fixed-p and fixed-λ models below are too simple to be applicable to many situations—and frankly, they often are. However, they can be used
as building blocks to produce more descriptive models. Section 8.2.1 presents a
set of linear regression models, which handle the snowflake problem well, but
throw out the probability framework presented in this chapter. But we can solve
the snowflake problem and still make use of simple probability models by em-
bedding a linear model inside a distribution: let p_i differ among each element i,
and estimate p_i using a linear combination of element i's characteristics. Page 288
covers examples of the many possibilities provided by models-within-models.

Statistics and their estimators

- The catalog in this section includes the three most typical items one would want
  from a distribution: a random number generator (*RNG*) that would produce data
  with the given distribution, a *probability density function* (*PDF*), and a *cumulative
  density function* (*CDF*).[7]

---

[7]This is for the case when the independent variable is continuous. When it is discrete, the PDF is named a
probability mass function (*PMF*) and the CDF a *cumulative mass function* (*CMF*). For every result or statement
about PDFs, there is an analogous result about PMFs; for every result or statement about CDFs, there is an
analogous result about CMFs.

- The catalog also includes the expected value and variance of these distributions, which are distinct from the means and variances to this point in a key manner: given a data set, the mean is a statistic—a function of data of the form $f(x)$. Meanwhile, given a model, the mean and variance of a draw are functions of the parameters, e.g., given a Binomial distribution with parameters n and p, $E(x|n, p) = np$. Of course, we rarely know all the parameters, so we are left with estimating them from data, but our estimate of p, $\hat{p}$ is once again a function of the data set on hand. We will return to this back-and-forth between estimates from data and estimates from the model after the catalog of models.

- The full details of RNG use will be discussed in Chapter 11, but the RNG functions are included in the catalog here for easy reference; each requires a pointer to a `gsl_rng`, which will be named `r`.

THE BERNOULLI FAMILY   The first set of distributions are built around a narrative of drawing from a pool of events with a fixed probability. The most commonly-used example is flipping a coin, which is a single event that comes out *heads* with probability $p = 1/2$. But other examples abound: draw one recipient from a sales pitch out of a list of such people and check whether he purchased or did not purchase, or pull a single citizen from a population and see if she was the victim of a crime. For one event, a draw can take values of only zero or one; this is known as a *Bernoulli draw*.

Bernoulli   The Bernoulli distribution represents the result of one Bernoulli draw, meaning that $P(x = 1|p) = p$, $P(x = 0|p) = 1 - p$, and $P(x = \text{anything}$ else $|p) = 0$. Notice that $E(x|p) = p$, even though x can be only zero or one.

$$P(x, p) = p^x(1 - p)^{(1-x)}, x \in \{0, 1\}$$
$$= \texttt{gsl\_ran\_bernoulli\_pdf(x, p)}$$
$$E(x|p) = p$$
$$\text{var}(x|p) = p(1 - p)$$
$$RNG : \texttt{gsl\_ran\_bernoulli(r, p)}$$

Binomial   Now take n Bernoulli draws, so we can observe between 0 and n events. The output is now a count: how many people dunned by a telemarketer agree to purchase the product, or how many crime victims there are among a fixed population. The probability of observing exactly k events is $p(k) \sim \text{Binomial}(n, p)$.

Counting x successes out of n trials is less than trivial. For example, there are six ways by which two successes could occur over four trials: $(0, 0, 1, 1)$, $(0, 1, 0, 1)$,

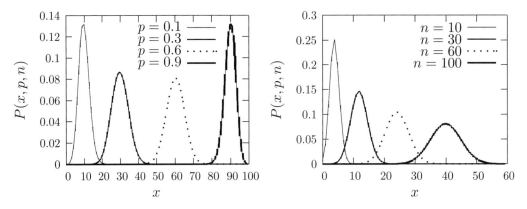

Figure 7.5 Left: The Binomial distribution with $n = 100$ and various values of p. Right: the Binomial distribution with $p = 0.4$ and various values of n. As n grows, the distribution approaches an $\mathcal{N}(np, \sqrt{np(1-p)})$.

$(0, 1, 1, 0)$, $(1, 1, 0, 0)$, $(1, 0, 0, 1)$, or $(1, 0, 1, 0)$, and the model underlying the Binomial model treats them all equally.

The form of the distribution therefore borrows a counting technique from combinatorics. The notation $\binom{n}{x}$ indicates n *choose* x, the number of unordered sets of x elements that can be pulled from n objects. The equation is

$$\binom{n}{x} = \frac{n!}{x!(n-x)!},$$

and the function is `gsl_sf_choose(n,x)` (in the GSL's Special Functions section). For example, we could get exactly thirty successful trials out of a hundred in $\binom{100}{30}$ ways ($\approx 2.94\text{e}25$).

Combinatorics also dictates the shape of the curve. There is only one way each to list four zeros or four ones—$(0, 0, 0, 0)$ and $(1, 1, 1, 1)$—and there are four ways to list one one—$(0, 0, 0, 1)$, $(0, 0, 1, 0)$, ...—and symmetrically for one zero. In order, the counts for zero through four ones are 1, 4, 6, 4, and 1. This simple counting scheme already produces something of a bell curve. Returning to coin-flipping, if $p = 1/2$ and the coin is flipped 100 times ($n = 100$), $p(50 \text{ heads})$ is relatively high, while $p(0 \text{ heads})$ or $p(100 \text{ heads})$ is almost nil.

In fact, as $n \to \infty$, the probability distribution approaches the familiar Normal distribution with mean np and variance $np(1 - p)$, as in Figure 7.5. Assuming that every telemarketer's probability of selling is equal, we expect that a plot of many months' telemarketer results will look like a Normal distribution, with many telemarketers successfully selling to np victims, and others doing exceptionally well or poorly. The assumption that every telemarketer is equally effective can even be tested, by checking for digression from the Normal distribution.

$$P(x, p, n) = \binom{n}{x} p^x (1-p)^{(n-x)}$$

$$= \texttt{gsl\_ran\_binomial\_pdf}(\texttt{x}, \texttt{p}, \texttt{n})$$

$$E(x|n, p) = np \tag{7.2.1}$$

$$\mathrm{var}(x|n, p) = np(1-p) \tag{7.2.2}$$

$$RNG : \texttt{gsl\_ran\_binomial}(\texttt{r}, \texttt{p}, \texttt{n})$$

- If $X \sim \text{Bernoulli}(p)$, then for the sum of n independent draws, $\sum_{i=1}^{n} X_i \sim \text{Binomial}(n, p)$.

- As $n \to \infty$, $\text{Binomial}(n, p) \to \text{Poisson}(np)$ or $\mathcal{N}(np, \sqrt{np(1-p)})$.

Since n is known and $E(x)$ and $\mathrm{var}(x)$ can be calculated from the data, Equations 7.2.1 and 7.2.2 are an oversolved system of two variables for one unknown, p. Thus, you can test for *excess variance*, which indicates that there are interactions that falsify that the observations were *iid* (independent and identically distributed) Bernoulli events.

A variant: one p from n draws   The statistic of interest often differs from that calculated in this catalog, but it is easy to transform the information here. Say that we multiply the elements of a set x by k. Then the mean goes from being $\mu_x \equiv \sum_i x_i/n$ to being $\mu_{kx} \equiv \sum_i (kx_i)/n = k\sum_i x_i/n = k\mu_x$. The variance goes from being $\sigma_x^2 \equiv \sum_i (x_i - \mu_x)^2/n$ to $\sigma_{kx}^2 \equiv \sum_i (kx_i - k\mu_x)^2/n = k^2 \sum_i (x_i - \mu_x)^2/n = k^2 \sigma_k^2$.

For example, we are often interested in estimating $\hat{p}$ from data with a Binomial-type story. Since $E(x) = np$ under a Binomial model, one could estimate $\hat{p}$ as $E(x/n)$. As for the variance, let k in the last paragraph be $1/n$; then the variance of x/n is the original variance ($\mathrm{var}(x) = n\hat{p}(1-\hat{p})$) times $1/n^2$, which gives $\mathrm{var}(\hat{p}) = \hat{p}(1-\hat{p})/n$.

Hypergeometric   Say that we have a pool of N elements, initially consisting of s successes and $f \equiv N - s$ failures. So $N = s + f$, and the Bernoulli probability for the entire pool is $p = s/N$. What are the odds that we get x successes from n draws *without replacement*? In this case, the probability of a success changes as the draws are made. The counting is more difficult, resulting in a somewhat more involved equation.

$$P(x,s,f,n) = \frac{\binom{s}{x}\binom{f}{n-x}}{\binom{N}{n}}$$

$$= \texttt{gsl\_ran\_hypergeometric\_pdf}(x,s,f,n)$$

$$E(x|n,s,f) = \frac{ns}{N}$$

$$\mathrm{var}(x|n,s,f) = \frac{n(\frac{s}{N})(1-\frac{s}{N})(N-n)}{(N-1)}$$

$$RNG : \texttt{gsl\_ran\_hypergeometric}(r,s,f,n)$$

- As $N \to \infty$, drawing with and without replacement become equivalent, so the Hypergeometric distribution approaches the Binomial.

Multinomial The Binomial distribution was based on having a series of events that could take on only two states: success/failure, sick/well, heads/tails, et cetera. But what if there are several possible events, like left/right/center, or Africa/Eurasia/Australia/Americas? The Multinomial distribution extends the Binomial distribution for such cases.

The Binomial case could be expressed with one parameter, p, which indicated success with probability p and failure with probability $1 - p$. The Multinomial case requires k variables, $p_1, \ldots, p_k$, such that $\sum_{i=1}^{k} p_i = 1$.

$$P(\mathbf{x},\mathbf{p},n) = \frac{n!}{x_1! \cdots x_k!} p_1^{x_1} \cdots p_k^{x_k}$$

$$= \texttt{gsl\_ran\_multinomial\_pdf}(k,p,n)$$

$$E\left(\begin{bmatrix} x_1 \\ x_2 \\ \vdots \\ x_n \end{bmatrix} \middle| n, \mathbf{p}\right) = n \begin{bmatrix} p_1 \\ p_2 \\ \vdots \\ p_n \end{bmatrix}$$

$$\mathrm{var}(\mathbf{x}|n,\mathbf{p}) = n \begin{bmatrix} p_1(1-p_1) & -p_1 p_2 & \cdots & -p_1 p_k \\ -p_1 p_2 & p_2(1-p_2) & \cdots & -p_2 p_k \\ \vdots & & \ddots & \vdots \\ -p_k p_2 & -p_k p_2 & \cdots & p_k(1-p_k) \end{bmatrix}$$

$$RNG : \texttt{gsl\_ran\_multinomial}(r,draws,k,p,out)$$

- There are two changes from the norm for the GSL's functions. First, p and n are arrays of doubles of size k. If $\sum_{i=1}^{k} \texttt{p[i]} \neq 1$, then the system normalizes the probabilities to make this the case. Also, most RNGs draw one number at a time, but this one draws K elements at a time, which will be put into the bins of the out array.

- You can verify that when $k = 2$, this is the Binomial distribution.

Normal   You know and love the bell curve, aka the Gaussian distribution. It is pictured for a few values of σ^2 in Figure 7.6.

As Section 9.1 will explain in detail, any set of means generated via iid draws will have a Normal distribution. That is,

- Draw K items from the population (which can have any nondegenerate distribution), $x_1, x_2, \ldots, x_k$. The Normal approximation works best when K is large.
- Write down the mean of those items, $\bar{x}_i$.
- Repeat the first two steps n times, producing a set $\mathbf{x} = \{\bar{x}_1, \bar{x}_2, \ldots, \bar{x}_n\}$.

Then $\mathbf{x}$ has a Normal distribution.

Alternatively, the Binomial distribution already produced something of a bell curve with $n = 4$ above; as $n \to \infty$, the Binomial distribution approaches a Normal distribution.

$$P(x, \mu, \sigma) = \frac{1}{\sigma\sqrt{2\pi}} \exp\left(-\frac{1}{2}\left[\frac{(x-\mu)}{\sigma}\right]^2\right)$$
$$= \texttt{gsl\_ran\_gaussian\_pdf(x, sigma)} + \texttt{mu}$$
$$E(x|\mu, \sigma) = \mu$$
$$\text{var}(x|\mu, \sigma) = \sigma^2$$
$$\int_{-\infty}^{x} \mathcal{N}(y|\mu, \sigma)dy = \texttt{gsl\_cdf\_gaussian\_P(x - mu, sigma)}$$
$$\int_{x}^{\infty} \mathcal{N}(y|\mu, \sigma)dy = \texttt{gsl\_cdf\_gaussian\_Q(x - mu, sigma)}$$
$$RNG : \texttt{gsl\_ran\_gaussian(r, sigma)} + \texttt{mu}$$

- If $X \sim \mathcal{N}(\mu_1, \sigma_1)$ and $Y \sim \mathcal{N}(\mu_2, \sigma_2)$ then $X + Y \sim \mathcal{N}(\mu_1 + \mu_2, \sqrt{\sigma_1^2 + \sigma_2^2})$.
- Because the Normal is symmetric, $X - Y \sim \mathcal{N}(\mu_1 - \mu_2, \sqrt{\sigma_1^2 + \sigma_2^2})$.
- Section 9.1 (p 297) discusses the Central Limit Theorem in greater detail.

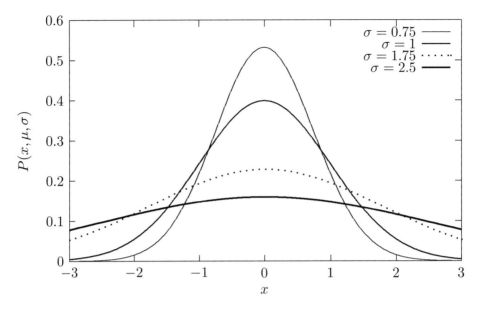

Figure 7.6 The Normal distribution, with $\mu = 0$.

Multivariate Normal   Just as the Normal distribution is the extension of the Binomial, the Multivariate Normal is the extension of the Multi-
nomial. Say that we have a data set $\mathbf{X}$ that includes a thousand observations and
seven variables (so $\mathbf{X}$ is a 1000×7 matrix). Let its mean be μ (a vector of length
seven) and the covariance among the variables be Σ (a seven by seven matrix).
Then the Multivariate Normal distribution that you could fit to this data is

$$P(\mathbf{X}, \boldsymbol{\mu}, \boldsymbol{\Sigma}) = \frac{\exp\left(-\frac{1}{2}(\mathbf{X} - \boldsymbol{\mu})'\boldsymbol{\Sigma}^{-1}(\mathbf{X} - \boldsymbol{\mu})\right)}{\sqrt{(2\pi)^n \det(\boldsymbol{\Sigma})}}$$

$$E(\mathbf{X}|\boldsymbol{\mu}, \boldsymbol{\Sigma}) = \boldsymbol{\mu}$$

$$\mathrm{var}(\mathbf{X}|\boldsymbol{\mu}, \boldsymbol{\Sigma}) = \boldsymbol{\Sigma}$$

• When $\mathbf{X}$ has only one column and $\boldsymbol{\Sigma} = [\sigma^2]$, this reduces to the univariate Normal
distribution.

Lognormal   The Normal distribution is apropos when the items in a sample are the
mean of a set of draws from a population, $\overline{x}_i = (s_1 + s_2 + \cdots + s_k)/k$.
But what if a data point is the *product* of a series of iid samples, $\tilde{x}_i = s_1 \cdot s_2 \cdots \cdots s_k$?
Then the log of $\tilde{x}_i$ is $\ln(\tilde{x}_i) = \ln(s_1) + \ln(s_2) + \cdots + \ln(s_k)$, so the log is a sum

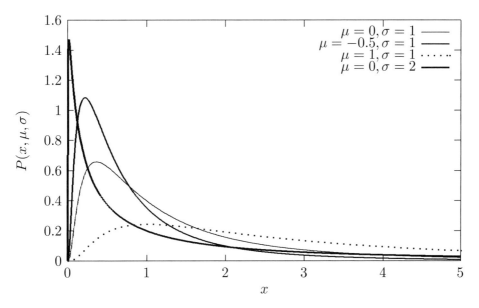

Figure 7.7 The Lognormal distribution.

of independent elements (i.e., n times a mean). Very broadly, when a point in the data set is produced by summing iid draws, it will be Normally distributed; when a point in the data set is produced by taking the product of iid draws, its log will be Normally distributed—i.e., it will have a lognormal distribution. The next section will present an example. Figure 7.7 shows some Lognormal distributions.

A notational warning: in the typical means of expressing the lognormal distribution, μ and σ refer to the mean of the Normal distribution that you would get if you replaced every element x in your data set with e^x, thus producing a standard Normal distribution. Be careful not to confuse this with the mean and variance of the data you actually have.

$$p(x, \mu, \sigma) = \frac{\exp\left(-(\ln x - \mu)^2/(2\sigma^2)\right)}{x\sigma\sqrt{2\pi}}$$
$$= \texttt{ran\_lognormal\_pdf}(x, \texttt{mu}, \texttt{sigma})$$
$$E(x|\mu, \sigma) = e^{\left(\mu + \frac{\sigma^2}{2}\right)}$$
$$\text{var}(x|\mu, \sigma) = (e^{\sigma^2} - 1)e^{(2\mu + \sigma^2)}$$
$$RNG : \texttt{ran\_lognormal}(\texttt{rng}, \texttt{mu}, \texttt{sigma})$$

Negative binomial   Say that we have a sequence of Bernoulli draws. How many failures will we see before we see n successes? If p percent of cars are illegally parked, and a meter reader hopes to write n parking tickets, the Negative binomial tells her the odds that she will be able to stop with $n + x$ cars.

The form is based on the Gamma function,

$$\Gamma(z) = \int_0^\infty x^{z-1} e^{-x} dx$$
$$= \texttt{gsl\_sf\_gamma(z)}.$$

You can easily verify that $\Gamma(z+1) = z\Gamma(z)$. Also, $\Gamma(1) = 1, \Gamma(2) = 1, \Gamma(3) = 2,$ $\Gamma(4) = 6$, and generally, $\Gamma(z) = (z-1)!$ for positive integers. Thus, if n and x are integers, formulas based on the Gamma function reduce to more familiar factorial-based counting formulas.

$$P(x, n, p) = \frac{\Gamma(n + x)}{\Gamma(x + 1)\Gamma(n)} p^n (1 - p)^x$$
$$= \texttt{gsl\_ran\_negative\_binomial\_pdf}(x, p, n)$$
$$E(x|n, p) = \frac{n(1 - p)}{p}$$
$$\text{var}(x|n, p) = \frac{n(1 - p)}{p^2}$$
$$RNG : \texttt{gsl\_ran\_negative\_binomial}(rng, p, n)$$

RATES   A Poisson process is very much like a Bernoulli draw, but the unit of measurement is continuous—typically a measure of time or space. It makes sense to have half of an hour, but not half of a coin flip, so the stories above based on Bernoulli draws are modified slightly to allow for a rate of λ events per hour to be applied to half an hour or a week.

Baltimore, Maryland, sees about 110 days of precipitation per year, somewhat consistently spaced among the months. But for how many days will it rain or snow in a single week? The Poisson distribution answers this question. We can also do a count of weeks: how often does it rain once in a week, twice in a week, et cetera? The Exponential distribution answers this question. Turning it around, if we want a week with three rainfalls, how long would we have to wait? The Gamma distribution answers this question.

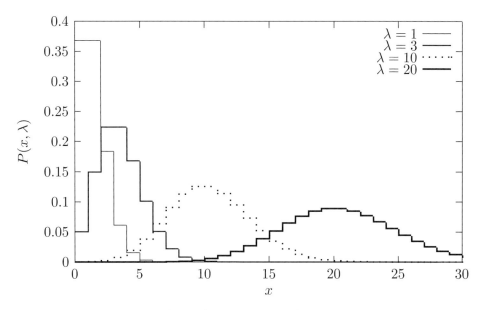

Figure 7.8 The Poisson distribution.

Poisson   Say that independent events (rainy day, landmine, bad data) occur at the
mean rate of λ events per span (of time, space, et cetera). What is the
probability that there will be x events in a single span?

We are assuming that events occur at a sufficiently even rate that the same rate
applies to different time periods: if the rate per day is λ_1, then the rate per week is
$7\lambda_1$, and the rate per hour is $\lambda_1/24$. See Figure 7.8.

$$P(x, \lambda) = \frac{e^{-\lambda}\lambda^x}{x!}$$
$$= \texttt{gsl\_ran\_poisson\_pdf(x, lambda)}$$
$$E(x|\lambda) = \lambda$$
$$\text{var}(x|\lambda) = \lambda$$
$$RNG : \texttt{gsl\_ran\_poisson(r, lambda)}$$

- As $n \to \infty$, Binomial$(n, p) \to$Poisson(np).
- If $X \sim$ Poisson(λ_1), $Y \sim$ Poisson(λ_2), and X and Y are independent, then
  $(X + Y) \sim$ Poisson$(\lambda_1 + \lambda_2)$.
- As $\lambda \to \infty$, Poisson$(\lambda) \to \mathcal{N}(\lambda, \sqrt{\lambda})$.

$Q_{7.7}$

- Calculate the Binomial-distributed probability of three rainfalls in seven days, given the probability of rain in one day of $p = (110/365)$.

- Calculate the Poisson-distributed probability of three rainfalls in seven days, given a one-day $\lambda = (110/365)$.

Gamma distribution   The Gamma *distribution* is so-named because it relies heavily on the Gamma *function*, first introduced on page 244. Along with the Beta distribution below, this naming scheme is one of the great notational tragedies of mathematics.

A better name in the statistical context would be 'Negative Poisson,' because it relates to the Poisson distribution in the same way the Negative binomial relates to the Binomial. If the timing of events follows a Poisson distribution, meaning that events come by at the rate of λ per period, then this distribution tells us how long we would have to wait until the nth event occurs.

The form of the Gamma distribution, shown for some parameter values in Figure 7.9, is typically expressed in terms of a shape parameter $\theta \equiv 1/\lambda$, where λ is the Poisson parameter. Here is the summary for the function in terms of both parameters:

$$P(x, n, \theta) = \frac{1}{\Gamma(n)\theta^n} x^{n-1} e^{-x/\theta}, x \in [0, \infty)$$

$$= \texttt{gsl\_ran\_gamma\_pdf(x, n, theta)}$$

$$P(x, n, \lambda) = \frac{1}{\Gamma(n)(\frac{1}{\lambda})^n} x^{n-1} e^{-\lambda x}, x \in [0, \infty) \ (7.2.3)$$

$$E(x|n, \theta \text{ or } \lambda) = n\theta = n/\lambda$$

$$\text{var}(x|n, \theta \text{ or } \lambda) = k\theta^2 = k/\lambda^2$$

$$\int_{-\infty}^{x} G(y|n, \theta)dy = \texttt{gsl\_cdf\_gamma\_P(x, theta)}$$

$$\int_{x}^{\infty} G(y|n, \theta)dy = \texttt{gsl\_cdf\_gamma\_Q(n, theta)}$$

$$RNG : \texttt{gsl\_ran\_gamma(r, n, theta)}$$

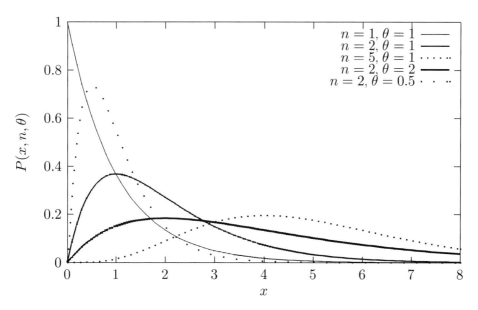

Figure 7.9 The Gamma distribution.

- With $n = df/2$ and $\theta = 2$, the Gamma distribution becomes a χ^2_{df} distribution (introduced on page 301).
- With $n = 1$, the Gamma distribution becomes an Exponential(λ) distribution.

Exponential distribution   The Gamma distribution found the time until n events occur, but consider the time until the first event occurs. $\Gamma(1) \equiv 1$, $1^\lambda = 1$ for all positive λ, and $x^0 = 1$ for all positive x, so at $n = 1$, Equation 7.2.3 defining the PDF of the Gamma distribution reduces to simply $e^{-\lambda x}$.

If we had a population of items, $\int_0^t e^{-\lambda x} dx$ percent would have had a first event between time zero and time t. If the event causes the item to leave the population, then one minus this percent are still in the population at time t. The form e^x is very easy to integrate, and doing so gives that the percent left at time $t = e^{-\lambda t}/\lambda$.

So we now have a story of a population where members leave via a Poisson process. Common examples include the stock of unemployed workers as some find a job every period, radioactive particles emanating from a block, or drug dosage remaining in a person's blood stream. Figure 7.10 shows a few examples of the Exponential distribution.

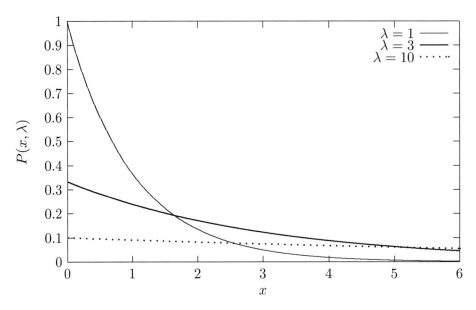

Figure 7.10 The Exponential distribution.

Since the exponent is $-\lambda$, this is sometimes called the *Negative exponential distribution*.

$$P(x,\lambda) = \frac{1}{\lambda}e^{\frac{-x}{\lambda}}$$
$$= \texttt{gsl\_ran\_exponential\_pdf(x,lambda)}$$
$$E(x|\lambda) = \lambda$$
$$\mathrm{var}(x|\lambda) = \lambda^2$$
$$\int_{-\infty}^{x} \mathrm{Exp}(\lambda)dy = \texttt{gsl\_cdf\_exponential\_P(x,lambda)}$$
$$\int_{x}^{\infty} \mathrm{Exp}(\lambda)dy = \texttt{gsl\_cdf\_exponential\_Q(x,lambda)}$$
$$RNG : \texttt{gsl\_ran\_exponential(r,lambda)}$$

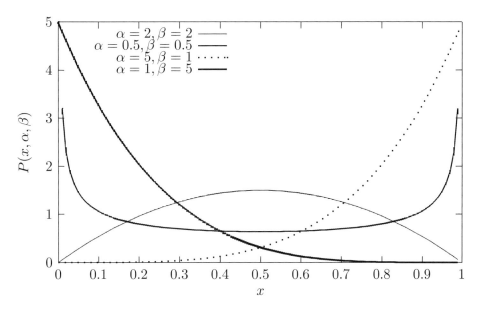

Figure 7.11 The Beta distribution.

DESCRIPTION   Here are a few more distributions that are frequently used in modeling to describe the shape of a random variable.

Beta distribution   Just as the Gamma distribution is named for the Gamma function, the Beta distribution is named after the Beta function—whose parameters are typically notated as α and β. This book will spell out Beta($\cdot$) for the Beta function and use $\mathcal{B}(\cdot, \cdot)$ for the Beta distribution.

The Beta function can be described via the following forms:

$$\begin{aligned}
\text{Beta}(\alpha, \beta) &= \int_0^1 x^{(\alpha-1)}(1 - x)^{(\beta-1)}dx \\
&= \frac{\Gamma(\alpha)\Gamma(\beta)}{\Gamma(\alpha + \beta)} \\
&= \texttt{gsl\_sf\_beta}(\texttt{alpha}, \texttt{beta}).
\end{aligned}$$

The Beta distribution is a flexible way to describe data inside the range $[0, 1]$. Figure 7.11 shows how different parameterizations could lead to a left-leaning, right-leaning, concave, or convex curve; see page 358 for more.

$$P(x, \alpha, \beta) = \text{Beta}(\alpha, \beta)x^{\alpha-1}(1 - x)^{\beta-1}$$
$$= \texttt{gsl\_ran\_beta\_pdf}(\texttt{x}, \texttt{alpha}, \texttt{beta})$$
$$E(x|\alpha, \beta) = \frac{\alpha}{\alpha + \beta}$$
$$\text{var}(x|\alpha, \beta) = \frac{\alpha\beta}{(\alpha + \beta)^2(\alpha + \beta + 1)}$$
$$\int_{-\infty}^{x} \mathcal{B}(y|\alpha, \beta)dy = \texttt{gsl\_cdf\_beta\_P}(\texttt{x}, \texttt{alpha}, \texttt{beta})$$
$$\int_{x}^{\infty} \mathcal{B}(y|\alpha, \beta)dy = \texttt{gsl\_cdf\_beta\_Q}(\texttt{x}, \texttt{alpha}, \texttt{beta})$$
$$RNG : \texttt{gsl\_ran\_beta}(\texttt{r}, \texttt{alpha}, \texttt{beta})$$

- If $\alpha < 1$ and $\beta < 1$, then the distribution is bimodal, with peaks at zero and one.
- If $\alpha > 1$ and $\beta > 1$, then the distribution is unimodal.
- As α rises, the distribution leans toward one; as β rises, the distribution leans toward zero; if $\alpha = \beta$, then the distribution is symmetric.
- If $\alpha = \beta = 1$, then this is the Uniform$[0, 1]$ distribution.

❋ *The Beta distribution and order statistics*   The first *order statistic* of a set of numbers **x** is the smallest number in the set; the second is the next-to-smallest, up to the largest order statistic, which is max(**x**).

Assume that the $\alpha + \beta - 1$ elements of **x** are drawn from a Uniform$[0, 1]$ distribution. Then the αth order statistic has a $\mathcal{B}(\alpha, \beta)$ distribution.

$Q_{7.8}$

- Write a function that takes in a `gsl_rng` and two integers a and b, produces a list of a+b-1 random numbers in $[0, 1]$, sorts them, and returns the ath order statistic.

- Write a function to call that function 10,000 times and plot the PDF of the returned data (using `apop_plot_histogram`). It helps to precede the plot output to Gnuplot with `set xrange [0:1]` to keep the range consistent.

- Write a `main` that produces an animation of the PDFs of the first through 100th order statistic for a set of 100 numbers.

- Replace the call to the draw-and-sort function with a draw from the $\mathcal{B}(\texttt{a}, \texttt{b})$ distribution, and re-animate the results.

Uniform distribution   What discussion of distributions would be complete without mention of the Uniform? It represents a belief that any value within $[\alpha, \beta]$ is equally possible.

$$P(x, \alpha, \beta) = \begin{cases} \frac{1}{\beta - \alpha} & x \in [\alpha, \beta] \\ 0 & x < \alpha, x > \beta \end{cases}$$

$$= \texttt{gsl\_ran\_flat\_pdf}(\texttt{x}, \texttt{alpha}, \texttt{beta})$$

$$E(x|\alpha, \beta) = \frac{\beta - \alpha}{2}$$

$$\text{var}(x|\alpha, \beta) = \frac{(\beta - \alpha)^2}{12}$$

$$\int_{-\infty}^{x} \mathcal{U}(y|\alpha, \beta) dy = \begin{cases} 0 & x < \alpha \\ \frac{x - \alpha}{\beta - \alpha} & x \in [\alpha, \beta] \\ 1 & x > \beta \end{cases}$$

$$= \texttt{gsl\_cdf\_flat\_P}(\texttt{x}, \texttt{alpha}, \texttt{beta})$$

$$\int_{x}^{\infty} \mathcal{U}(y|\alpha, \beta) dy = \texttt{gsl\_cdf\_flat\_Q}(\texttt{x}, \texttt{alpha}, \texttt{beta})$$

$$RNG, \text{general} : \texttt{gsl\_ran\_flat}(\texttt{r}, \texttt{alpha}, \texttt{beta})$$

$$RNG, \alpha = 0, \beta = 1 : \texttt{gsl\_rng\_uniform}(\texttt{r})$$

➤ Probability theorists through the ages have developed models that indicate that if a process follows certain guidelines, the data will have a predictable form.

➤ A single draw from a binary event with fixed probability has a Bernoulli distribution; from this, a wealth of other distributions can be derived.

➤ An event which occurs with frequency λ per period (or λ per volume, et cetera) is known as a Poisson process; a wealth of distributions can be derived for such a process.

➤ If $\bar{x}$ is the mean of a set of independent, identically distributed draws from *any* nondegenerate distribution, then the distribution of $\bar{x}$ approaches a Normal distribution. This is the Central Limit Theorem.

➤ The Beta distribution is useful for modeling a variety of variables that are restricted to $[0, 1]$. It can be unimodal, bimodal, lean in either direction, or can simply match the Uniform distribution.

7.3 **USING THE SAMPLE DISTRIBUTIONS**   Here are some examples of how
                                                you could use the distributions
described above to practical benefit.

LOOKING UP FIGURES   If I have fifty draws from a Bernoulli event with probability
                           .25, what is the likelihood that I will have more than twenty
         successes?

Statistics textbooks used to include an appendix listing tables of common distributions, but those tables are effectively obsolete, and more modern textbooks refer the reader to the appropriate function in a stats package. For those who long for the days of grand tables, the code supplement includes `normaltable.c`, code for producing a neatly formatted table of CDFs for a set of Normal distributions (the p-value often reported with hypothesis tests is one minus the listed value).

The code is not printed here because it is entirely boring, but the tables it produces provide another nice way to get a feel for the distributions.

Alternatively, Apophenia's command-line program `apop_lookup` will look up a quick number for you.

GENERATING DATA FROM A DISTRIBUTION   Each distribution neatly summarizes
                                            an oft-modeled story, and so each
         can be used as a capsule simulation of a process, either by itself or as a building block for a larger simulation.

Listing 7.12 gives a quick initial example. It is based on work originated by Gibrat (1931) and extended by many others, including Axtell (2006), regarding *Zipf's law*, that the distribution of the sizes of cities, firms, or other such agglomerations tends toward an Exponential-type form. In the model here, this comes about because agents' growth rates are assumed to be the mean of a set of iid random shocks, and so are Normally distributed.

- First, the program produces a set of agents with one characteristic: size, stored in a `gsl_vector`. The `initialize` function draws agent sizes from a Uniform[0, 100] distribution. To do this, it requires a `gsl_rng`, which `main` allocates using `apop_-rng_alloc` and passes to `initialize`. See Chapter 11 for more on using random number generators.
- Each period, the firms grow by a Normally distributed rate (via the `grow` function). That is, the `grow` function randomly draws g from a `gsl_rng_gaussian`, and then reassigns the firm size to $size \leftarrow size * \exp(g)$. The most likely growth rate is therefore $\exp(0) = 1$. When $g < 0$, $\exp(g) < 1$; and when $g > 0$, $\exp(g) > 1$.

```
#include <apop.h>

int agentct = 5000;
int periods = 50;
int binct = 30;
double pauselength = 0.6;
gsl_rng *r;

void initialize(double *setme){
    *setme = gsl_rng_uniform(r)*100;
}

void grow(double *val){
    *val *= exp(gsl_ran_gaussian(r,0.1));
}

double estimate(gsl_vector *agentlist){
    return apop_vector_mean(agentlist);
}

int main(){
  gsl_vector *agentlist = gsl_vector_alloc(agentct);
    r = apop_rng_alloc(39);
    apop_vector_apply(agentlist, initialize);
    for (int i=0; i< periods; i++){
        apop_plot_histogram(agentlist, binct, NULL);
        printf("pause %g\n", pauselength);
        apop_vector_apply(agentlist, grow);
    }
    fprintf(stderr, "the mean: %g\n", estimate(agentlist));
}
```

Listing 7.12 A model of Normally distributed growth. Online source: `normalgrowth.c`.

Also, $\exp(g) * \exp(-g) = 1$, and by the symmetry of the Normal distribution, g and $-g$ have equal likelihood, so it is easy for an agent to find good luck in period one countered by comparable bad luck in period two, leaving it near where it had started.

• The output is a set of Gnuplot commands, so use `./normalgrowth | gnuplot`. With a `pause` between each histogram, the output becomes an animation, showing a quick transition from a Uniform distribution to a steep Lognormal distribution, where most agents are fast approaching zero size, but a handful have size approaching 1,000.[8]

---

[8]Here, the x-axis is the firm size, and the y-axis is the number of firms. Typically, Zipf-type distributions are displayed somewhat differently: the x-axis is the *rank* of the firm, 1st, 2nd, 3rd, et cetera, and the y-axis is the size of the so-ranked firm. Converting to this form is left as an exercise to the reader. (*Hint*: use `gsl_vector_-sort`.)

• The last step is a model estimation, to which we will return in a few pages. Its output is printed to `stderr`, aka the screen, so that the pipe to Gnuplot is not disturbed.

SIMULATION   Fein *et al.* (1988) found that their depressive patients responded well to a combination of Lithium and a monoamine oxidase inhibitor (MAOI). But both types of drug require careful monitoring: Lithium overdoses are common and potentially damaging, while the combination of MAOIs and chocolate can be fatal.

```
#include <apop.h>

double find_lambda(double half_life){
    double lambda = −half_life/log(1/2.);
    return gsl_cdf_exponential_Q(1, lambda);
}

int main(){
    double li = 0, maoi = 0;
    int days = 10;
    gsl_matrix *d = gsl_matrix_alloc(days*24,4);
    double hourly_decay1 = find_lambda(20.); //hours; lithium carbonate
    double hourly_decay2 = find_lambda(11.); //hours; phenelzine
        for (size_t i=0; i < days*24; i ++){
            li *= hourly_decay1;
            maoi *= hourly_decay2;
            if (i % 24 == 0)
                li+= 600;
            if ((i+12) % 24 == 0)
                maoi+= 45;
            APOP_MATRIX_ROW(d, i, onehour);
            apop_vector_fill(onehour, i/24., li/10., maoi, maoi/li*100.);
        }
    printf("plot 'maoi.out' using 1:2 with lines title 'Li/10', \
                'maoi.out' using 1:3 with lines title 'MAOI', \
                'maoi.out' using 1:4 with lines title 'MAOI/Li, pct\n");
    remove("maoi.out");
    apop_matrix_print(d, "maoi.out");
}
```

Listing 7.13 A simulation of the blood stream of a person taking two drugs. Online source: `maoi.c`.

Listing 7.13 simulates a patient's blood stream as she follows a regime of Lithium carbonate (average half life: about 20 hours, with high variance) and an MAOI named phenelzine (average half life: 11 hours). As per the story on page 247, when the drug leaves the blood stream via a Poisson process, the amount of a drug remaining in the blood is described by an Exponential distribution.

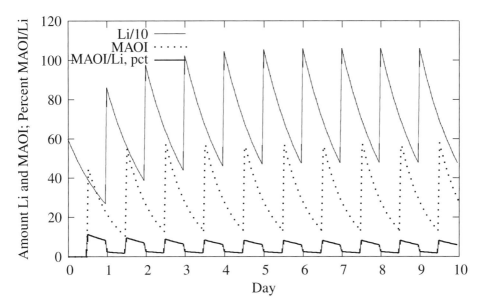

Figure 7.14 The typical sawtooth pattern of decay and renewal.

- The first step is to convert from the half life commonly used by pharmacists to the λ parameter in the exponential distribution. The `find_lambda` function does this.

- Given λ, `gsl_cdf_exponential_Q(1, lambda)` answers the question of what percentage of a given initial level is remaining after one hour.

- The main simulation is a simple hourly `for` loop, that decrements the amount of drug in the blood stream by the amount calculated above, then checks whether it is time for our patient to take one of her meds, and then records the various levels on her chart.

- The ungainly `printf` statement at the end plots the result. Gnuplot does not save data, so it needs to reread the data file three times to plot the three columns.

Figure 7.14 shows the density in blood as our subject takes 600 mg of Lithium at midnight every day, and 45 mg of MAOI at noon every day. For convenience in scaling of the plot, the amount of Lithium in the blood stream is divided by ten. In the later days, after the subject's system has reached its dynamic equilibrium, she takes in 600 mg of Li per day, and loses about 600 mg per day; similarly for the 45 mg of MAOI. The ratio of MAOI/Li jumps constantly over the range of 187% to 824%.

Q7.9 | Derive or verify that the `find_lambda` function correctly converts between half life and the Exponential distribution's λ parameter.

$Q_{7.10}$ Let there be two stocks: employed and unemployed. Let the half-life of employment (i.e., transition to unemployment) be 365 days, and the half-life of unemployment (i.e., finding a job) be 3 weeks (21 days). Modify the Lithium/MAOI program to model the situation. For each period, calculate the loss from both the employment and the unemployment stocks, and then transfer the appropriate number of people to the other stock. What is the equilibrium unemployment rate?

FITTING EXISTING DATA   The common goal throughout the book is to estimate the parameters of the model with data, so given a data set, how can we find the parameters for the various distributions above?

You can see above that almost every parameter can be solved—sometimes over-solved—using the mean and variance. For example, Equations 7.2.1 and 7.2.2 (describing the parameters of a Binomial distribution) are a system of two equations in two unknowns:

$$\mu = np$$
$$\sigma^2 = np(1-p)$$

It is easy to calculate estimates of μ and σ from data, $\hat{\mu}$ and $\hat{\sigma}$, and we could plug those estimates into the above system of equations to find the parameters of the distribution. You can verify that for these two equations we would have

$$\hat{n} = \frac{\hat{\mu}^2}{\hat{\mu} - \hat{\sigma}^2}$$
$$\hat{p} = 1 - \frac{\hat{\sigma}^2}{\hat{\mu}}.$$

This is *method of moments* estimation (see, e.g., Greene (1990, pp 117*ff*)). To summarize the method, we write down the parameter estimates as functions of the mean, variance, skew, and kurtosis, then we find estimates of those parameters from the data, and use those parameter estimates to solve for the parameter estimates of the model itself.

But problems easily crop up. For example, we can just count observations to find the value of n for our data set, so given n, $\hat{\mu}$, and $\hat{\sigma}^2$, our system of equations is now two equations with only one unknown (p). The Poisson distribution had a similar but simpler story, because its single parameter equals two different moments:

$$\mu = \lambda$$
$$\sigma^2 = \lambda.$$

So if our data set shows $\hat{\mu} = 1.2$ and $\hat{\sigma}^2 = 1.4$, which do we use for $\hat{\lambda}$? Apophenia doesn't fret much about this issue and just uses $\hat{\mu}$, because this is also the maximum likelihood estimator (MLE) of λ (where MLE will be discussed fully in Chapter 10).

For the Uniform, the method of moments doesn't work either: the expression $(\beta - \alpha)$ is oversolved with the two equations, but there is no way to solve for α or β alone. However, a few moments' thought will show that the most likely value for (α, β) given data $\mathbf{x}$ is simply $(\min(\mathbf{x}), \max(\mathbf{x}))$.

Most of the above distributions have an `apop_model` associated (`apop_normal`, `apop_gamma`, `apop_uniform`, et cetera), and if you have a data set on hand, you can quickly estimate the above parameters:

```
apop_data *d = your_data;
apop_model *norm = apop_estimate(d, apop_normal);
apop_model *beta = apop_estimate(d, apop_beta);
apop_model_show(norm);
apop_model_show(beta);
apop_model_show(apop_estimate(d, apop_gamma));
```

Q7.11
Listing 7.12 produces a data set that should be Zipf distributed. Add an estimation in the `estimate` function to see how well it fits.

Better still, run a tournament. First, declare an array of several models, say the Lognormal, Zipf, Exponential, and Gamma. Write a `for` loop to estimate each model with the data, and fill an array of confidence levels based on log-likelihood tests. [Is such a tournament valid? See the notes on the multiple testing problem on 316.]

The method of moments provides something of a preview of the working of the various model-based estimations in the remainder of the book. It took in data, and produced an estimate of the model parameters, or an estimate of a statistic using the estimate of the model parameters that were produced using data.

As the reader may have noticed, all these interactions between data, model parameters, and statistics create many opportunities for confusion. Here are some notes to bear in mind:

- The expected value, variance, and other such measures of a *data set*, when no model is imposed, is a function of the data. [E.g., $E(\mathbf{x})$.]
- The expected value, variance, and other such measures of a *model* are functions of the ideal parameters, not any one data set. [E.g., $E(\mathbf{x}|\beta)$ is only a function of β.]
- Our *estimate* of model parameters given a data set is a function of the given data set (and perhaps any known parameters, if there are any on hand). For example, the Normal parameter μ is a part of the model specification, but the estimate of μ,

which we write as $\hat{\mu}$, is a function of the data. Any variable with a hat, like $\hat{p}$, could be notated as a function of the data, $\hat{p}(\mathbf{x})$.

- We will often have a statistic like $E(\mathbf{x})$ that is a function of the data—in fact, we define a *statistic* to be *a function of data*. But models often have data-free analogues to these statistics. Given a probability distribution $P(x, \beta)$, the expected value $E(f(x)|\beta) = \int_{\forall x} f(x)P(x, \beta)dx$, meaning that we integrate over all x, and so $E(f(x)|\beta)$ is a function of only β. The model in which β lives is almost always taken as understood by context, and many authors take the parameters as understood by context as well, leaving the expected value to be written as $E(f(x))$, even though this expression is a function of β, not x.

BAYESIAN UPDATING   The definition of a conditional probability is based on the statement $P(A \cap B) = P(A|B)P(B)$; in English, the likelihood of A and B occurring at the same time equals the likelihood of A occurring given that B did, times the likelihood that B occurs. The same could be said reversing A and B: $P(A \cap B) = P(B|A)P(A)$. Equating the two complementary forms and shunting over $P(B)$ gives us the common form of Bayes's rule:

$$P(A|B) = \frac{P(B|A)P(A)}{P(B)}.$$

Now to apply it. Say that we have a prior belief regarding a parameter, such as that the distribution of the mean of a data set is $\sim \mathcal{N}(0, 1)$; let this be $Pri(\beta)$. We gather a data set $\mathbf{X}$, and can express the likelihood that we would have gathered this data set given any haphazard value of β, $P(\mathbf{X}|\beta)$. Let $\mathbb{B}$ be the entire range of values that β could take on. We can then use Bayes's rule to produce a *posterior distribution*:

$$Post(\beta|\mathbf{X}) = \frac{P(\mathbf{X}|\beta)Pri(\beta)}{P(\mathbf{X})}$$

So on the right-hand side, we had a prior belief about β's value expressed as a distribution, and a likelihood function $P(\mathbf{X}|\beta)$ expressing the odds of observing the data we observed given any one parameter. On the left-hand side, we have a new distribution for β, which takes into account the fact that we have observed the data $\mathbf{X}$. In short, this equation used the data to update our beliefs about the distribution of β from $Pri(\beta)$ to $Post(\beta)$.

The numerator is relatively clear, and requires only local information, but we can write $P(\mathbf{X})$ in its full form—

$$Post(\beta|\mathbf{X}) = \frac{P(\mathbf{X}|\beta)Pri(\beta)}{\int_{\forall B \in \mathbb{B}} P(\mathbf{X}|B)Pri(B)dB}$$

—to reveal that the denominator is actually global information, because calculating it requires covering the entire range that β could take on. Local information is easy and global information is hard (see pages 325 *ff*), so Bayesian updating is often described via a form that just ignores the global part:

$$Post(\beta|\mathbf{X}) \propto P(\mathbf{X}|\beta)Pri(\beta).$$

That is, the posterior equals the amount on the right-hand side times a fixed amount (the denominator above) that does not depend on any given value of β. This is already enough to compare ratios like $Post(\beta_1|\mathbf{X})/Post(\beta_2|\mathbf{X})$, and given the right conditions, such a ratio is already enough for running likelihood ratio tests (as discussed in Chapter 10).

Computationally, there are two possibilities for moving forward given the problem of determining the global scale of the distribution. First, there are a number of *conjugate distribution* pairs that can be shown to produce an output model that matches the prior in form but has updated parameters. In this case, the `apop_-update` function simply returns the given model and its new parameters; see the example below.

Chapter 11 will present a computationally-intensive method of producing a posterior distribution when the analytic route is closed (i.e., *Monte Carlo Maximum Likelihood*). But for now we can take `apop_update` as a black box that takes in two models and outputs an updated conjugate form where possible, and an empirical distribution otherwise. We could then make draws from the output distribution, plot it, use it as the prior to a new updating procedure·when a new data set comes in, et cetera.

An example: Beta ♡ Binomial   For now, assume that the likelihood that someone has a tattoo is constant for all individuals, regardless of age, gender, ... (we will drop this clearly false assumption in the section on multilevel modeling, page 288). We would like to know the value of that overall likelihood. That is, the statistic of interest is $p \equiv$ (count of people who have tattoos)/(total number of people in the sample).

Because we have weak knowledge of p, we should describe our beliefs about its value using a distribution: p has small odds of being near zero or one, a reasonable chance of being about 10%, and so on. The Beta distribution is a good way to describe the distribution, because it is positive only for inputs between zero and one. Let $\mathcal{B}$ indicate the Beta distribution; then $\mathcal{B}(1,1)$ is a Uniform(0, 1) distribution, which is a reasonably safe way to express a neutral prior belief about p. Alternatively, setting $Pri(p)$ to be $\mathcal{B}(2,2)$ will put more weight around $p = 1/2$ and less at the extremes, and raising the second parameter a little more will bring the mode of our beliefs about p below $1/2$ [See Figure 7.11].

Given p, the distribution of the expected count of tattooed individuals is Binomial. For each individual, there is a p chance of having a tattoo—a simple Bernoulli draw. The overall study makes $n = 500$ such draws, and thus fits the model underlying the Binomial distribution perfectly. But we do not yet know p, so this paragraph had to begin by taking p as given. That is, the Binomial distribution describes $P(\text{data}|p)$.

It so happens that the Beta and Binomial distributions are conjugate. This means that, given that $Pri(p)$ is a Beta distribution and $P(\text{data}|p)$ is a Binomial distribution, the posterior $Post(p|\text{data})$ is a Beta distribution, just like the prior. Tables of other such conjugate pairs are readily available online.

However, the parameters are updated to accommodate the new information. Let x be the number of tattoos observed out of n subjects, and the prior distribution be $\mathcal{B}(\alpha, \beta)$. Then the posterior is a $\mathcal{B}(\alpha + x, \beta + n - x)$ distribution. The discussion of the prior offered possibilities like $\alpha = \beta = 1$ or $\alpha = \beta = 2$. But the survey has 500 subjects; the count of tattooed individuals alone dwarfs $\alpha = 2$. Therefore, we can approximate the posterior as simply $\mathcal{B}(x, n - x)$.

The catalog above listed the expected value of a Beta distribution as $\frac{\alpha}{\alpha+\beta}$. With $\alpha = x$ and $\beta = n - x$, this reduces simply to x/n. That is, the expected posterior value of p is the percentage of people in our sample who have tattoos ($\hat{p}$). Bayesian updating gave us a result exactly as we would expect.

The variance of a Beta distribution is

$$\frac{\alpha\beta}{(\alpha + \beta)^2(\alpha + \beta + 1)}.$$

Again, with N around 500, the 1 in the denominator basically disappears. Filling in $\alpha = x$ and $\beta = n - x$, we get

$$\frac{\hat{p}(1 - \hat{p})}{n}.$$

Again, this is what we would get from the Binomial distribution.

We call a $\mathcal{B}(\alpha, \beta)$ distribution with small α and β a *weak prior*, by which we mean that a moderately-sized data set entirely dwarfs the beliefs we expressed in the prior. So what is the point of the updating process? First, we could use a stronger prior, like $\alpha = 200, \beta = 300$, which would still have some effect on the posterior distribution even after updating with the data set.

Second, the system provides a consistent mechanism for combining multiple data sets. The posterior distribution that you have after accounting for a data set will have a form appropriate for use as a prior distribution to be updated by the next data set. Thus, Bayesian updating provides a natural mechanism for running metastudies.

Q7.12

Verify that apop_update using a Beta prior, a Binomial likelihood function, and the tattoo data does indeed produce the estimated mean and variance as the simple x/n estimate. Gather the data from the column tattoos.'ego has tattoos', which is coded 1=yes, 2=no, and calculate $\hat{\mu}$ and $\hat{\sigma}^2$ using the formulæ in the above few paragraphs.

What results do you get when you assume a stronger prior, like $\mathcal{B}(200, 800)$ or $\mathcal{B}(800, 200)$?

7.4 NON-PARAMETRIC DESCRIPTION

Say that we have a data set and would like to know the distribution from which the data was drawn. To this point, we assumed the form of the distribution (Normal, Binomial, Poisson, et cetera) and then had only to estimate the parameters of the distribution from data. But without assuming a simple parametric form, how else could we describe the distribution from which the data was drawn?

The simplest answer would be a plain old histogram of the drawn data. This is often sufficient. But especially for small data sets, the histogram has dissatisfactory features. If we make four draws, and three have value 20 and one has value 22, does this mean that 21 has probability zero, or we just didn't have the luck of drawing a 21 this time?

Thus, a great deal of nonparametric modeling consists of finding ways to smooth a histogram based on the claim that the actual distribution is not as lumpy as the data.

The histogram   The histogram is the most assumption-free way to describe the likelihood distribution from which the data was drawn. Simply lay down a row of bins, pile each data point into the appropriate bin, normalize the bins to sum to one if so desired, and plot. Because the most common use of a histogram (after just plotting it) is using it to make random draws, the full discussion of histogram production will appear in the chapter on random draws, on page 361.

The key free variable in a histogram is the *bandwidth*—the range over the x-axis that goes into each data point. If the bandwidth is too small, then the histogram will have many slices and generally be as spiky as the data itself. A too-large bandwidth oversmooths—at an infinite bandwidth, the histogram would have only one bar, which is not particularly informative. Formally, there is a bias-variance trade-off between the two extremes, but most of us just try a few bandwidths until we get something that looks nice. See Givens & Hoeting (2005, ch 11) for an extensive discussion of the question in the context of data smoothing.

Moving average   The simplest means of smoothing data is a moving average, replacing each data point with the mean of the adjacent b data points
(where b is the bandwidth). You could use this for histograms or for any other
series. For example, `movingavg.c` in the online code supplement plots the temperature deviances as shown in `data-climate.db`, and a moving average that
replaces each data point with the mean deviation over a two-year window, based
on the `apop_vector_moving_average` function.

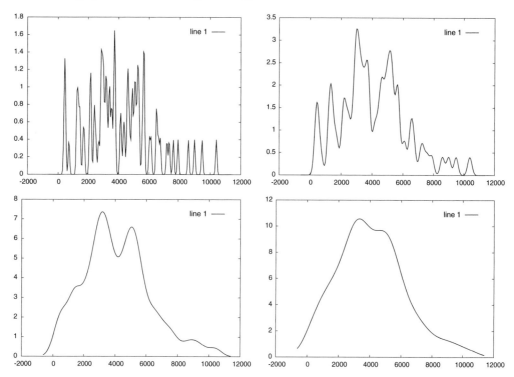

Figure 7.15 A series of density plots. As h rises, the kernel density smooths out and has fewer peaks.

Kernel smoothing   The *kernel density estimate* is based on this function:

$$\hat{f}(t, X, h) = \frac{\sum_{i=1}^{n} \mathcal{N}((t - X_i)/h)}{n \cdot h},$$

where $X_1, X_2, \ldots X_n \in \mathbb{R}$ are the n data points observed, $\mathcal{N}(y)$ is a Normal$(0, 1)$
density function evaluated at y, and $h \in \mathbb{R}^+$ is the bandwidth. Thus, the overall
curve is the sum of a set of subcurves, each centered over a different data point.
Figure 7.15 shows the effect of raising h on the shape of a set of fitted curves.[9]
When h is very small, the Normal distributions around each data point are sharp
spikes, so there is a mode at every data point. As h grows, the spikes spread out
and merge, until the sum of subdistributions produces a single bell curve. See page

[9]The data is the male viewership for 86 TV specials, from Chwe (2001).

376 for more on how these plots were generated; see also Silverman (1985).

As usual, there is a simple form for code to produce a default kernel density from a data set, and a more extensive form that allows more control. Try Listing 7.16, which plots the histogram of precipitation figures and the kernel-density smoothed version based on a $\mathcal{N}(0, 0.1)$ kernel. Also try $\sigma = 0.001$, 0.01, and 0.2 to see the progression from the data's spikes to a smooth bell curve.

```
#include <apop.h>

int main(){
    apop_db_open("data−climate.db");
    apop_data *data = apop_query_to_data("select pcp from precip");
    apop_model *h = apop_estimate(data, apop_histogram);
    apop_histogram_normalize(h);
    remove("out.h"); remove("out.k");
    apop_histogram_print(h, "out.h");
    apop_model *kernel = apop_model_set_parameters(apop_normal, 0., 0.1);
    apop_model *k = apop_model_copy(apop_kernel_density);
    Apop_settings_add_group(k, apop_kernel_density, NULL, h, kernel, NULL);
    apop_histogram_print(k, "out.k");
    printf("plot 'out.h' with lines title 'data', 'out.k' with lines title 'smoothed'\n");
}
```

Listing 7.16 A histogram before and after smoothing via kernel densities. Run via `smoothing | gnuplot`. Online source: `smoothing.c`.

Plot the `data-tv` set using:

- a histogram, using 40 and 100 bins,

- a smoothed version of the 40-bin histogram, via a moving average of bandwidth four,

- the 40-bin histogram smoothed via a Normal(x, 100.0) kernel density,

- the 40-bin histogram smoothed via a Uniform($x - 500.0$, $x + 500.0$) kernel density.

LINEAR PROJECTIONS

Our weakness forbids our considering the entire universe and makes us cut it up into slices.

—Poincaré (1913, p 1386)

This chapter covers models that make sense of data with more dimensions than we humans can visualize. The first approach, taken in Section 8.1 and known as principal component analysis (PCA), is to find a two- or three-dimensional subspace that best describes the fifty-dimensional data, and flatten the data down to those few dimensions.

The second approach, in Section 8.2, provides still more structure. The model labels one variable as the dependent variable, and claims that it is a linear combination of the other, independent, variables. This is the ordinary least squares (OLS) model, which has endless variants. The remainder of the chapter looks at how OLS can be applied by itself and in combination with the distributions in the prior chapter.

One way to characterize the two projection approaches is that both aim to project N-dimensional data onto the best subspace of significantly fewer than N dimensions, but they have different definitions of *best*. The standard OLS regression consists of finding the one-dimensional line that minimizes the sum of squared distances between the data and that line, while PCA consists of finding the few dimensions where the variance of the projection of the data onto those dimensions is maximized.

8.1 ❋ **PRINCIPAL COMPONENT ANALYSIS**   *PCA* is closely related to factor analysis, and in some fields is known as *spectral decomposition*. The first phase (calculating the eigenvalues) is sometimes called the *singular value decomposition*. It is a purely descriptive method. The idea is that we want a few dimensions that will capture the most variance possible—usually two, because we can plot two dimensions on paper.

After plotting the data, perhaps with markers for certain observations, we may find intuitive descriptions for the dimensions on which we had just plotted the data. For example, Poole & Rosenthal (1985) projected the votes cast by all Congressmen in all US Congresses, and found that 90% of the variance in vote patterns could be explained by two dimensions.[1] One of these dimensions could be broadly described as 'fiscal issues' and the other as 'social issues.' This method stands out because Poole & Rosenthal did not have to look at bills and place them on either scale—the data placed itself, and they just had to name the scales.

Shepard & Cooper (1992) asked their sighted subjects questions regarding color words (red, orange, ...), and did a principal component analysis on the data to place the words in two dimensions, where they formed a familiar color wheel. They did the same with blind subjects, and found that the projection collapsed to a single dimension, with violet, purple, and blue on one end of the scale, and yellow and gold on the other. Thus, the data indicates that the blind subjects think of colors on a univariate scale ranging from *dark colors* to *bright colors*.

It can be shown that the best n axes, in the sense above, are the n eigenvectors of the data's covariance matrix with the n largest associated eigenvalues.

The programs discussed below query three variables from the US Census data: the population, median age, and males per 100 females for each US state and commonwealth, the District of Columbia and Puerto Rico. They do a factor analysis and then project the original data onto the space that maximizes variance, producing the plot in Figure 8.1.

The programs also display the eigenvectors on the screen. They find that the first eigenvector is approximately $(0.06, -1, 0.03)$. among the three variables given, the second term—population—by itself describes the most variance in the data. The X-axis in the plot follows population[2]

The eigenvalues for the Y-axis are $(0.96, 0.05, -0.29)$, and are thus a less one-sided combination of the first variable (males per female) and the last (median age). That said, how can we interpret the Y axis? Those familiar with US geography will observe that the states primarily covered by cities (at the extreme, DC) are high on

---

[1] They actually did the analysis using an intriguing maximum likelihood method, rather than the eigenvector method here. Nonetheless, the end result and its interpretation are the same.

[2] As of the 2000 census, California≈ 33.8 million, Texas≈ 20.8, New York≈ 19.0, Florida≈ 16.0, et cetera.

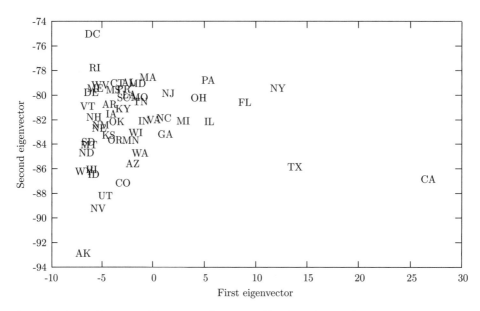

Figure 8.1 States decomposed into population on the X axis, and a combination of median age and gender balance (urban-ness?) on the Y axis.

the Y-axis, while states that are far more rural than urban (at the extreme, Alaska) are toward the bottom. Thus, the principal component analysis indicates that we could plausibly interpret the variation in median age and male/female balance by a single variable representing a state's urban–rural balance. Because interpreting the meaning of an artificial axis is a subjective exercise, other stories are also possible; for example, one could also argue that this second axis instead represents a state's East-to-West location.

✳ **CODING IT** As with many algorithms, the process of coding is straightforward, but involves a number of details. This section will show you the computation of a principal component analysis on two levels. The first goes through the steps of calculating eigenvectors yourself; the second is a single function call.

- The input and output functions are identical for both programs, so the redundancy-minimizing method of implementing these functions is via a separate file, Listing 8.2, and a corresponding header, which is too simple to be repeated here but is included in the online code supplement as `eigenbox.h`.

- The `query_data` function is self-explanatory. With the clause `select geo_-names as row_names`, Apophenia will use the state names as row names rather than plain data.

- These programs are intended to be run via pipe to Gnuplot, like `eigeneasy | gnuplot -persist`. So if we want additional information that Gnuplot will not

```
1   #include "eigenbox.h"
2
3   apop_data *query_data(){
4       apop_db_open("data−census.db");
5       return apop_query_to_data(" select postcode as row_names, \n\
6                           m_per_100_f, population/1e6 as population, median_age \n\
7                           from geography, income,demos,postcodes \n\
8                           where income.sumlevel= '040' \n\
9                           and geography.geo_id = demos.geo_id \n\
10                          and income.geo_name = postcodes.state \n\
11                          and geography.geo_id = income.geo_id ");
12  }
13
14  void show_projection(gsl_matrix *pc_space, apop_data *data){
15      apop_opts.output_type = 'p';
16      apop_opts.output_pipe = stderr;
17      fprintf(stderr,"The eigenvectors:\n");
18      apop_matrix_print(pc_space, NULL);
19      apop_data *projected = apop_dot(data, apop_matrix_to_data(pc_space), 0, 0);
20      printf("plot '−' using 2:3:1 with labels\n");
21      apop_data_show(projected);
22  }
```

Listing 8.2 The tools used below, including the query and a method for displaying labeled points.
Online source: `eigenbox.c`.

understand, we need to send it to a different location. Thus, lines 15–18 send output to `stderr`, so the eigenvectors are printed to screen instead of sent down the `stdout` pipeline to Gnuplot.

- The `plot` command on line 20 is `with labels`, meaning that instead of points, we get the two-letter postal codes, as seen in Figure 8.1.

- As for the actual math, Listing 8.3 shows every step in the process.[3] The only hard part is finding the eigenvalues of $\mathbf{X'X}$; the GSL saw us coming, and gives us the `gsl_eigen_symm` functions to calculate the eigenvectors of a symmetric matrix. The `find_eigens` function shows how one would use that function. The GSL is too polite to allocate large vectors behind our backs, so it asks that we pass in pre-allocated workspaces when it needs such things. Thus, the `findeigens` function allocates the workspace, calls the eigenfunction, then frees the workspace. It frees the matrix whose eigenvalues are being calculated at the end because the matrix is destroyed in the calculations, and should not be referred to again. To make sure future tests in the way of `if (!subject)...` work, the last line sets the pointer to `NULL`.

---

[3]There is one cheat: the `apop_sv_decomposition` function would use the `apop_normalize_for_-svd(xpx->matrix)` function to ensure that for each row, $\mathbf{x'x} = 1$. You can erase line 28 of `eigenhard.c`, input this SVD-specific normalization function after the `apop_dot` step, and look for subtle shifts.

```
1    #include "eigenbox.h"
2
3    void find_eigens(gsl_matrix **subject, gsl_vector *eigenvals, gsl_matrix *eigenvecs){
4        gsl_eigen_symmv_workspace * w = gsl_eigen_symmv_alloc((*subject)−>size1);
5        gsl_eigen_symmv(*subject, eigenvals, eigenvecs, w);
6        gsl_eigen_symmv_free (w);
7        gsl_matrix_free(*subject); *subject = NULL;
8    }
9
10   gsl_matrix *pull_best_dims(int ds, int dims, gsl_vector *evals, gsl_matrix *evecs){
11       size_t indexes[dims], i;
12       gsl_matrix *pc_space = gsl_matrix_alloc(ds,dims);
13           gsl_sort_vector_largest_index(indexes, dims, evals);
14           for (i=0; i<dims; i++){
15               APOP_MATRIX_COL(evecs, indexes[i], temp_vector);
16               gsl_matrix_set_col(pc_space, i, temp_vector);
17           }
18       return pc_space;
19   }
20
21   int main(){
22       int dims = 2;
23       apop_data *x = query_data();
24       apop_data *cp = apop_data_copy(x);
25       int ds = x−>matrix−>size2;
26       gsl_vector *eigenvals = gsl_vector_alloc(ds);
27       gsl_matrix *eigenvecs = gsl_matrix_alloc(ds, ds);
28           apop_matrix_normalize(x−>matrix, 'c', 'm');
29           apop_data *xpx = apop_dot(x, x, 1, 0);
30           find_eigens(&(xpx−>matrix), eigenvals, eigenvecs);
31           gsl_matrix *pc_space = pull_best_dims(ds, dims, eigenvals, eigenvecs);
32           show_projection(pc_space, cp);
33   }
```

Listing 8.3 The detailed version. Online source: `eigenhard.c`.

- If the space of the data has full rank, then there will be three eigenvectors for three-dimensional data. The `pull_best_dimensions` function allocates a new matrix that will have only `dims` dimensions, and the best eigenvectors are included therein. Again, the GSL saw us coming, and provides the `gsl_sort_-vector_largest_index` function, which returns an array of the indices of the largest elements of the `evals` vector. Thus, `indexes[0]` is the index of the largest eigenvalue, `indexes[1]` is the point in the vector of values with the second largest eigenvector, et cetera. Given this information, it is an easy `for` loop (lines 14–17) to copy columns from the set of all eigenvectors to the `pc_space` matrix.

- Finally, after the principal component vectors have been calculated, `show_projec-tion` produces a graphical representation of the result. It projects the data onto the space via the simple dot product data·pc_space, and then produces Gnuplottable output as per the tricks discussed above.

• Given all these functions, the main routine is just declarations and function calls to implement the above procedure: pull the data, calculate $\mathbf{X}'\mathbf{X}$, send the result to the eigencalculator, project the data, and show the results.

```
#include "eigenbox.h"

int main(){
    int dims = 2;
    apop_data *x = query_data();
    apop_data *cp = apop_data_copy(x);
    apop_data *pc_space = apop_matrix_pca(x−>matrix, dims);
        fprintf(stderr, "total explained: %Lf\n", apop_sum(pc_space−>vector));
        show_projection(pc_space−>matrix, cp);
}
```

Listing 8.4 The easy way: just let `apop_sv_decomposition` do the work. Online source: `eigeneasy.c`.

Listing 8.4 presents the same program using the `apop_matrix_pca` function to do the singular value decomposition in one quick function call. That function simply does the normalization and bookkeeping necessary to use the `gsl_linalg_SV_-decomp` function. All that is left for the `main` function in Listing 8.4 is to query the input data to a matrix, call the SVD function, and display the output results.

8.1

> A matrix is *positive definite* (PD) iff all of its eigenvalues are greater than zero, and *positive semidefinite* (PSD) iff all of its eigenvalues are greater than or equal to zero. Similarly for *negative definite* (ND) and *negative semidefinite* (NSD).
>
> These are often used as multidimensional analogues to positive or negative. For example, just as $x^2 \geq 0, \forall x \in \mathbb{R}$, $\mathbf{X} \cdot \mathbf{X}$ is PSD for any $\mathbf{X}$ with real elements. An extremum of $f(x)$ is a maximum when the second derivative $f''(x)$ is negative, and a minimum when $f''(x)$ is positive; $f(\mathbf{x})$ is a maximum when the matrix of second partial derivatives is NSD, and a minimum when the matrix of second partials is PSD (otherwise it's a saddle point representing a maximum along one direction and a minimum along another, and is thus a false extremum).   ≫

≫

Write a function `matrix_is_definite` that takes in a matrix and outputs a single character, say 'P', 'p', 'N', or 'n', to indicate whether the matrix is one of the above types (and another character, say 'x', if it is none of them). Write a test function that takes in any data matrix **X** (your favorite data set or a randomly-generated matrix) and checks that **X** · **X** is PSD.

Σ

➤ Principal component analysis projects data of several dimensions onto the dimensions that display the most variance.

➤ Given the data matrix **X**, the process involves finding the eigenvalues of the matrix $\underline{\mathbf{X}'\mathbf{X}}$ associated with the largest eigenvalues, and then projecting the data onto the space defined by those eigenvectors.

➤ `apop_matrix_pca` runs the entire process for the efficiently lazy user.

8.2 OLS AND FRIENDS

Assume that our variable of interest, **y**, is described by a linear combination of the explanatory variables, the columns of **X**, plus maybe a Normally-distributed error term, ϵ. In short, $\mathbf{y} = \mathbf{X}\beta + \epsilon$, where β is the vector of parameters to be estimated. This is known as the ordinary least squares (*OLS*) model, for reasons discussed in the introduction to this chapter

To a great extent, the OLS model is the null prior of models: it is the default that researchers use when they have little information about how variables interrelate. Like a good null prior, it is simple, it is easy for the computer to work with, it flexibly adapts to minor digressions from the norm, it handles nonlinear subelements (despite often being called linear regression), it is exceptionally well-studied and understood, and it is sometimes the case that **y** really is a linear combination of the columns of **X**.

OLS is frequently used for solving the snowflake problem from the last chapter: we had a series of very clean univariate models that stated that the outcome of interest is the result of a series of identical draws with equal probability, but in most real-world situations, each draw is slightly different—in the terminology of OLS, we need to control for the other characteristics of each observation. The term *control for* is an analogy to controlled experiments where nuisance variables are fixed, but the metaphor is not quite appropriate, because adding a column to **X** representing the control leads to a new projection onto a new space (see below), and may completely change the estimate of the original OLS parameters.[4] Later sections will present other options for surmounting the snowflake problem when using the distributions from the last chapter.

Because linear models are so well-studied and documented, this book will only briefly skim over them.[5] This chapter puts OLS in the context from the first page of this book: a model that claims a relationship between $\mathbf{X}$ and $\mathbf{y}$ and whose parameters can be estimated with data, whose computational tools provide several conveniences. The next chapter briefly covers OLS for hypothesis testing.

Unlike the models to this point, OLS implicitly makes a causal claim: the variables listed in $\mathbf{X}$ cause $\mathbf{y}$ to take the values they do. However, there is no true concept of causality in statistics. The question of when statistical evidence of causality is valid is a tricky one that will be left to the volumes that cover this question in detail.[6] For the purposes here, the reader should merely note the shift in descriptive goal, from fitting distributions to telling a causal story.

A brief derivation   Part of OLS's charm is that it is easy (and instructive) to derive $\hat{\beta}$ for the OLS case. We seek the parameter estimate $\hat{\beta}$ that minimizes the squared error, $\epsilon'\epsilon$, where $\mathbf{y} = \mathbf{X}\hat{\beta} + \epsilon$.

This is smallest when the error term ϵ is orthogonal to the space of $\mathbf{X}$, meaning that $\mathbf{X}'\epsilon = \mathbf{0}$. If $\mathbf{X}$ and $\epsilon = \mathbf{y} - \mathbf{X}\hat{\beta}$ were not orthogonal, then we could always reduce the size of ϵ by twiddling $\hat{\beta}$ by an iota to either $\hat{\beta} + \iota$ or $\hat{\beta} - \iota$.

✴ Proof: After adding ι to $\hat{\beta}$, the new equation would be $\mathbf{y} = \mathbf{X}\hat{\beta} + \mathbf{X}\iota + (\epsilon - \mathbf{X}\iota)$, meaning that the new error term is now $\epsilon_n \equiv \epsilon - \mathbf{X}\iota$, and

$$\epsilon_n'\epsilon_n = (\epsilon - \mathbf{X}\iota)'(\epsilon - \mathbf{X}\iota)$$
$$= \epsilon'\epsilon - 2\iota'\mathbf{X}'\epsilon + \iota'\mathbf{X}'\mathbf{X}\iota.$$

The last term, $\iota'\mathbf{X}'\mathbf{X}\iota$, is always a non-negative scalar, just as $x \cdot x$ is non-negative for any real value of x. So the only way that $\epsilon_n'\epsilon_n$ can be smaller than $\epsilon'\epsilon$ is if $2\iota'\mathbf{X}'\epsilon > 0$. But if $\mathbf{X}'\epsilon = \mathbf{0}$, then this is impossible.

The last step of the proof is to show that if $\iota'\mathbf{X}'\epsilon$ is not zero, then there is always a way to pick ι such that $\epsilon_n'\epsilon_n < \epsilon'\epsilon$. ℚ: Prove this. (*Hint*: if $\iota'\mathbf{X}'\epsilon \neq 0$, then the

---

[4]That is, regression results can be unstable, so never trust a single regression.

[5]Economists are especially dependent on linear regression, because it is difficult to do controlled studies on the open economy. Thus, econometrics texts such as Maddala (1977), Kmenta (1986), or Greene (1990) are a good place to start when exploring linear models.

[6]See Perl (2000) or any book on structural equation modeling. The typical full causal model is a directed acyclic graph representing the causal relationships among nodes, and the data can reject or fail to reject it like any other model.

same is true for $\iota_d = 2\iota$ and $\iota_h = \iota/2$.)   ◆

Projection   So we seek $\hat{\beta}$ that will lead to an error such that $\mathbf{X}'\boldsymbol{\epsilon} = 0$. Expanding $\boldsymbol{\epsilon}$
to $\mathbf{y} - \mathbf{X}\hat{\beta}$, we can solve for $\hat{\beta}$:

$$\mathbf{X}'[\mathbf{y} - \mathbf{X}\hat{\beta}] = 0$$
$$\mathbf{X}'\mathbf{y} = \mathbf{X}'\mathbf{X}\hat{\beta}$$
$$(\mathbf{X}'\mathbf{X})^{-1}\mathbf{X}'\mathbf{y} = \hat{\beta}$$

This is the familiar form from page 3.

Now consider the *projection matrix*, which is defined as

$$\mathbf{X}^P \equiv \mathbf{X}(\mathbf{X}'\mathbf{X})^{-1}\mathbf{X}'.^7$$

It is so named because, as will be demonstrated below, $\mathbf{X}^P\mathbf{v}$ projects the vector $\mathbf{v}$
onto the space of $\mathbf{X}$.

Start with $\mathbf{X}^P\mathbf{X}$. Writing this out, it reduces instantly: $\mathbf{X}(\mathbf{X}'\mathbf{X})^{-1}\mathbf{X}'\mathbf{X} = \mathbf{X}$. So
the projection matrix projects $\mathbf{X}$ onto itself.

The expression $\mathbf{X}^P\boldsymbol{\epsilon}$ also simplifies nicely:

$$\begin{aligned}
\mathbf{X}^P\boldsymbol{\epsilon} &= \mathbf{X}(\mathbf{X}'\mathbf{X})^{-1}\mathbf{X}'[\mathbf{y} - \mathbf{X}\hat{\beta}] \\
&= \mathbf{X}(\mathbf{X}'\mathbf{X})^{-1}\mathbf{X}'[\mathbf{y} - \mathbf{X}(\mathbf{X}'\mathbf{X})^{-1}\mathbf{X}'\mathbf{y}] \\
&= \mathbf{X}(\mathbf{X}'\mathbf{X})^{-1}\mathbf{X}'\mathbf{y} - \mathbf{X}(\mathbf{X}'\mathbf{X})^{-1}\mathbf{X}'\mathbf{X}(\mathbf{X}'\mathbf{X})^{-1}\mathbf{X}'\mathbf{y} \\
&= \mathbf{X}(\mathbf{X}'\mathbf{X})^{-1}\mathbf{X}'\mathbf{y} - \mathbf{X}(\mathbf{X}'\mathbf{X})^{-1}\mathbf{X}'\mathbf{y} \\
&= \mathbf{0}.
\end{aligned}$$

The projection matrix projects $\boldsymbol{\epsilon}$ onto the space of $\mathbf{X}$, but $\boldsymbol{\epsilon}$ and $\mathbf{X}$ are orthogonal,
so the projected vector is just $\mathbf{0}$.

What does the projection matrix have to do with OLS? Since $\hat{\beta} = (\mathbf{X}'\mathbf{X})^{-1}\mathbf{X}'\mathbf{y}$,
$\mathbf{X}\hat{\beta} = \mathbf{X}(\mathbf{X}'\mathbf{X})^{-1}\mathbf{X}'\mathbf{y} = \mathbf{X}^P\mathbf{y}$. Thus, the OLS estimate of $\mathbf{y}$, $\hat{\mathbf{y}} \equiv \mathbf{X}\hat{\beta}$, is the
projection of $\mathbf{y}$ onto the space of $\mathbf{X}$: $\hat{\mathbf{y}} = \mathbf{X}^P\mathbf{y}$.

And that, in a nutshell, is what OLS is about: the model claims that $\mathbf{y}$ can be
projected onto the space of $\mathbf{X}$, and finds the parameters that achieve that with least
squared error.

---

[7]This is sometimes called the *hat matrix*, because (as will be shown), it links $\mathbf{y}$ and $\hat{\mathbf{y}}$.

A sample projection   At this point in the narrative, most statistics textbooks would include a picture of a cloud of points and a plane onto which they are projected. But if you have a Gnuplot setup that lets you move plots, you can take the data and projected points in your hands, and get a feel for how they move.

```
1   #include "eigenbox.h"
2   gsl_vector *do_OLS(apop_data *set);
3
4   gsl_vector *project(apop_data *d, apop_model *m){
5       apop_data *d2 = apop_data_copy(d);
6       APOP_COL(d2, 0, ones);
7       gsl_vector_set_all(ones, 1);
8       return apop_dot(d2, m−>parameters, 0, 'v')−>vector;
9   }
10
11  int main(){
12      apop_data *d = query_data();
13      apop_model *m = apop_estimate(d, apop_ols);
14      d−>vector = project(d, m);
15      //d−>vector = do_OLS(d);
16      d−>names−>rowct = 0;
17      d−>names−>colct = 0;
18      apop_data_print(d, "projected");
19      FILE *cmd = fopen("command.gnuplot", "w");
20      fprintf(cmd, "set view 20, 90\n\
21              splot 'projected' using 1:3:4 with points, 'projected' using 2:3:4\n");
22  }
```

Listing 8.5 Code to project data onto a plane via OLS. Compile with `eigenbox.c` from earlier. Online source: `projection.c`.

Listing 8.5 queries the same data as was plotted in Figure 8.1, does an OLS projection, and produces a Gnuplot command file to plot both the original and projected data.

- Ignore lines 2 and 15 for now; they will allow us to do the projection manually later on.

- Line 13 does the OLS regression. The `apop_estimate` function estimates the parameters of a model from data. It takes in a data set and a model, and returns an `apop_model` structure with the parameters set.

- The `project` function makes a copy of the data set and replaces the dependent variable with a column of ones, thus producing the sort of data on the right-hand side of an OLS regression. Line seven calculates $X\beta$.

- When the data set is printed on line 18, the first column will be the $X\beta$ vector just calculated, the second will be the original y data, and the third and fourth columns will be the non-constant variables in **X**.

- Gnuplot has some awkwardness with regards to replotting data (see page 170). Lines 16–18 write the data to a file (with the row and column names masked out), then lines 19–20 write a two-line command file. You can run it from the command line via `gnuplot command.gnuplot -`, and should then have on your screen a plot that you can move about.

The view set on line 20 is basically a head-on view of the plane onto which the data has been projected, and you can see how the data (probably in green) shifts to a projected value (red). It is a somewhat different picture from the PCA in Listing 170. [Q: Add the `with label` commands to put labels on the plot.] Spinning the graph a bit, to `set view 83, 2`, shows that all of the red points are indeed on a single plane, while in another position, at `set view 90, 90`, you can verify that the points do indeed match on two axes.[8]

THE CATALOG Because OLS is so heavily associated with hypothesis testing, this section will plan ahead and present both the estimates of β produced by each model, its expected value, and its variance. This gives us all that we need to test hypotheses regarding elements of our estimate of β, $\hat{\beta}$.

OLS The model:

- $\mathbf{y} = \mathbf{X}\beta + \epsilon$
- n = the number of observations, so $\mathbf{y}$ and ϵ are $n \times 1$ matrices.
- k = the number of parameters to be estimated. $\mathbf{X}$ is $n \times k$; β is k by 1.
- Many results that will appear later assume that the first column of β is a column of ones. If it isn't, then you need to replace every non-constant column $\mathbf{x}_i$ below with $\mathbf{x}_i - \overline{\mathbf{x}}_i$, the equivalence of which is left as an exercise for the reader.[9] See below for the actual format to use when constructing your `apop_data` set.

Assumptions:

- $E(\epsilon) = 0$.
- $\text{var}(\epsilon_i) = \sigma^2$, a constant, $\forall\, i$.

---

[8]If your Gnuplot setup won't let you spin the plot with a pointing device, then you can modify the Gnuplot command file to print three separate static graphs in three files via the `set view` commands here, or you can try a command like `set view 83, 2; replot` at the Gnuplot command prompt.

[9]If you would like to take the route of normalizing each column to have mean zero, try `apop_matrix_norm-alize(dataset, 'm')`. This will normalize a **1** column to **0**, so after calling the normalize function, you may need to do something like `APOP_COL(your_data, 0, onecol); gsl_vector_set_all(onecol, 1)`.

- $\text{cov}(\epsilon_i, \epsilon_j) = 0$, $\forall\, i \neq j$. Along with the above assumption, this means that the $n \times n$ covariance matrix for the observations' errors is $\Sigma \equiv \sigma^2 \mathbf{I}$.
- The columns of $\mathbf{X}$ are not collinear (i.e., $\det(\mathbf{X}'\mathbf{X}) \neq 0$, so $(\mathbf{X}'\mathbf{X})^{-1}$ exists).[10]
- $n > k$.

Notice that we do not assume that ϵ has a Normal distribution, but that assumption will be imposed in the next chapter, when we apply t tests to $\hat{\beta}$. When all of that holds, then

$$\hat{\beta}_{\mathrm{OLS}} = (\mathbf{X}'\mathbf{X})^{-1}(\mathbf{X}'\mathbf{y})$$
$$E(\hat{\beta}_{\mathrm{OLS}}) = \beta$$
$$\text{var}(\hat{\beta}_{\mathrm{OLS}}) = \sigma^2(\mathbf{X}'\mathbf{X})^{-1}$$

Almost anything can have a variance, which often creates confusion. A column of data has a variance, the column of errors ϵ has a variance, the estimate $\hat{\beta}_{\mathrm{OLS}}$ has a covariance matrix, and (if so inclined) you could even estimate the variance of the variance estimate $\hat{\sigma}^2$. The variance listed above is the $K \times K$ covariance matrix for the estimate $\hat{\beta}_{\mathrm{OLS}}$, and will be used for testing hypotheses regarding β in later chapters. It is a combination of the other variances: the variance of the error term ϵ is σ^2, and the various columns of the data set have covariance $\mathbf{X}'\mathbf{X}$, so the variance of $\hat{\beta}_{\mathrm{OLS}}$ is the first times the inverse of the second.

INSTRUMENTAL VARIABLES   The proofs above gave us a guarantee that we will calculate a value of $\hat{\beta}$ such that $\mathbf{X}$ will be uncorrelated to $\hat{\epsilon} = \mathbf{y} - \mathbf{x}\hat{\beta}$.

Our hope is to estimate a 'true' model, claiming that $\mathbf{y} = \mathbf{X}\beta + \epsilon$, where $\mathbf{X}$ is uncorrelated to ϵ, and so on. But if it is the case that a column of $\mathbf{X}$ really is correlated to the error in this model, then there is no way that $\hat{\beta}$ (which guarantees that the estimated error and the columns of $\mathbf{X}$ are not correlated) could match β (which is part of a model where error and a column of $\mathbf{X}$ are correlated). This creates major problems.

For example, say that one column of the data, $\mathbf{x}_i$, is measured with error, so we are actually observing $\mathbf{x}_i + \epsilon_i$. This means that the error term in our equation is now

---

[10]If this assumption is barely met, so $\det(\mathbf{X}'\mathbf{X})$ is not zero but is very small, then the resulting estimates will be unstable, meaning that very small changes in $\mathbf{X}$ would lead to large changes in $(\mathbf{X}'\mathbf{X})^{-1}$ and thus in the parameter estimates. This is known as the problem of *multicollinearity*. The easiest solution is to simply exclude some of the collinear variables from the regression, or do a principal component analysis to find a smaller set of variables and then regress on those. See, e.g., Maddala (1977, pp 190–194) for further suggestions.

the true error joined with an offset to the measurement error, $\hat{\epsilon} = \epsilon - \epsilon_i$. If the true x_i and the true ϵ have no correlation, then $x_i + \epsilon_i$ and $\epsilon - \epsilon_i$ almost certainly do.

As above, the OLS estimate of β is $\hat{\beta}_{OLS} = (X'X)^{-1}X'y$, or taking a step back in the derivation, $(X'X)\hat{\beta}_{OLS} = X'y$. Also, the true β is defined to satisfy $y = X\beta + \epsilon$. With a few simple manipulations, we can find the distance between β and $\hat{\beta}$:

$$y = X'\beta + \epsilon$$
$$X'y = X'X\beta + X'\epsilon$$
$$(X'X)\hat{\beta}_{OLS} = X'X\beta + X'\epsilon$$
$$(X'X)(\hat{\beta}_{OLS} - \beta) = X'\epsilon$$
$$\hat{\beta}_{OLS} - \beta = (X'X)^{-1}X'\epsilon \qquad (8.2.1)$$

If $X'\epsilon = 0$, then the distance between $\hat{\beta}_{OLS}$ and β is zero—$\hat{\beta}_{OLS}$ is a consistent estimator of β. But if X and ϵ are correlated, then $\hat{\beta}_{OLS}$ does not correctly estimate β. Further, unless we have a precise measure of the right-hand side of Equation 8.2.1, then we don't know if our estimate is off by a positive, negative, large or small amount. Still further, every column of X affects every element of $(X'X)^{-1}$, so mismeasurement of one column can throw off the estimate of the OLS coefficient for every other column as well.

The solution is to replace the erroneously-measured column of data with an *instrument*, a variable that is not correlated to ϵ but is correlated to the column of data that it will replace. Let x_i be the column that is measured with error (or is otherwise correlated to ϵ), let z be a column of alternate data (the instrument), and let Z be the original data set X with the column x_i replaced by z. If $\text{cov}(z, \epsilon) = 0$, then the following holds:

$$\hat{\beta}_{IV} = (Z'X)^{-1}(Z'y)$$
$$E(\hat{\beta}_{IV}) = \beta$$
$$\text{var}(\hat{\beta}_{IV}) = \sigma^2(Z'X)^{-1}Z'Z(X'Z)^{-1}$$

Whe $\det(Z'X)$ is small, $(Z'X)^{-1}$—and thus the variance—is large; this happens when $\text{cov}(z, x_i) \to 0$. We want the variance of our estimator to be as small as possible, which brings us to the usual rule for searching for an instrument: find a variable that can be measured without significant error, but which is as well-correlated as possible to the original variable of interest.

GLS   Generalized Least Squares generalizes OLS by allowing $\epsilon'\epsilon$ to be a known matrix Σ, with no additional restrictions. Note how neatly plugging $\sigma^2 \mathbf{I}$ in to the estimator of β and its variance here reduces the equations to the OLS versions above.

$$
\begin{aligned}
\hat{\beta}_{\mathrm{GLS}} &= (\mathbf{X}'\Sigma^{-1}\mathbf{X})^{-1}(\mathbf{X}'\Sigma^{-1}\mathbf{y}) \\
E(\hat{\beta}_{\mathrm{GLS}}) &= \beta \\
\mathrm{var}(\hat{\beta}_{\mathrm{GLS}}) &= (\mathbf{X}'\Sigma^{-1}\mathbf{X})^{-1}
\end{aligned}
$$

But there is a computational problem with GLS as written here. For a data set of a million elements, Σ is $10^6 \times 10^6 = 10^{12}$ (a trillion) elements, and it is often the case that all but a million of them will be zero. A typical computer can only hold tens of millions of `doubles` in memory at once, so simply writing down such a matrix is difficult—let alone inverting it. Thus, although this form is wonderfully general, we can use it only when there is a special form of Σ that can be exploited to make computation tractable.

WEIGHTED LEAST SQUARES   For example, let Σ be a diagonal matrix. That is, errors among different observations are uncorrelated, but the error for each observation itself is different. This is *heteroskedastic* data. The classic example in Econometrics is due to differences in income: we expect that our errors in measurement regarding the consumption habits of somebody earning \$10,000/year will be about a thousandth as large as our measurement errors regarding consumption by somebody earning \$10 million/year.

It can be shown (e.g., Kmenta (1986)) that the optimum for this situation, where σ_i is known for each observation i, is to use the GLS equations above, with Σ set to zero everywhere but the diagonal, where the ith element is $\frac{1}{\sigma_i^2}$.

The GLS equations about β now apply directly to produce Weighted Least Squares estimates—and there is a trick that lets us do computation using just a vector of diagonal elements of Σ, instead of the full matrix. Let $\sqrt{\sigma}$ be a vector where the ith element is the square root of the ith diagonal element of Σ. For WLS, the ith element of $\sqrt{\sigma}$ is thus $\frac{1}{\sigma_i}$. Now let $\mathbf{y}_\Sigma$ be a vector whose ith element is the ith element of $\mathbf{y}$ times the ith element of $\sqrt{\sigma}$, and $\mathbf{X}_\Sigma$ be the column-wise product of $\mathbf{X}$ and $\sqrt{\sigma}$. That is,

```
void columnwise_product(gsl_matrix *data, gsl_vector *sqrt_sigma){
    for (size_t i=0; i< data->size2; i++){
        Apop_matrix_col(data, i, v);
        gsl_vector_mul(v, sqrt_sigma);
    }
}
```

Then you can verify that $\mathbf{X}'_\Sigma \mathbf{X}_\Sigma = \mathbf{X}'\Sigma\mathbf{X}$, $\mathbf{X}'_\Sigma \mathbf{y}_\Sigma = \mathbf{X}'\Sigma\mathbf{y}$, and so

$$\hat{\beta}_{\mathrm{WLS}} = \left(\mathbf{X}'_\Sigma \mathbf{X}_\Sigma\right)^{-1}\left(\mathbf{X}'_\Sigma \mathbf{y}_\Sigma\right)$$
$$E(\hat{\beta}_{\mathrm{WLS}}) = \beta$$
$$\mathrm{var}(\hat{\beta}_{\mathrm{WLS}}) = \sigma^2\left(\mathbf{X}'_\Sigma \mathbf{X}_\Sigma\right)^{-1}$$

Thus, we can solve for the Weighted Least Squares elements without ever writing down the full Σ matrix in all its excess. This is the method used by `apop_wls` (which is what you would use in practice, rather than calculating $\mathbf{X}_\Sigma$ yourself).

FITTING IT   If you already have a data matrix in `apop_data *set`, then you can estimate an OLS model in one line:

```
apop_estimate_show(apop_estimate(set, apop_ols));
```

If your data set has a non-NULL `weights` vector, then you could replace `apop_ols` in the above with `apop_wls`.

If you would like more control over the details of the estimation routine, see page 145 on the format for changing estimation settings, and the online references for the detailed list of options. The `apop_IV` model requires the settings setting treatment, because that model requires that the settings' `instruments` element is set.

Q8.2   In the exercise on page 232, you found a relationship between males per female (the dependent variable) and both density and male–female wage ratios (the independent variables). Produce an appropriate data set and send it to `apop_estimate(`*your_data*`, apop_ols)` to check the coefficients when both independent variables are included.

If the assumptions of OLS do not fit, then you will need to be ready to modify the innards of the OLS estimation to suit, so the remainder of this section goes back to the linear algebra layer of abstraction, to go over the steps one would take to estimate β.

Listing 8.6 extends the example in Listing 8.5 (page 273) by doing every step in the linear algebra of OLS. Uncomment Line 15 in that code, link it with this (by adding `projectiontwo.o` to the `OBJECTS` line in the makefile), and re-run to produce the new projection, which should be identical to the old.

```
1   #include <apop.h>
2
3   gsl_vector *do_OLS(apop_data *set){
4       apop_data *d2 = apop_data_copy(set);
5       APOP_COL(d2, 0, firstcol);
6       apop_data *y_copy = apop_vector_to_data(apop_vector_copy(firstcol));
7       gsl_vector_set_all(firstcol, 1);
8
9       apop_data *xpx = apop_dot(d2,d2,'t',0);
10      gsl_matrix *xpxinv = apop_matrix_inverse(xpx->matrix); //(X'X)⁻¹
11      apop_data *second_part = apop_dot(apop_matrix_to_data(xpxinv), d2,0,'t');
12
13      apop_data *beta = apop_dot(second_part, y_copy, 0, 0); //(X'X)⁻¹X'y
14      strcpy(beta->names->title, "The OLS parameters");
15      apop_data_show(beta);
16
17      apop_data *projection_matrix = apop_dot(d2, second_part,0,0); //X(X'X)⁻¹X'
18      return apop_dot(projection_matrix, y_copy, 0,0)->vector;
19  }
```

Listing 8.6 The OLS procedure and its use for projection, spelled out in great detail. Online source: `projectiontwo.c`.

- Typically, the **y** values are the first column of the data matrix, so the first step is to extract the data into a separate vector. Lines 4–7 do this, and end up with one `apop_data` set named `y_copy` which copies off the first column of the input matrix, and a second set named d2, which is a copy of the input matrix with the same first column set to **1**.
- Lines 9–11 find $(\mathbf{X'X})^{-1}\mathbf{X'}$. If you like, you can use the debugger to check the value of any of the intermediate elements all along the calculation.
- Now that we have $(\mathbf{X'X})^{-1}\mathbf{X'}$, Lines 13–15 do the single additional dot product to find $\beta = (\mathbf{X'X})^{-1}\mathbf{X'y}$, and display the result.
- Line 17 produces the projection matrix $\mathbf{X}(\mathbf{X'X})^{-1}\mathbf{X'}$, and given the projection matrix, another dot product projects **y** onto it.

The GSL has a function to solve equations of the type $\mathbf{A}\beta = \mathbf{C}$ using Householder transformations, and this happens to be exactly the form we have here—$(\mathbf{X}'\mathbf{X})\beta = (\mathbf{X}'\mathbf{y})$. Given apop_data sets for $\mathbf{X}'\mathbf{X}$, $\mathbf{X}'\mathbf{y}$, and a gsl_vector allocated for beta, this line would fill beta with the solution to the equation:

```
gsl_linalg_HH_solve (xpx−>matrix, xpy−>vector, *beta);
```

Q8.3   In practice, it is almost necessary to take the inverse to solve for OLS parameters, because the covariance matrix of $\hat{\beta}$ is $\sigma^2(\mathbf{X}'\mathbf{X})^{-1}$. But the Householder method manages to avoid explicitly finding the inverse of $\mathbf{X}'\mathbf{X}$, and may thus be quicker in situations like the Cook's distance code (page 133) where a for loop solves for thousands of OLS parameter sets.
Rewrite projectiontwo.c to use the Householder transformation to find beta. Check that the results using this alternate method match the beta found via inversion.

$\sum$

➤ The Ordinary Least Squares model assumes that the dependent variable is an *affine projection* of the others. Given this and other assumptions, the likelihood-maximizing parameter vector is $\beta_{\mathrm{OLS}} = (\mathbf{X}'\mathbf{X})^{-1}(\mathbf{X}'\mathbf{Y})$.

➤ If $\Sigma \neq \mathbf{I}$, then $\beta_{\mathrm{GLS}} = (\mathbf{X}'\Sigma\mathbf{X})^{-1}(\mathbf{X}'\Sigma\mathbf{Y})$. Depending on the value of Σ, one can design a number of models.

➤ In most cases, you can use apop_estimate. But if need be, coding these processes is a simple question of stringing together lines of linear algebra operations from Chapter 4.

8.3 DISCRETE VARIABLES   To this point, the regression methods have been assuming that both $\mathbf{y}$ and the elements of $\mathbf{X}$ are all continuous variables $\in \mathbb{R}$. If they are discrete variables $\in \{0, 1, 2, \dots\}$, then we need to make modifications.

There are a number of distinctions to be made. First, the approaches to handling columns of discrete variables in $\mathbf{X}$ are different, and simpler, than approaches to handling discrete values of the outcome variable $\mathbf{y}$. If $\mathbf{y}$ only takes on integer values, or worse, is just zero or one, then there can't possibly be a β and a Normally distributed ϵ that satisfy $\mathbf{y} = \mathbf{X}\beta + \epsilon$. There are convolutions like stipulating that for observations where $\mathbf{x}\beta < 0$ we force ϵ such that $\mathbf{x}\beta + \epsilon = 0$ and where $\mathbf{x}\beta > 1$ we force $\mathbf{x}\beta + \epsilon = 1$, but then ϵ is non-Normal, which makes a mess of the hypothesis tests we'll run in later chapters. More generally, such an approach lacks grace, and requires a cascade of little fixes (Greene, 1990, p 637).

How we proceed in either case depends on the type of discrete variable at hand:

- A discrete binary variable takes on exactly two values, such as male/female or case/control, and can be rewritten as simply either zero or one.
- Ordered, discrete data are typically a count, such as the number of children or cars a person owns.
- Most qualitative categorizations lead to multi-valued and unordered data. For example, a Metro station could be on the Red, Blue, Green, Orange, and Yellow line, or a voter could align himself with a Red, Blue, Green, or Other party.

DUMMY VARIABLES   For a column of zero–one data in the independent data $\mathbf{X}$, we don't really have to modify OLS at all, but we change the interpretation slightly: taking zero as observations in the baseline and one as the observations in the treatment group, the OLS coefficient on this zero–one *dummy variable* can indicate how effective the treatment is in changing the outcome. Tests for the significance of the dummy variable can be shown to be equivalent to ANOVA tests of a category's significance.

As an extension, if the categories are ordered but discrete, like the number of children, then we again don't have to do anything: if a jump from $x = 0$ to $x = 1$ induces a shift of size β in the outcome, then a jump from $x = 1$ to $x = 2$ would do the same under a linear model, and a jump from $x = 0$ to $x = 2$ would produce a jump of size 2β. If it is natural to presume this, then the model does no violence to reality.[11]

But if the categories are discrete and unordered, then we can't use one variable with the values $\{0, 1, 2, \dots\}$, because the implied linear relationship makes no sense. With 0=Green, 1=Blue, 2=Red, does it make sense that a shift from Green to Blue will have exactly the same effect as a shift from Blue to Red, and that a shift from Green to Red would have exactly twice that effect? In this case, the solution is to have a separate variable for all but one of the choices, and thus to produce $n - 1$ coefficients for n options. The excluded option is the baseline, like the zero option in the binary case. Here, the function `apop_data_to_dummies` saves the day: it takes the ith column from the data set and outputs a table with $n-1$ binary dummy variables; see the example below for usage.

Given multiple categories, you could even produce *interaction terms* to represent membership in multiple categories: e.g., control $\times$ male, control $\times$ female,

---

[11]If the perfectly linear form is implausible, it may be sensible to transform the input variable, to the square of x or $\sqrt{x}$.

treatment $\times$ male, treatment $\times$ female. The easiest way to do this is to simply create a column with two other columns mashed together: in SQLite, the query `select 'left' || 'right'` produces the string `leftright`; in mySQL, `concat('left', 'right')` does the same. Then break the column down into $n-1$ dummy variables as above.

The underlying claim with zero–one dummy variables is $\mathbf{y} = \mathbf{X}\boldsymbol{\beta} + \mathbf{k}$, where $\mathbf{k}$ indicates a constant value that gets added to the controls but not the cases, for example. But say that we expect the slopes to differ from cases to controls; for cases the equation is $\mathbf{y} = \mathbf{x}_1\beta_1 + \mathbf{x}_2\beta_2$ and for controls it is $\mathbf{y} = \mathbf{x}_1(\beta_1+k) + \mathbf{x}_2\beta_2$. The way to get a standard regression to produce k in this case would be to produce a data set which has the appropriate value $\mathbf{x}_1$ for each control, but zero for each case. Then the regression would be equivalent to $\mathbf{y} = \mathbf{x}_1\beta_1 + \mathbf{x}_1 k + \mathbf{x}_2\beta_2$ for controls and $\mathbf{y} = \mathbf{x}_1\beta_1 + \mathbf{x}_2\beta_2$ for cases, as desired.

```
1   #include <apop.h>
2
3   apop_model * dummies(int slope_dummies){
4       apop_data *d = apop_query_to_mixed_data("mmt", "select riders, year−1977, line \
5               from riders, lines \
6               where riders.station=lines.station");
7       apop_data *dummified = apop_data_to_dummies(d, 0, 't', 0);
8       if (slope_dummies){
9           Apop_col(d, 1, yeardata);
10          for(int i=0; i < dummified−>matrix−>size2; i ++){
11              Apop_col(dummified, i, c);
12              gsl_vector_mul(c, yeardata);
13          }
14      }
15      apop_data *regressme = apop_data_stack(d, dummified, 'c');
16      apop_model *out = apop_estimate(regressme, apop_ols);
17      apop_model_show(out);
18      return out;
19  }
20
21  #ifndef TESTING
22  int main(){
23      apop_db_open("data−metro.db");
24      printf("With constant dummies:\n"); dummies(0);
25      printf("With slope dummies:\n"); dummies(1);
26  }
27  #endif
```

Listing 8.7 Two regressions with dummy variables. Online source: `dummies.c`.

Listing 8.7 runs a simple regression of ridership at each station in the Washington Metro on a constant, the year, and a set of dummy variables for Green, Orange,

Red, and Yellow lines (meaning the Blue line is the baseline).

- The query on line four pulls two numeric columns (ridership and year) and one text column.

- Line 7 converts text column zero into an `apop_data` set consisting of dummy variables. If you ask the debugger to display the `dummified` data set, you will see that every row has a 1 in at most a single column.

- Line 15 stacks the dummies to the right of the other numeric data, at which point the data set is in the right form to be regressed, as on line 16. Again, it is worth asking the debugger to show you the data set in its final, regressable form.

- We will test the claim that these two tests are equally likely on page 354; the `#ifndef` wrapper on lines 21 and 27 will let us read this file into the testing program. Here, the wrapper is innocuous.

- The `slope_dummies` switch determines whether the dummies will be constant dummies or slope dummies. For constant dummies, we need only zeros or ones, so as above, `apop_data_to_dummies` gives us what we need. For slope dummies, we need replace every 1 with the current year, which is what the column-by-column vector multiplication achieves over lines 9–13.

PROBIT AND LOGIT Now we move on to the question of discrete outcome variables. For example, a person either buys a house or does not, which makes house purchasing a binary variable. Say that the value of a house is a linear sum of the value of its components: an extra bedroom adds so many dollars, a nearby highly-rated school adds so many more, each square meter is worth a few thousand dollars, et cetera. That is, one could write down a linear model that total value $U = \beta_1 \cdot \text{cost} + \beta_2 \cdot \text{bedrooms} + \beta_3 \cdot \text{school quality} + \beta_4 \cdot \text{square meters} + \cdots + \epsilon$, where β_1 is typically normalized to -1 and ϵ is the usual error term.[12]

To phrase the model briefly: $U = \mathbf{x}\beta + \epsilon$. But rather than leaving the model in the standard linear form, we add the rule that a person buys iff $U > 0$. Thus, the input is the same set of characteristics $\mathbf{x}$ (of consumers or of houses) and weightings β, and the output is a binary variable: the person acted or did not; the buyer consumed or did not; the geyser erupted or did not, et cetera.

From here, the details of the decision model depend on whether the outcome is binary, multivalued but unordered, or ordered.

- Logit with two choices: The first alternative, the *logit model* or *logistic model*, assumes that $\epsilon \sim$ a *Gumbel distribution*. The Gumbel is also known as the Type I

---

[12] Real estate economists call this the *hedonic pricing model*, whose implementation typically differs in some respects from the logit model discussed here. See Cropper *et al.* (1993) for a comparison.

Extreme Value distribution, meaning that it expresses the distribution of the statistic $\max(\mathbf{x})$ under appropriate conditions. McFadden (1973) won a Nobel prize partly by showing that the above assumptions, including the rather eccentric error term, reduce to the following simple form:

$$P(0|\mathbf{x}) = \frac{1}{1 + \exp(\mathbf{x}\boldsymbol{\beta}_1)}$$

$$P(1|\mathbf{x}) = \frac{\exp(\mathbf{x}\boldsymbol{\beta}_1)}{1 + \exp(\mathbf{x}\boldsymbol{\beta}_1)}$$

There is no $\boldsymbol{\beta}_0$ because is it normalized to $\mathbf{0}$, so $\exp(\mathbf{x}\boldsymbol{\beta}_0) = 1$.

- Probit: The *probit model* has a similar setup to the logit, but $\epsilon \sim \mathcal{N}(0,1)$. The model can be rephrased slightly. Say that $U(\mathbf{x})$ is deterministic, but the probability of consuming given some utility is random. To describe such a setup, let $U = -\mathbf{x}\boldsymbol{\beta}$ rather than $-\mathbf{x}\boldsymbol{\beta} + \epsilon$ (the introduction of a minus sign will be discussed in detail below), and let the probability that the agent acts be

$$P(\mathbf{x}|\boldsymbol{\beta}) = \int_{-\infty}^{U(-\mathbf{x}\boldsymbol{\beta})} \mathcal{N}(y|0,1)dy,$$

where $\mathcal{N}(y|0,1)$ indicates the Normal PDF at y with $\mu = 0$ and $\sigma = 1$, so the integral is the CDF up to $U(-\mathbf{x}\boldsymbol{\beta})$.
As you can see, this is a much more ornery form, especially since the CDF of the Normal (what engineers call the *error function*, or more affectionately, *erf*) does not simplify further and is not particularly easy to calculate. This partly explains the prevalence of the logit form in existing journal articles, but is not much of an issue now that maximum likelihood estimation is cheap, even when erf is involved.

- With k different options, the logit model generalizes neatly for the case of multiple values, to the *multinomial logit*:

$$P(0|B) = \frac{1}{1 + \sum_{i=1}^{k} \exp(\mathbf{x}\boldsymbol{\beta}_i)} \text{ and}$$

$$(8.3.1)$$

$$P(k|B) = \frac{\exp(\mathbf{x}\boldsymbol{\beta}_x)}{1 + \sum_{i=1}^{k} \exp(\mathbf{x}\boldsymbol{\beta}_i)}, k \neq 0$$

where B is the set of choices from which x is chosen, and $\boldsymbol{\beta}_k$ is a different vector of coefficients for each option $k \neq 0$. You can see that $\boldsymbol{\beta}_0$ is once again normalized to $\mathbf{0}$, and given only options zero and one, the form here reduces to the binomial form above.[13]

- Ordered multinomial probit: Presume that there is an underlying continuous variable $Y^* = -\mathbf{X}\boldsymbol{\beta}$ such that the true $Y_i = 0$ if $Y_i^* \leq 0$; $Y_i = 1$ if $0 < Y_i^* \leq A_1$; $Y_i = 2$ if $A_1 < Y_i^* \leq A_2$; and so on. That is,

---

[13]The generalization is so clean that Apophenia doesn't even include separate binary and multinomial logit models: the `apop_logit` model just counts the options and acts accordingly.

$$P(Y_i = 0) = \int_{-\infty}^{A_1 - \mathbf{X}\boldsymbol{\beta}} \mathcal{N}(y|0, 1)dy$$

$$P(Y_i = 1) = \int_{A_1 - \mathbf{X}\boldsymbol{\beta}}^{A_2 - \mathbf{X}\boldsymbol{\beta}} \mathcal{N}(y|0, 1)dy$$

$$P(Y_i = 2) = \int_{A_2 - \mathbf{X}\boldsymbol{\beta}}^{\infty} \mathcal{N}(y|0, 1)dy$$

Rather than finding a single cutoff between acting and not acting, we estimate several cutoffs, between choosing zero, one, two, …items; estimating `apop_multinomial_probit` will therefore return the usual vector of parameters β, plus a vector of $A_1, A_2, \ldots, A_{k-1}$ for k options.

Parameter overload   Very reasonable models can produce an overwhelming number of parameters. For the multinomial logit model above, with k options and n elements in each observation $\mathbf{x}$, there are $n \times (k - 1)$ parameters to be estimated. Your goals will dictate upon which parameters you should focus, and how you get meaning from them.

If the sole goal is to prove that A causes B, then the univariate test that the coefficient on A is significant is all we will look at from one of the above tests.

If we want to predict how comparable people will act in the future, then we need the model's estimate for the odds that each individual will select any one choice, and the predicted most likely selection. Perhaps you will want the full catalog, or just the average probabilities via `apop_data_summarize(`*outmodel*`.expected)`.

Or, you may want to know the dynamic shifts: what happens to B when A rises by 1%? Economists call this the *marginal change*, though virtually every field has some interest in such questions. This is closely related to the question of whether the parameter on A is statistically significant: you will see in the chapter on maximum likelihood estimates that their variance (used to measure confidence for a hypothesis tests) is literally the inverse of the second derivative (used to measure change in outcome variable given change in income variable).

For a linear model, the marginal change is trivial to calculate: we proposed that $y = \mathbf{x}\boldsymbol{\beta}$, meaning that $\partial y / \partial x_i = \beta_i$. That is, a 1% change in x_i leads to a β% change in y.

Interpreting the marginal changes in the above models can be more difficult. For the probit, the cutoff between draws that induce a choice of option zero and those that lead to option one was $-\mathbf{x}\boldsymbol{\beta}$. As $+\mathbf{x}\boldsymbol{\beta}$ grows (either because of a marginal expansion in an element of $\mathbf{x}$ or of $\boldsymbol{\beta}$), the cutoff falls slightly, and the set of draws that lead to option one grows. This fits our intuition, because we characterized the elements of the linear sum $\mathbf{x}\boldsymbol{\beta}$ as increasing the proclivity to choose option one.

For the ordered probit, there are multiple cutoffs. A marginal change in β will perhaps raise the cutoff between options two and three, lowering the odds of choosing three, but at the same time raise the cutoff between options three and four, raising the odds of choosing three. See Greene (1990, p 672) or your estimation program's documentation for more on making productive use of the estimated model.

IIA   Return to the multinomial logit formula, Equation 8.3.1, and consider the ratio of the probability of two events, E_1 and E_2.

$$\frac{P(E_1|x, B)}{P(E_2|x, B)} = \frac{\frac{\exp(\mathbf{x}\boldsymbol{\beta}_1)}{\sum_{y \in B} \exp(\mathbf{x}\boldsymbol{\beta}_y)}}{\frac{\exp(\mathbf{x}\boldsymbol{\beta}_2)}{\sum_{y \in B} \exp(\mathbf{x}\boldsymbol{\beta}_y)}}$$

$$= \frac{\exp(\mathbf{x}\boldsymbol{\beta}_1)}{\exp(\mathbf{x}\boldsymbol{\beta}_2)}$$

The key point in this simple result is that this ratio does not depend on what else is in the set of options. The term for this property is *independence of irrelevant alternatives*, or *IIA*. The term derives from the claim that the choice between one option and another in no way depends upon the other options elsewhere.

The standard example is commuters choosing among a red bus, a blue bus, and a train, where the red bus and blue bus are identical except for their paint job. We expect that the ratio of P(red)/P(blue) = 1, and let P(red)/P(train) = P(blue)/P(train) = $\frac{1}{x}$. If the train were out of service, then P(red)/P(blue) would still be 1, because as many former train riders would now pick the blue bus as would pick the red, but if the blue bus were knocked out of commission, then everybody who was riding it would likely just switch to the red bus, so without the blue bus, P(red)/P(train) = $\frac{2}{x}$. In short, the train option is irrelevant to the choice of red or blue bus, but the blue bus option is not irrelevant in the choice between train and red bus.

In the code supplement, you will find `data-election.db`, which includes a small selection of survey data regarding the United States's 2000 election, from National Election Studies (2000). The survey asked respondents their opinion of various candidates (including the two candidates for Vice President) on a scale from 0 to 100; in the database, you will find the person that the respondent most preferred.[14]

---

[14]*Duverger's law* tells us that a first-past-the-post system such as that of the United States tends to lead to a stable two-party system, due to the various strategic considerations that go into a vote. We see this in the data: question 793 of the survey (not included in the online supplement) asked the respondent for whom he or she expects to vote, and the totals were: (Gore, 704), (Bush, 604), (everybody else, 116). The favorite from the thermometer score avoids the strategic issues involved in vote choice, and produces a much broader range of preferences: (Gore, 711), (Bush, 524), (Nader, 157), (Cheney, 68), et cetera. ℚ: Write a query to get the complete count for each candidate from the survey data.

The IIA condition dictates that the logit estimates of the relative odds of choosing candidates won't change as other candidates enter and exit, though intuitively, we would expect some sort of shift. The examples also suggest a two-level model, wherein people first choose a category (bus/train, Republican/Democrat/Green), and then having chosen a category, select a single item (such as red bus/blue bus, Buchanan/McCain). Such a model weakens IIA, because a choice's entry or exit may affect the choice of category. These hierarchies of class and item are a type of *multilevel model*; the next section will give an overview of multilevel modeling possibilities, including some further discussion of the nested logit.

The theoretical IIA property is irrefutable—there is not much room for error in the two lines of algebra above. But many theoretical models can be pushed to produce mathematically clean results that are just not relevant in the application of the model to real data. Does IIA have a strong influence on real-world logit estimates?

We would test IIA by running one unconstrained logit, and another logit estimation restricted so that one option is missing; this fits the likelihood ratio (LR) form which will appear on page 351, and a few variants on this test exist in the literature. Cheng & Long (2007) built a number of artificial data sets, some designed around agents who make choices conforming to IIA, and some designed to not have the IIA property. They found that the LR tests were ineffective: the unconstrained and constrained odds ratios sometimes did not differ (demonstrating IIA) and sometimes did (against IIA), but there was no sufficiently reliable relationship between when the underlying data had the IIA property and when the parameter estimates demonstrated IIA. Fry & Harris (1996) used a less broad method that had slightly more optimistic results, but still encountered glitches, such as problems with power and results that were sensitive to which option was removed.

The probit model matches the logit save for the type of bell curve that describes the error term, so one expects the parameters and predicted odds to be similar. In fact, Amemiya (1981) points out that logit parameters typically match probit parameters, but that they are scaled by a factor of about 1.6. That is, for each option i, $\beta_i^L \approx 1.6\beta_i^P$. Yet, the logit parameters have the IIA property, while the probit parameters do not. This is hard to reconcile, unless we accept that the IIA property of the logit is theoretically absolute but empirically weak.

➤ For discrete explanatory variables, you can use the standard family of OLS models, by adding dummy variables.

➤ The standard linear form $\mathbf{X}\beta$ can be plugged in to a parent function, like the Probit's comparison of $\mathbf{X}\beta$ to a Normal distribution or the logit's comparison to a Gumbel distribution, to generate models of discrete choice.   ⋙

➤ The Logit model has the property that the ratio of the odds of selecting two choices, e.g., $p(A)/p(B)$, does not depend on what other choices exist.

➤ The computational pleasantries that OLS demonstrates are no longer applicable, and we usually need to do a maximum likelihood search to find the best parameters.

8.4 MULTILEVEL MODELING   Retake the snowflake problem from page 236: the models to this point all assumed that each observation is independently and identically distributed relative to the others, but this is frequently not the case.

One way in which this may break is if observations fall into clusters, such as families, classrooms, or geographical region. A regression that simply includes a family/classroom/region dummy variable asserts that each observation is iid relative to the others, but its outcome rises or falls by a fixed amount depending on its group membership. But the distribution of errors for one family may be very different from that of another family, the unobserved characteristics of one classroom of students is likely to be different from the unobserved characteristics of another, and so on.

A better alternative may be to do a model estimation for each group separately. At the level of the subgroup, unmodeled conditions are more likely to be constant for all group members. Then, once each group has been estimated, the parameters can be used to fit a model where the observational unit is the group.

The other type of multilevel model is that where there are no distinct subclusters, but we would like to give more detail about the derivation of the parameters. For example, say that we feel that the propensity to consume a good is Normally distributed, but we know that the likelihood of consumption is also based on a number of factors. We just saw this model—it is the probit, which asserted that the likelihood of consuming is simply the CDF of a Normal distribution up to a parameter, and that parameter has the form $\mathbf{X}\beta$. The logit asserted that the probability of consuming was $\exp(\theta)/(1 + \exp(\theta))$, where θ is of the form $\mathbf{X}\beta$.

These are examples of *multilevel models*. There is a parent model, which is typically the primary interest, but instead of estimating it with data, we estimate it using the results from subsidiary models. To this point, we have broadly seen two types of model: simple distributions, like the Normal(μ, σ) or the Poisson(λ), and models asserting a more extensive causal structure, like the OLS family. Either form could be the parent in a multilevel model, and either form could be the child.

Clustering   A classroom model may assume a distribution of outcomes for each
classroom and estimate μ_i and σ_i for each class, and then assert a linear
form that the outcome variable is $\beta_1\mu_i + \beta_2\sigma_i$. Unemployment models are typically
modeled as Poisson processes, so one could estimate λ_i for each region i, but then
link those estimates together to assert that the λ's have a Normal distribution.

```
1    #include <apop.h>
2
3    void with_means(){
4        apop_data *d2 = apop_query_to_data("select avg(riders), year−1977 \
5                from riders, lines \
6                where riders.station=lines.station\
7                group by lines.line, year");
8        apop_model_show(apop_estimate(d2, apop_ols));
9    }
10
11   void by_lines(){
12     apop_data *lines = apop_query_to_text("select distinct line from lines");
13     int linecount = lines−>textsize[0];
14     apop_data *parameters = apop_data_alloc(0, linecount, 2);
15       for(int i=0; i < linecount; i ++){
16           char *color = lines−>text[i][0];
17           apop_data *d = apop_query_to_data("select riders, year−1977 from riders, lines\n\
18                where riders.station = lines.station and lines.line = '%s'", color);
19           apop_model *m = apop_estimate(d, apop_ols);
20           APOP_ROW(parameters, i, r);
21           gsl_vector_memcpy(r, m−>parameters−>vector);
22       }
23     apop_data_show(parameters);
24     apop_data *s = apop_data_summarize(parameters);
25     apop_data_show(s);
26   }
27
28   int main(){
29       apop_db_open("data−metro.db");
30       printf("Regression parent, Normal child:\n\n"); with_means();
31       printf("Normal parent, regression child:\n\n"); by_lines();
32   }
```

Listing 8.8 Two model estimations: with a regression parent and with a Normal parent. Online
source: `ridership.c`.

Listing 8.8 estimates two models of the change in ridersip of the Washington Metro
over time. You already saw two models of this data set, in the form of two dummy-
variable regressions, on page 282. Here, we take a multilevel approach: first with
a regression parent and distribution child, and then the other way around.

Having SQL on hand pays off immensely when it comes to clustering, because it is so easy to select data where its group label has a certain value, or calculate aggregates using a group by clause. It will not be apparent in this example using just under 2,000 data points, but because SQL tables can be indexed, grouping and aggregation can be much faster than sifting through a table line by line.

The first model is a regression parent, distribution child: we estimate a Normal model for each Metro line for each year, and then regress on the set of $\hat{\mu}$'s thus estimated. Of course, finding $\hat{\mu}$ is trivial—it's the mean, which even standard SQL can calculate, so the output from the query on line 4 is already a set of statistics from a set of Normal distribution estimations. Then, the regression estimates the parameters of a regression on those statistics.

The second model is a distribution parent, regression child: we run a separate regression for every Metro line and then find the mean and variance of the OLS parameters. Line 12 queries a list of the five colors (Blue line, Red line, ...), and then lines 15–22 are a for loop that runs a separate regression for each color, and writes the results in the parameters matrix. Then, line 24 finds $\hat{\mu}$ and $\hat{\sigma}$ for the set of parameters.

The estimations tell different stories, and produce different estimates for the slope of ridership with time. Notably, the Green line added some relatively unused stations in the 1990s, which means that the slope of the Green line's ridership with time is very different from that of the other lines. This is very clear in the second case where we run a separate regression for every line, but is drowned out when the data set includes every line.

Notice also that the size of the parent model's data set changes with different specifications: in the first model, it was 150; in the second, it was 5. Thus, the first model had greater confidence that the slope was different from zero.[15] Gelman & Hill (2007) point out that we test parameters only on the parent model, which means that if n is very small for some of the clusters, then this should have no effect on the parent—even $n = 1$ for some clusters is not a problem. Clusters with small n should show large natural variation in child parameter estimates, and that would be reflected in the quality of the estimation of the parent parameters. But since we neither make nor test claims about the child parameters, there is no need to concern ourselves directly with the 'quality' of the child parameters.

---

[15] In the second specification, with five data points, one has a negative slope and four a positive slope. Nonetheless, the mean is still about three times $\hat{\sigma}/\sqrt{n}$, giving us about 99% confidence that the mean is significant.

To get a better handle on what differs among the lines and within the overall regression, draw some graphs:

- total ridership per year

- average ridership per year

- total ridership for each line.

 8.4

- average ridership for each line.

The difference between the total and average ridership is based on the fact that the number of stations is not constant over time—produce a crosstab of the data to see the progression of new station additions.

The plots for each line are best written via a `for` loop based on the one beginning on line fifteen of Listing 8.8. How do the graphs advise the statistics calculated by the two dummy and two multilevel models?

An example: multi-level logit   We can break down the process of choosing a presidential candidate into two steps: first, the voter decides whether to choose a Democrat, Republican, or Green candidate, and then chooses among the offered candidates in the chosen party. The probability of selecting a candidate is thus $P(\text{candidate}) = P(\text{candidate|party})P(\text{party})$. We would thus need to do a few logit estimations: one for the Democratic candidates, one for the Republican candidates, and one for the party choice.[16] The IIA property would continue to hold within one party, but among parties, the change in candidates could shift $P(\text{party})$, so the proportional odds of choosing a given Democrat versus choosing a given Republican may shift. Listing 8.9 does some of these logit estmations along the way to the multilevel model.

- Once again, SQL saves the day in doing these multilevel estimations. In this case, the explanatory variables won't change from model to model, so they are fixed in the `logit` function, but the outcome variable will change. Thus, the function takes in the varying tail of the query and appends it to the fixed list of explanatory variables to pull out consistent data.

- The tails of the queries, in `main`, are straightforward: get all candidates, get all Democrats, get all Republicans, get only party names.

- Out of the overwhelming amount of data, the `logit` function displays on screen only the mean odds of selecting each option.

---

[16]There was only one Green candidate, Ralph Nader, so $P(\text{Nader}) = P(\text{Green})$.

```
1    #include <apop.h>
2
3    apop_model * logit(char *querytail){
4        apop_data *d = apop_query_to_mixed_data("mmmmmt", "select 1, age, gender, \
5                illegal_immigrants, aid_to_blacks, %s", querytail);
6        apop_data_text_to_factors(d, 0, 0);
7        apop_model *m = apop_estimate(d, apop_logit);
8        apop_data *ev = apop_expected_value(d, m);
9        apop_data_show(apop_data_summarize(ev));
10       return m;
11   }
12
13   int main(){
14       apop_db_open("data−nes.db");
15       logit("favorite from choices");
16       logit("favorite from choices c, party p where p.person = c.favorite and p.party = 'R'");
17       logit("favorite from choices c, party p where p.person = c.favorite and p.party = 'D'");
18       logit("p.party from choices c, party p where p.person = c.favorite");
19   }
```

Listing 8.9 Logit models with and without grouping of candidates by party. Online source: `candidates.c`.

Q_{8.5}

> The example is not quite complete: we have the probability of choosing a candidate given a party and the probability of choosing a party, but not the product of the two. Fill in the remainder of the code by finding the predicted odds for each candidate for each observation.
> You can count how often a person chose the candidate that the model says the person is most likely to pick. Did the multilevel logit do a better job of picking candidates than the standard one-level logit?

The simple concept of nesting together models is akin to McFadden's nested logit (McFadden, 1978). In fact, if these aren't enough possibilities for you, see Gelman & Hill (2007), who offer several more.

Describing snowflakes   Now consider the multilevel model as used to model parameters that are used as inputs to other models. This is the typical form for when we want to use one of the stories from Chapter 7, like the one about the Negative Binomial model or the one about the Poisson, but we want the parameter to vary according to per-observation conditions. As above, the probit model from the last section fits this description, as each agent has a cutoff based on a linear model and that agent's characteristics (i.e., the cutoff for person i is $x_i\beta$), and that cutoff is then fed into a parent Normal distribution.

Because many of Apophenia's model-handling functions can work with a model
that has only a log likelihood specified, the process of writing a log likelihood for
such a model is supremely simple:

- Find the linear estimates of the parameters for each observation, probably using
  `apop_dot` on the input matrix and input parameters.
- Use the stock `log_likelihood` function to find the log likelihood of the outcome
  data given the parameter estimates from the last step.

Listing 8.10 implements such a process. The first step is building the model, which
basically consists of reusing existing building blocks. The log likelihood function
is simply the two steps above: find $\mathbf{X}\beta$, then run a `for` loop to estimate a separate
Poisson model for each row, based on the ith outcome variable and the ith Poisson
parameter. The `main` program declares the model, pulls the data, and then runs the
data through both the new model and the standard OLS model.

```
#include <apop.h>

double pplc_ll(apop_data *d, apop_model *child_params){
    apop_data *lambdas = apop_dot(d, child_params->parameters, 0);
    apop_data *smallset = apop_data_alloc(0, 1, 1);
    double ll = 0;
    for(size_t i=0; i < d->vector->size; i ++){
        double lambda = apop_data_get(lambdas, i, -1);
        apop_model *pp = apop_model_set_parameters(apop_poisson, lambda);
        apop_data_set(smallset, 0,0, apop_data_get(d, i, -1));
        ll += pp->log_likelihood(smallset, pp);
    }
    return ll;
}

int main(){
    apop_model pplc = {"Poisson parent, linear child", -1,
                        .log_likelihood= pplc_ll, .prep=apop_probit.prep};
    apop_db_open("data-tattoo.db");
    apop_data *d = apop_query_to_data("select \
                tattoos.'ct tattoos ever had' ct, tattoos.'year of birth' yr, \
                tattoos.'number of days drank alcohol in last 30 days' booze \
                from tattoos \
                where yr+0.0 < 97 and ct+0.0 < 10 and booze notnull");
    apop_data *d2 = apop_data_copy(d);
    apop_model_show(apop_estimate(d, pplc));
    apop_model_show(apop_estimate(d2, apop_ols));
}
```

Listing 8.10 A multilevel model. Online source: `probitlevels.c`.

In this case, you can interpret the parameters in a similar manner to the discussion of the probit and logit cases above. The parent parameter λ is calculated as $\mathbf{X}\beta$, so a 1% shift in x_i leads to a $\beta_i\%$ shift in λ. Thus, after checking whether the parameters are significantly different from zero, you can directly compare the relative magnitudes of the parameters to see the relative effect of a 1% shift in the various inputs.

Statistics is a subjective field   At this point, you have seen quite a range of models: you saw the distribution models from Chapter 7, the linear models from earlier in this chapter, and here you can see that you can embed any one model into any other to form an infinite number of richer models. Plus, there are the simulation and agent-based models, standing alone or nested with a different type of model. The Metro data has already been modeled at least four different ways (depending on how you count), and the male-to-female ratio among counties in still more ways. Which model to choose?

A model is a subjective description of a situation, and many situations afford multiple perspectives. This is rare advice from a statistics textbook, but *be creative.* We're not living in the 1970s anymore, and we have the tools to tailor the model to our perceptions of the world, instead of forcing our beliefs to fit a computationally simple model. Try as many models as seem reasonable. Having many different perspectives on the same situation only raises the odds that you will come away with a true and correct understanding of the situation.

We also have some more objective tools at our disposal for selecting a model: Chapter 10 will demonstrate a means of testing which of two disparate models has a better fit to the data. For example, running the code in Listing 8.10 here reveals that the likelihood of the alternative model is higher than the log likelihood of the OLS model. You can use the test on page 353 to test whether the difference in likelihood is truly significant.

$\sum$

➤ We can create models where the data used to estimate the parameters in one model is generated by estimation in another model.

➤ One common use is clustering: develop a model for individual groups (states, classrooms, families) and then discuss patterns among a set of groups (the country, school, or community).

➤ We can also use multiple levels to solve the snowflake problem, specifying a different probability of an event for every observation, but still retaining a simple basic model of events. For example, the probit is based on plugging the output from a standard linear form into a Normal CDF.

HYPOTHESIS TESTING WITH THE CLT

I'm looking for the patterns in static: They start to make sense
the longer I'm at it.

—Gibbard (2003)

The purpose of descriptive statistics is to say something about the data you have. The purpose of hypothesis testing is to say something about the data you don't have.

Say that you took a few samples from a population, maybe the height of several individuals, and the mean of your sample measurements is $\hat{\mu} = 175$ cm. If you did your sums right, then this is an indisputable, certain fact. But what is the mean height of the population from which you drew your data set? To guess at the answer to this question, you need to make some assumptions about how your data set relates to the population from which it was drawn.

Statisticians have followed a number of threads to say something about data they don't have. Each starts with a data set and some assumptions about the environment and data generation method, and concludes with an output distribution that can be compared to the data. Here is a list of some common assumptions. It is impossible for it to be comprehensive, and many of the categories overlap, but it offers a reasonable lay of the discussion in the following chapters.

- *Classical methods*: Claim that the data was produced via a process that allows application of the Central Limit Theorem.

- *Maximum likelihood estimation*: Write down a likelihood function for any given data/parameter pairing, and find the most likely parameter given the data.

- *Bayesian analysis*: Claim a distribution expressing prior beliefs about the parameters and a likelihood function for the data on hand, then combine them to produce a posterior distribution for the parameters.

- *Resampling methods*: Claim that random draws from the data are comparable to random draws from the population (the *bootstrap principle*), then generate a distribution via random draws from the data.

- *Kernel/smoothing methods*: Claim that the histogram of the existing data is a lumpy version of the true distribution; smooth the data to produce the output distribution.

All of these approaches will be discussed over the remainder of this book. This chapter will focus on the first: making inferences about the population via use of the Central Limit Theorem (*CLT*). The CLT describes the distribution of the sample mean, $\bar{x}$, and works regardless of the form of the underlying data. That is, no matter the true distribution of the data, the distribution of the sample mean has a very specific form—as long as n is large enough. For relatively small n, another of the above methods of inference, such as the Monte Carlo methods discussed in Chapter 11, may be preferable.

Metadata

Metadata is data about data. Any statistic is a function of data, so it is by definition metadata. Be careful not to confuse the characteristics of the data and metadata; for example, the variance of the mean is almost always smaller than the variance of the base data. Like many hypothesis tests, the Central Limit Theorem is primarily concerned not with the distribution of the base data set, but the distribution of the mean of the data set.

The CLT gives us a basis for the Normal distribution; we can then produce variants based on the Normal. The square of a Normally distributed variable x will have a Chi squared distribution (which is written as $x^2 \sim \chi_1^2$, and read as: the statistic is distributed as a Chi squared with one degree of freedom). Dividing a Normal distribution by a transformed χ^2 distribution produces another distribution (the t distribution), and the ratio of two χ^2's produces an F distribution. Because all of this is rooted in the CLT, the statements are true regardless of the vagaries of the underlying population that the statistics describe.

Having found a means of describing the distribution of the unobservable population parameter β, Section 9.3 will then look a number of simple tests regarding β. They are direct applications of the above distributions, and so are often given names like the t test, χ^2 test, and F test.

The remainder of the chapter applies these building blocks in more complex structures to test hypotheses about more elaborate statistics. For example, if two independent statistics β_1 and β_2 are $\sim \chi_1^2$, then $\beta_1 + \beta_2 \sim \chi_2^2$. So if the squared distance between a histogram segment and a hypothesized distribution is $\sim \chi_1^2$, then the total distance between a thousand such segments and the hypothesized distribution is $\sim \chi_{1000}^2$, and that total distance could thus be used to test the aggregate claim that the data histogram is close to the distribution.

9.1 THE CENTRAL LIMIT THEOREM

The CLT is the key piece of magic for this chapter. Make a series of n *independent, identically distributed* draws, $x_1, x_2, \ldots x_n$, from a fixed underlying population. The underlying population may have *any* nondegenerate distribution.[1] Let the mean of this sequence of draws be $\bar{x}$, and the true mean of the overall population be μ. Then as $n \to \infty$,

$$\sqrt{n}\frac{(\bar{x} - \mu)}{\sigma} \sim \mathcal{N}(0, 1). \tag{9.1.1}$$

That is, no matter the underlying population, the distribution of a mean of a series of draws will approach a Normal distribution.

Put another way, if we repeated the procedure and drew k independent data sets from the population and plotted a histogram of $\bar{x}_1, \ldots, \bar{x}_k$, we would eventually see a familiar bell curve.

Because it is about the distribution of $\bar{x}$, the CLT embodies a two-stage procedure: we first produce the means of a series of k sets of draws—metadata from the base distribution—and then seek the distribution of the metadata (the k means), not the data itself. Listing 9.1 demonstrates exactly how this two-stage procedure works, and is worth understanding in detail.

- On line four (and the first panel of Figure 9.2), you can see the data from which the program makes draws. It is nowhere near a bell curve: everything is either ≤ 11 or ≥ 90.

- The inner loop of the `make_draws` function (the j-indexed loop) takes ct draws from the CDF, and adds them to `total`. When `total` is divided by ct in the line after the loop, it becomes the mean of ct draws from the distribution. The outer loop (the i-indexed loop) records drawct such means. Line 23 plots the distribution of this set of several means.

  This double-loop is the base of the CLT, and is reflected in the assumptions about data sets below. Say that we have drawct data points in our data set, and we are

---

[1] By 'nondegenerate' I mean that more than one outcome has positive probability. If your data is fifty items that all have value 10, the mean of your samples will be 10 no matter how you slice them. However, if your sample takes as few as two distinct values, the CLT will eventually provide you with a bell curve. You can verify this by modifying the code in Listing 9.1. There are also theoretical distributions with infinite variance, which also cause problems for the CLT, but this is of course not an issue for finite data sets.

```
1    #include <apop.h>
2
3    int drawct = 10000;
4    double data[] = {1, 2, 3, 10, 11, 10, 11, 90, 91, 90, 91};
5
6    gsl_vector *make_draws(int ct, gsl_rng *r){
7        double total;
8        gsl_vector *out = gsl_vector_alloc(drawct);
9          for(int i=0; i< drawct; i++){
10             total = 0;
11             for(int j=0; j< ct; j++)
12                 total += data[gsl_rng_uniform_int(r, sizeof(data)/sizeof(data[0]))];
13             gsl_vector_set(out, i, total/ct);
14         }
15         return out;
16   }
17
18   int main(){
19       gsl_rng *r = apop_rng_alloc(23);
20         for (int ct=1; ct<= 1018; ct+=3){
21             printf("set title 'Mean of %i draws'\n", ct);
22             gsl_vector *o =make_draws(ct, r);
23             apop_plot_histogram(o, 200, NULL);
24             gsl_vector_free(o);
25             printf("pause 0.6\n");
26         }
27   }
```

Listing 9.1 Take the mean of an increasing number of draws. The distribution of the means approaches a Normal distribution. Online source: `cltdemo.c`.

claiming that they are Normally distributed. We presume that they are Normally distributed (and not just constant) because a multitude of events have affected each data point in some sort of haphazard way. That is, each individual data point went through a process like the inner loop in lines 11–12, absorbing a large number of random shocks. After all the little shocks, we gathered a single data point, as in line 13.

- The `main` function is intended to show that the CLT works best when each data point is the mean of several draws from the base distribution. Line 22 repeatedly calls `make_draws`. At the first call, `ct==1`, so `make_draws` makes 10,000 draws from the distribution itself. The next call produces 10,000 data points where each is the mean of four draws, and so on up to each data point being the mean of 1018 draws. The program dumps plots of the histograms to STDOUT, so run the program via `./cltdemo | gnuplot`.

Figure 9.2 shows a few frames of output from the program. The first frame of the animation is simply a set of spikes representing the base data. The second frame, where each data point is the mean of four draws, has a series of humps, because some draws have all large numbers, some have three large numbers and one small, some have two of each, and so on. In the third frame, there are more combinations possible, so there are more humps. As the frames progress, the humps merge together to form a familiar bell curve. This is a re-telling of the counting story on page 237, which explained why the Binomial distribution approaches a bell curve as well.

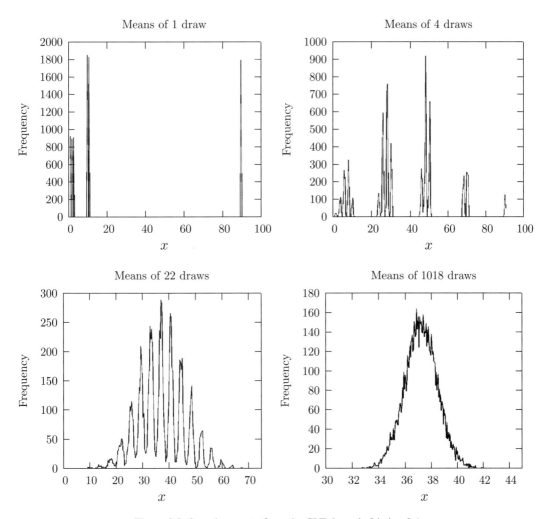

Figure 9.2 Sample outputs from the CLT demo in Listing 9.1

Finally, notice the x-axes of the snapshots: the original data was plotted from 0–100, but the scale in the fourth frame only goes from 30 to 45.[2] So not only does the distribution of $\bar{x}$ approach a bell curve, but it approaches a rather narrow bell curve.

Modify line 4 to try different base distributions from which the system will draw. [Thanks to the creative use of `sizeof` on line 12, you don't need to specify the size of the array. But see the footnote on page 125 on why this could be bad form.] Deleting the elements $\{1, 2, 3\}$ produces some especially interesting patterns. What sort of data sets lead quickly to a bell curve, and what data sets require averaging together thousands of elements before achieving a decent approximation to the Normal?

Equation 9.1.1 put the story more formally: if we have a data set x with n elements, true mean μ, and variance σ^2, then as $n \to \infty$, $(\bar{x} - \mu) / \frac{\sigma}{\sqrt{n}}$ approaches a $\mathcal{N}(0, 1)$ distribution. From that regularity of nature, we can derive all of the distributions to follow.

Variants on variance   There is often confusion about when to use σ^2, σ, or $\sigma/\sqrt{n}$, so it is worth a quick review.

- For any data set or distribution, the variance is notated as σ^2. The formula for the variance of a data set is $\sum_{i=1}^{n}(x_i - \mu)^2/n$, so it makes sense that its symbol would have a square included.

- For the Normal distribution, the square root of the variance is known as the standard deviation, σ, and is used to describe the 'width' of the distribution. For example, just over 95% of a Normal distribution is within plus or minus two standard deviations of the mean. Outside of the Normal distribution, σ is rarely used.

- Let us say that we have a *data set* x whose variance is σ^2. Then the variance of the *mean of* x, $\text{var}(\bar{x})$, is σ^2/n, and the standard deviation of $\bar{x} = \sigma/\sqrt{n}$. Once again, it is important to bear in mind whether you are dealing with data or metadata.

➤ The Central Limit Theorem states that if each observation $\bar{x}_i$ is the mean of some draws from an iid distribution, then as $n \to \infty$, the distribution of $\bar{x}_i$ follows Equation 9.1.1.

➤ That is, if μ is the overall mean and σ is the square root of the variance of the set of $\bar{x}_i$'s, then $(\bar{x}_i - \mu)/(\sigma/\sqrt{n})$ approaches a $\mathcal{N}(0, 1)$ distribution.

---

[2]If the shift in x-axis bothers you, you could ask Gnuplot to hold a constant scale by adding a line like `printf("set xrange [0:100]\n");` at the top of `main`.

9.2   MEET THE GAUSSIAN FAMILY   With the exception of the Normal, the distributions below are distinct from the distributions of Section 7.2. The distributions there are typically used to describe data that we have observed in the real world. The distributions here are aimed at describing metadata, such as the means and variances of model parameters.

NORMAL   The Normal distribution, presented on page 241, will also be used to describe some of the parameters derived below. The big problem with the Normal distribution is that it depends on σ, an unknown. It also depends on μ, but we are frequently testing a claim that μ has some fixed value, so we assume rather than derive it. Thus, much of the trickery in this section involves combining distributions in ways such that the unknown σ's cancel out.

χ^2 DISTRIBUTION   The square of a variable with distribution $\mathcal{N}(0,1)$ has a χ^2 distribution with one degree of freedom, and the sum of n independent χ^2-distributed variables is also $\sim \chi^2$, with n degrees of freedom. Figure 9.3 shows the distribution for a few different degrees of freedom.

- If $X \sim \mathcal{N}(0,1)$, then $X^2 \sim \chi_1^2$.
- If $X_i \sim \mathcal{N}(0,1)$ for $i = 1, \ldots, n$, then $X_1^2 + \cdots + X_n^2 \sim \chi_n^2$.
- If $X_i \sim \chi_n^2$, then $E(X_i) = n$.

The summation property is immensely useful, because we often have sums of variables to contend with. The most common case is the sample variance, which is a sum of squares. Being a sum of squares of Normally-distributed elements, it is easy to show that (Snedecor & Cochran, 1976, p 74)

$$\frac{\left[\sum_n (x - \bar{x})^2\right]}{\sigma^2} \sim \chi_{n-1}^2. \tag{9.2.1}$$

The numerator is the estimate of the sample variance times $n - 1$, so we can use this to test that the sample variance equals a given σ^2, or establish a confidence interval around an estimate of the variance. But we will see that it is useful for much more than just describing variance estimates.

The sample variance is $\sim \chi_{n-1}^2$, not χ_n^2, because given the first $n - 1$ data points and the mean, the last one can actually be calculated from that data, meaning that we effectively have the sum of $n - 1$ variables $\sim \chi_1^2$, plus a no longer stochastic constant. For more on degrees of freedom, see the sidebar on page 222.

It is worth mentioning the origin of the χ^2 distribution as a common form. Pearson (1900) did a Taylor expansion of errors from what we now call a Multinomial

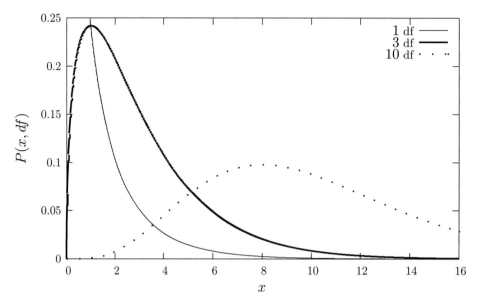

Figure 9.3 The χ^2 distribution for 3, 5, and 10 degrees of freedom. Being a sum of squares, it is always greater than zero. As more elements are added in to the sum, the distribution becomes less lopsided, and approaches a Normal distribution.

distribution, of the form $k_1\chi + k_2\chi^2 + k_3\chi^3 + \cdots$, where the k_i's are appropriate constants. He found that one can get satisfactory precision using only the χ^2 term of the series. The ANOVA family of tests is based on this approximation, because those tests claim that the data are random draws that fit the story of the Multinomial distribution (as on page 240), so a sum of such distributions leads to a χ^2 distribution as well.

STUDENT'S t DISTRIBUTION   Let $\mathbf{x}$ be a vector of data (such as the error terms in a regression). Then

$$\frac{\overline{\mathbf{x}} - \mu}{\hat{\sigma}/\sqrt{n}} \sim t_{n-1},$$

where $\hat{\sigma} = \mathbf{x}'\mathbf{x}/n$ (a scalar). It might look as though this is just an approximation of the Normal, with $\hat{\sigma}$ replacing σ, but it is not. To see where the form of the t distribution came from, consider dividing the CLT equation (Equation 9.1.1),

$$\frac{\overline{\mathbf{x}} - \mu}{\frac{\sigma}{\sqrt{n}}} \sim \mathcal{N}(0,1),$$

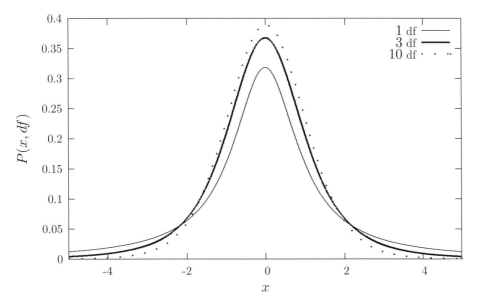

Figure 9.4 The t distribution for 1, 3, and 10 degrees of freedom. For one degree of freedom, the distribution has especially heavy tails—the variance is infinite—but as df grows, the distribution approaches a $\mathcal{N}(0,1)$.

by

$$\sqrt{\frac{\hat{\sigma}^2}{\sigma^2}} \sim \sqrt{\frac{\chi^2_{n-1}}{n-1}}.$$

Then

$$\left(\frac{\overline{\mathbf{x}} - \mu}{\frac{\sigma}{\sqrt{n}}} \middle/ \sqrt{\frac{\hat{\sigma}^2}{\sigma^2}} \right) = \frac{\overline{\mathbf{x}} - \mu}{\hat{\sigma}/\sqrt{n}}.$$

The key stumbling block, the unknown value of σ, cancels out from the numerator and denominator. This is a work of genius by Mr. Student, because he could calculate the exact shape of the distribution through straightforward manipulation of the Normal and χ^2 tables.[3] Some t distributions for various degrees of freedom are pictured in Figure 9.4.

- The t_1 distribution (i.e., $n = 2$) is called a Cauchy distribution.
- As $n \to \infty$, the t_n distribution approaches the $\mathcal{N}(0,1)$ distribution.

---

[3] *Student* is actually Mr. William Sealy Gosset, who published the t distribution in 1908 based on his work as an employee of the Guinness Brewery.

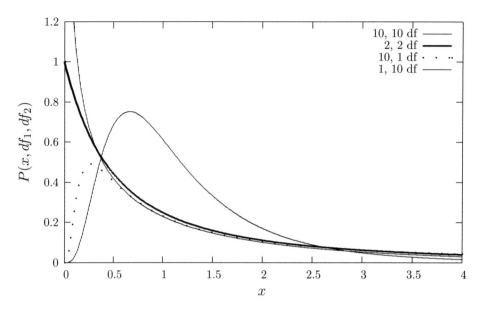

Figure 9.5 The F distribution for various pairs of numerator/denominator degrees of freedom.

F **DISTRIBUTION**   Instead of a ratio of an $\mathcal{N}$ and a $\sqrt{\chi^2}$, you could also take the ratio of two χ^2-distributed variables. The σ's in both denominators would again cancel out, leaving a distribution that could be calculated from the χ^2 tables. This is the derivation and definition of the F distribution:

$$[\chi_m^2/m]/[\chi_n^2/n] \sim F(m, n).$$

See Figure 9.5 for some sample F distributions.

Also, consider the square of a t distribution. The numerator of a t_n distribution is a Normal distribution, so its square is a χ_1^2; the denominator is the square root of a χ_n^2 distributed variable, so its square is a χ_n^2. Thus, the square of a t_n-distributed variable has an $F_{1,n}$ distribution as well.

The F distribution allows us to construct tests comparing one χ^2-distributed variable in the numerator to another χ^2-distributed variable in the denominator, and either of these χ^2 variables could be the sum of an arbitrary number of elements. We can thus use the F distribution to construct comparisons among relatively complex statistics.

LOOKUP TABLES   There are three things that cover most of what you will be doing with a distribution: finding values of its PDF, values of its CDF, and occasionally values of its inverse CDF. For a single quick number, see the command-line program `apop_lookup`. For getting the the value of the PDFs at a given point from inside a program, here are the headers for the GSL's PDF lookup functions:

```
double gsl_ran_gaussian_pdf (double x, double sigma);
double gsl_ran_tdist_pdf (double x, double df);
double gsl_ran_chisq_pdf (double x, double df);
double gsl_ran_fdist_pdf (double x, double df1, double df2);
```

- The prefix `gsl_ran` indicates that these functions are from the random number generation module (`#include <gsl/gsl_randist.h>`). Random number generation itself will be delayed to Chapter 11.
- The mean for the Normal function is fixed at zero, so modify X accordingly, e.g., if X is drawn from a $\mathcal{N}(7, 1)$, then ask the GSL for `gsl_ran_gaussian_pdf(X-7, 1)`.

The next most common distribution calculation found in tables in the back of statistics texts is calculating the CDF above or below a point. The P-functions below calculate the CDF below a point, i.e. $\int_{-\infty}^{X} f(y)dy$, while the Q-functions calculate the CDF above a point, i.e. $\int_{X}^{\infty} f(y)dy$. These sum to one, so you can express any area in terms of whichever function is clearest.

Here is the list of functions:

```
double gsl_cdf_gaussian_P (double x, double sigma);
double gsl_cdf_tdist_P (double x, double df);
double gsl_cdf_chisq_P (double x, double df);
double gsl_cdf_fdist_P (double x, double df1, double df2);
```

... plus all of these with the P replaced by a Q.

These will be used to test *hypotheses*, which in this context are claims like $\mu > 0$. If you are shaky with hypothesis tests, see the next section. But if you are well-versed with the traditional notation for hypothesis tests, notice that the overuse of the letter P can easily lead to confusion. The `gsl_cdf_gaussian_P` function gives what is known as the p-value for the one-tailed test that the mean is less than zero, and the `gsl_cdf_gaussian_Q` function gives the infrequently-used q-value for the same hypothesis. Put another way, if we find that the mean of our Normally distributed data is 2.2 standard deviations below zero, then we reject the one-tailed hypothesis that the mean is less than or equal to zero with confidence `1 - gsl_cdf_gaussian_P(2.2, 1)` == `gsl_cdf_gaussian_Q(2.2, 1)`.

For a hypothesis that $\mu > 0$, everything is reversed. Here is a table that strives to clarify which function goes with the confidence with which the null is rejected and which goes with the p-value, and when:

	$H_0 : \mu > 0$	$H_0 : \mu < 0$
confidence	gsl_...._P	gsl_...._Q
p-value	gsl_...._Q	gsl_...._P

For a centered two-tailed test, the p-value takes the form

```
2 * GSL_MIN(gsl_ran_gaussian_P(mu, sigma), gsl_ran_gaussian_Q(mu, sigma))
// or equivalently,
2 * gsl_ran_gaussian_Q(fabs(mu), sigma)
```

The confidence with which we fail to reject the two-tailed null is one minus this.

In the other direction, we may want to know where we will need to be to reject a hypothesis with 95% certainty. For example, a value-at-risk oriented regulator will want to know the worst one-day loss a bank can expect over a month. To formalize the question, what is the value of the 1-in-20, or 5%, point on the CDF? Assuming a Normal(μ, σ) distribution of profit and loss,[4] the bank will report a value at risk of `gsl_cdf_gaussian_Pinv` (0.05, σ) + μ. Here are the requisite function declarations:

```
double gsl_cdf_gaussian_Pinv (double p, double sigma);
double gsl_cdf_tdist_Pinv (double p, double df);
double gsl_cdf_chisq_Pinv (double p, double df);
double gsl_cdf_fdist_Pinv (double p, double df1, double df2);
```

...plus all of these with the `Pinv`s replaced by `Qinv`s.

Q9.2   The *power* of a test is the likelihood of successfully rejecting the null hypothesis if there really is an effect in the data and the null should be rejected (see page 335 for more). When designing an experiment, you will need to estimate the power of a given design so you can decide whether to gather ten samples or a million.
I expect the mean of my Normally-distributed data to be 0.5, and $\hat{\sigma}$ to be 1.1. Given these assumptions, what must n be to reject the null hypothesis $\mu = 0$ with 99.9% confidence? What if the Normal approximation assumption is deemed inapplicable, so the data is taken to be t-distributed?

---

[4]This assumption is false. Securities typically have leptokurtic (fat-tailed) returns; see page 230.

➤ The square of a Normal distribution is a χ^2 distribution.

➤ Both of these distributions rely on an unknown variance σ^2. We can guess at σ^2, but then our confidence intervals are mere approximations as well.

Σ

➤ The ratio of a Normal over the square root of a transformed χ^2 distribution is a t distribution. By taking the ratio of the form of the two distributions, the unknown σ's cancel out, so a valid confidence region can be constructed from a finite amount of data.

➤ The ratio of two χ^2 distributions is an F distribution. Again, the unknown σ's cancel out.

9.3 TESTING A HYPOTHESIS The chapter to this point has discussed how certain manners of gathering data and aggregating it into a statistic, such as taking its mean or taking the sum of squares, lead to certain known distributions. Thus, if we have a statistic produced in such a manner, we can evaluate the confidence with which a claim about that statistic is true. For example, the mean of a data set can be transformed to something having a t distribution (assuming the CLT holds). Similarly for the difference between the means for two data sets, so a precise probability can be placed on claims about the difference in two means.

CLAIMING A FIXED MEAN This test is sometimes called a z-test, but see the footnote below. The claim is that the mean of a column of data is equal to μ_H. The procedure to the test:

• Find the mean of the data $\hat{\mu}$.
• Given the variance of the data set $\hat{\sigma}_d^2$, estimate the standard deviation of the mean via $\hat{\sigma}_m = \hat{\sigma}_d/\sqrt{n}$.
• For a one-tailed test, find the percentage of the t_{n-1} distribution that is over $|\mu_H - \hat{\mu}|/(\hat{\sigma}/\sqrt{n})$, i.e., `gsl_cdf_tdist_Q(fabs(`$\mu_H - \hat{\mu}$`)/`$\hat{\sigma}_m$`, n-1)`. Report this as the p-value.
• For a two-tailed test, report twice the calculated number as the p-value.

Q9.3

Can we reject the claim H_0: *The typical Puerto Rican county has over a 50% poverty rate*? Use the county-level info from the `poverty_pct_all` column from the `income` table of `data-census.db` as your data set. Given that the US Census Bureau defines poverty by the cost of living in the main part of the United States, how would you interpret your result?

Consider the parameters $\hat{\beta}$ from the OLS regression, which you will recall takes the form $\hat{\beta} = (\mathbf{X}'\mathbf{X})^{-1}\mathbf{X}'\mathbf{y}$. This is a more complex expression than a simple mean, but each element of the $\hat{\beta}$ vector, $\hat{\beta}_0$, $\hat{\beta}_1$, ..., can still be shown to have a simple t distribution. Thus, the standard test that a regression parameter β_i is significantly different from zero is a variant of the test here regarding a t-distributed scalar. The full details will be given in the section on regression tests below.

COMPARING THE MEAN OF TWO SETS   Among the most common and simplest questions is: are two sets of observations from the same process? Chapter 3 (page 109) already showed how to do a t test to test this claim.[5] You are encouraged to reread that section with an eye toward the test procedure.

Reporting confidence

There is some tradition of reporting only whether the p value of a test is greater than or less than some artificial threshold, such as $p > 0.05$ or $p < 0.05$. But Gigerenzer (2004) cites Fisher (1956) as stating:

> ...no scientific worker has a fixed level of significance at which from year to year, and in all circumstances, he rejects hypotheses; he rather gives his mind to each particular case in the light of his evidence and his ideas.

Based on this observation, it would be better form to list the actual confidence calculated, and allow the reader to decide for herself whether the given value provides small or great confidence in the results. The error bars from Chapter 5 provide a convenient way to do this.

The paired t test is a common variant to the standard t test. Say that the data are paired in the sense that for each element in the first set, there is an element in the second set that is related; put another way, this means that for each element a_i, there is a corresponding element b_i such that the difference $a_i - b_i$ makes real-world sense. For example, we could look at student scores on a test before a set of lessons and scores by the same students after the lessons. Then, rather than looking at the t distribution for the *before* data and comparing it to the t distribution for

---

[5]There is no standardized naming scheme for tests. A test basically consists of three components:
1. the context,
2. the statistic to be calculated, and
3. the distribution that the statistic is compared to.

Thus, there are tests with names like the paired data test (using component 1), sum of squares test (component 2), or F test (component 3).

There is no correct way to name a procedure, but you are encouraged to avoid approach #3 where possible. First, there are really only about four distributions (Normal, χ^2, t, F) that are used in most real-world applications, which means that approach #3 gives us only four names for myriad tests. Two people could both be talking about *running a chi-squared test* and find that they are talking about entirely different contexts and statistics.

There is an odd anomaly regarding naming customs for the Normal distribution: rather than calling the statistic to be compared to the Normal a *normal statistic* or *Gaussian statistic*, it is typically called a z statistic. There is a z distribution, but it has nothing to do with the z test: it is one half the log of an F distribution, and is no longer commonly used because the F is slightly more convenient.

Finally, which distribution to use depends on the context and data: if a statistic has a t distribution for small n, then it approaches Normal as $n \to \infty$, so we could easily find ourselves in a situation where we are looking up the statistic for what is called a t test on the Normal distribution tables, or looking up a z statistic in the t distribution tables.

the *after* data, we could look at the vector of differences $a_i - b_i$ and find the confidence with which zero falls in the appropriate t distribution. This is generally a more powerful test, meaning that we are more likely to reject the null hypothesis of no difference between the two vectors, and therefore the paired t test is generally preferable over the unpaired version (when it is applicable). Apophenia provides the `apop_paired_t_test` function to run this test where appropriate.

χ^2**-BASED TESTS**   One quick application of the χ^2 distribution is for testing whether a variance takes on a certain value. We posit that the denominator of Equation 9.2.1 is a fixed number, and then check the χ^2 tables for the given degrees of freedom. This is a relatively rare use of the distribution.

A more common use is to take advantage of the summation property to combine individual statistics into more elaborate tests. Any time we have statistics of the form (observed $-$ expected)^2/expected, where (observed $-$ expected) should be Normally distributed, we can use Pearson's Taylor series approximation to piece together the statistics to form a χ^2 test. There are examples of this form in the section on ANOVA and goodness-of-fit testing below.

F**-BASED TESTS**   Because of all the squaring that goes into a χ^2 distributed statistic, x and $-x$ are indistinguishable, and so it becomes difficult to test one-tailed claims of the form $a > b$. We could use the t test for a one-tailed claim about a single variable, or an F test for a combination of multiple variables.

Let H_0 be the claim that $\mathbf{Q}'\boldsymbol{\beta} = \mathbf{c}$. This is a surprisingly versatile hypothesis. For example, say that $\boldsymbol{\beta}$ is a vector with three elements, $\begin{bmatrix} \beta_1 \\ \beta_2 \\ \beta_3 \end{bmatrix}$, $\mathbf{Q} = \begin{bmatrix} 1 \\ 0 \\ 0 \end{bmatrix}$, and $\mathbf{c} = [7]$.

Then H_0 is $\beta_1 = 7$. Or, $\mathbf{Q} = \begin{bmatrix} 1 \\ -1 \\ 0 \end{bmatrix}$ and $\mathbf{c} = [0]$ gives $H_0 : \beta_1 = \beta_2$. Or, say we want to test $H_0 : \beta_2 = 2\beta_3$. Then let $\mathbf{Q} = \begin{bmatrix} 0 \\ 1 \\ -2 \end{bmatrix}$ and $\mathbf{c} = 0$. In ANOVA terminology, a hypothesis about a linear combination of coefficients is known as a *contrast*.

To test all three hypotheses at once, simply stack them, one hypothesis to a row:

$$\mathbf{Q}' = \begin{bmatrix} 1 & 0 & 0 \\ 1 & -1 & 0 \\ 0 & 1 & -2 \end{bmatrix} \quad \mathbf{c} = \begin{bmatrix} 7 \\ 0 \\ 0 \end{bmatrix}. \tag{9.3.1}$$

Any linear (or affine) hypothesis having to do with a parameter β can be fit into this form.

Define q to be the number of constraints (rows of $\mathbf{Q}'$), n the sample size, and k the number of parameters to be estimated (β). As before, let $\underline{\mathbf{X}}$ represent $\mathbf{X}$ normalized so that each column but the first has mean zero, and the first column is the constant vector $\mathbf{1}$. Now, if H_0 is true and β was estimated using OLS, then $\mathbf{Q}'\beta \sim \mathcal{N}(\mathbf{c}, \sigma^2 \mathbf{Q}'(\underline{\mathbf{X}}'\underline{\mathbf{X}})^{-1}\mathbf{Q})$,[6] and we can construct a χ^2-distributed linear combination of the square of q standard Normals via

$$\frac{(\mathbf{Q}'\hat{\beta} - \mathbf{c})'[\mathbf{Q}'(\underline{\mathbf{X}}'\underline{\mathbf{X}})^{-1}\mathbf{Q}]^{-1}(\mathbf{Q}'\hat{\beta} - \mathbf{c})}{\sigma^2} \sim \chi_q^2. \qquad (9.3.2)$$

Alternatively, say that we are testing the value of the variance of the regression error, and $\epsilon \sim \mathcal{N}$; then

$$\frac{\epsilon'\epsilon}{\sigma^2} \sim \chi_{n-k}^2. \qquad (9.3.3)$$

As above, we can divide scaled versions of Equation 9.3.2 by Equation 9.3.3 to give us a statistic with an F distribution and no unknown σ^2 element:

$$\frac{n-k}{q} \frac{(\mathbf{Q}'\hat{\beta} - \mathbf{c})'[\mathbf{Q}'(\underline{\mathbf{X}}'\underline{\mathbf{X}})^{-1}\mathbf{Q}]^{-1}(\mathbf{Q}'\hat{\beta} - \mathbf{c})}{\epsilon'\epsilon} \sim F_{q,n-k}. \qquad (9.3.4)$$

If you have read this far, you know how to code all of the operations in Equation 9.3.4. But fortunately, Apophenia includes a function that will calculate it for you. To do this, you will need to feed the function an estimate of β and an `apop_data` set indicating the set of contrasts you wish to test, whose vector element is $\mathbf{c}$ and matrix element is $\mathbf{Q}'$. As in Equation 9.3.1, each *row* of the input matrix represents a hypothesis, so to test all three equality constraints at once, you could use a vector/matrix pair like this:

```
double line = {
    7, 1, 0, 0
    0, 1, −1, 0
    0, 0, 1, −2};
apop_data *constr = apop_line_to_data(line, 3, 3, 3);
```

---

[6]For any other method, the form of the variance is $\mathbf{Q}'$(the variance from Section 8.2)$\mathbf{Q}$. See, e.g., Amemiya (1994).

The final [vector|matrix] form for the constraint matches the form used for constraints on pp 152–153, but in this case the constraints are all equalities.

Listing 9.6 presents a full example.

- It runs a regression on the males-per-female data from page 267, so link this code together with that code.
- The constraint is only one condition: that $\beta_3 = 0$.
- The `apop_F_test` function takes in a set of regression results, because the F test as commonly used is so closely married to OLS-type regressions.

```
#include "eigenbox.h"

int main(){
    double line[] = {0, 0, 0, 1};
    apop_data *constr = apop_line_to_data(line, 1, 1, 3);
    apop_data *d = query_data();
    apop_model *est = apop_estimate(d, apop_ols);
    apop_model_show(est);
    apop_data_show(apop_f_test(est, constr));
}
```

Listing 9.6 Run an F-test on an already-run regression. Online source: `ftest.c`.

Here is a useful simplification. Let R^2 be the coefficient of determination (defined further below), n be the number of data points, k be the number of parameters (including the parameter for the constant term), and ϕ be the F-statistic based on $\mathbf{Q} = \mathbf{I}$ and $\mathbf{c} = \mathbf{0}$. Then it can be shown that

$$\frac{(n-k)R^2}{k(1-R^2)} = \phi. \tag{9.3.5}$$

$\mathbb{Q}_{9.4}$ Verify the identity of Equation 9.3.5 using Equation 9.3.4 and these definitions (from page 228):

$$\mathbf{y}_{\text{est}} \equiv \mathbf{X}\boldsymbol{\beta} \text{ (the estimated value of } \mathbf{y}),$$
$$SSR \equiv \sum (\mathbf{y}_{\text{est}} - \overline{\mathbf{y}})^2,$$
$$SSE \equiv \boldsymbol{\epsilon}'\boldsymbol{\epsilon}, \text{ and}$$
$$R^2 \equiv \frac{SSR}{SSE}.$$

Statistical custom is based on the availability of computational shortcuts, so the F statistic of Equation 9.3.5 often appears in the default output of many regression packages.[7] It is up to you to decide whether this particular test statistic is relevant for the situation you are dealing with, but because it is a custom to report it, Apophenia facilitates this hypothesis test by assuming it as the default when you send in NULL variables, as in apop_F_test(estimate, NULL).

$\mathbb{Q}_{9.5}$ Verify the identity of Equation 9.3.5 by running a linear regression on the data set you produced for the exercise on page 278, then passing the apop_model thus produced to apop_F_test to find the F statistic and Apophenia's R^2-finding function to find the SSE and SSR.

Σ

➤ The simplest hypothesis test regarding the parameters of a model is the t test. It claims that the mean of a data set is different from a given value of μ. A special case is the claim that the mean of two data sets differ.

➤ The χ^2 test allows the comparison of linear combinations of allegedly Normal parameters. But since everything is squared to get the χ^2 parameter, it can not test asymmetric one-tailed hypotheses.

➤ The F test provides full generality, and can test both one-tailed and two-tailed hypotheses, and claims that several contrasts are simultaneously true.

9.4 ANOVA

ANOVA is a contraction for *analysis of variance*, and is a catchall term for a variety of methods that aim to decompose a data set's variance into subcategories. Given a few variables and a few groups, is there more variation between groups or within groups? Can the variation in a dependent variable be attributed primarily to some independent variables more than others?

The descriptive portion of ANOVA techniques was covered back on pages 224–227. This section covers the hypothesis testing part of ANOVA.

You may want to re-run metroanova.c, which first appeared as Listing 7.2 on page 226. It produces an ANOVA table that includes several sums of squared errors, and the ratio between them. At this point, you will recognize a sum of squared errors as having a χ^2 distribution (assuming the errors are Normally distributed), and the df-weighted ratio of two sums of squared errors as being F-distributed.

---

[7]Scheffé (1959) parodies the singular focus on this form by calling it *"the" F test* throughout the book.

Thus, the traditional ANOVA table includes an F test testing the claim that the among-group variation is larger than the within-group variation, meaning that the grouping explains a more-than-random amount of variation in the data.

Independence   The crosstab represents another form of grouping, where the rows divide observations into the categories of one grouping, and the columns divide the observations into categories of another. Are the two groupings independent?

To give a concrete example, say that we have a two-by-two array of events, wherein 178 people chose between up/down and left/right:

	Left	Right	Σ
Up	30	86	116
Down	24	38	62
Σ	54	124	178

Is the incidence of Up/Down correlated to the incidence of Left/Right, or are the two independent? Draws from the four boxes should follow a Multinomial distribution: if Up/Down were a Bernoulli draw with probabilities p_U and p_D, and Left/Right were a separate, independent Bernoulli draw with probabilities p_L and p_R, then the expected value of Up/Left would be $E_{UL} = np_Up_L$, and similarly for E_{DL}, E_{UR}, and E_{DR}. Notating the actual incidence of Up/Left as $O_{UL} = 30$, we can use the fact (from page 301) that the χ^2 is a reasonable approximation of errors from a Multinomial distribution to say that the observed variance over the expected value $(O_{UL} - E_{UL})^2/E_{UL} \sim \chi^2$. Similarly for the other three cells, so the sum

$$\frac{(O_{UL} - E_{UL})^2}{E_{UL}} + \frac{(O_{UR} - E_{UR})^2}{E_{UR}} + \frac{(O_{DL} - E_{DL})^2}{E_{DL}} + \frac{(O_{DR} - E_{DR})^2}{E_{DR}} \sim \chi_1^2.$$

(9.4.1)

This expression has one degree of freedom because the horizontal set has two elements and one mean $\Rightarrow$ one df; similarly for the vertical set; and 1 df $\times$ 1 df = 1 df. If there were three rows and six columns, there would be $2 \times 5 = 10$ df.

Listing 9.7 calculates this, once the long way and twice the short way. The `calc_-chi_squared` function calculates Equation 9.4.1, using the `one_chi_sq` function to calculate each individual term. Finally, `main` gathers the data and calls the above functions. After all that, it also calls `apop_test_anova_independence`, which does all this work for you on one line.

The distribution of means of a series of Binomial draws will approach a Normal as $n \to \infty$, but for many situations, n is closer to around ten. For such a case, Fisher

```
#include <apop.h>

double one_chi_sq(apop_data *d, int row, int col, int n){
    APOP_ROW(d, row, vr);
    APOP_COL(d, col, vc);
    double rowexp = apop_vector_sum(vr)/n;
    double colexp = apop_vector_sum(vc)/n;
    double observed = apop_data_get(d, row, col);
    double expected = n * rowexp * colexp;
    return gsl_pow_2(observed − expected)/expected;
}

double calc_chi_squared(apop_data *d){
    double total = 0;
    int n = apop_matrix_sum(d−>matrix);
        for (int row=0; row <d−>matrix−>size1; row++)
            for (int col=0; col <d−>matrix−>size2; col++)
                total += one_chi_sq(d, row, col, n);
        return total;
}

int main(){
    double dataline[] = { 30,86,
                          24,38 };
    apop_data *data = apop_line_to_data(dataline, 0, 2, 2);
    double stat, chisq;
    stat = calc_chi_squared(data);
    chisq = gsl_cdf_chisq_Q(stat, (data−>matrix−>size1 − 1)* (data−>matrix−>size2 − 1));
    printf("chi squared statistic: %g; p, Chi−squared: %g\n", stat,chisq);
    apop_data_show(apop_test_anova_independence(data));
    apop_data_show(apop_test_fisher_exact(data));
}
```

Listing 9.7 Pearson's χ^2 test and Fisher's Exact test. Online source: `fisher.c`.

(1922) calculated the probability of a given table using direct combinatorial computation. The equations for the Fisher exact test are a mess, but the story is the same as above—its null hypothesis is that Up/Down and Left/Right are independent—and its calculation is trivial: just call `apop_test_fisher_exact`, as in the last line of Listing 9.7.

※ *Scaling*   How would the calculation be affected if we replicated every count in the data set into k counts, so $O'_{UL} = kO_{UL}$ and $E'_{UL} = kE_{UL}$? Then $(O'_{UL} - E'_{UL})^2/E'_{UL} = k(O_{UL} - E_{UL})^2/E_{UL}$. That is, scaling the data set by k scales the test statistic by k as well. For almost any data set, there exists a k for which the null hypothesis will be rejected.

Across data sets, the scale can easily be different, and statistical significance will be easier to achieve in the set with the larger scale. Generally, it is tenuous to assert that a data set with a test statistic in the 96th percentile of the χ^2 distribution diverges from independence less than a data set whose test statistic is in the 99.9th percentile. Use the test to establish whether the data rejects the null hypothesis, then use other methods (a simple covariance will often do) to establish the magnitude of the difference.

For comparison, prior tests involving the mean are not as sensitive to scale. Notably, consider the ratio upon which the Central Limit Theorem is based, after every element of the data vector $\mathbf{x}$ is scaled by k:

$$\frac{\text{mean}}{\sqrt{\text{var}/n}} = \frac{\sum(kx - k\overline{\mathbf{x}})/n}{\sqrt{\sum(kx - k\overline{\mathbf{x}})^2/n^2}}$$

$$= \frac{\sum(x - \overline{\mathbf{x}})}{\sqrt{\sum(x - \overline{\mathbf{x}})^2}}.$$

All else equal, the ratio of the mean to $\sqrt{\hat{\sigma}^2/n}$ (often written $\hat{\sigma}/\sqrt{n}$) is not affected by the scale of $\mathbf{x}$, or even the number of elements in the data set, the way the χ^2 statistic above was affected by rescaling.

9.5 REGRESSION   In the prior chapter, we used the linear regression model for purely descriptive purposes, to find the best way to project $\mathbf{y}$ onto $\mathbf{X}$. If we add the assumption that ϵ is Normally distributed, then we can also test hypotheses regarding the parameter estimates. Given this assumption, it can be shown that the coefficients on the independent variables (the $\boldsymbol{\beta}$ vector) have a t distribution, and therefore the confidence with which an element of $\boldsymbol{\beta}$ differs from any constant can be calculated (see, e.g., Greene (1990, p 158)).

The covariance matrix of $\boldsymbol{\beta}_{\text{OLS}}$ is $\Sigma = \sigma^2(\mathbf{X}'\mathbf{X})^{-1}$, where σ^2 is the variance of the regression's error term: if ϵ is the vector of errors, and there are n data points and k regressors (including any constant column $\mathbf{1}$), then $\epsilon'\epsilon/(n-k)$ provides an unbiased estimate of σ^2.[8] The estimated variance of β_1 is the first diagonal element, Σ_{11}; the estimated variance of β_2 is the second diagonal element, Σ_{22}; and so on for all β_i.

As is typical for a test of a statistic of the data, the count of degrees of freedom is data points minus constraints; specifically, for n data points and k regression parameters (including the one attached to the constant column), $df = n - k$.

---

[8]The form of the variance of the error term analogizes directly with the basic one-variable unbiased estimate of variance, $\sum_{i=1}^{n}(x_i - \overline{\mathbf{x}})^2/(n-1)$. First, the setup of OLS guarantees that $\overline{\epsilon} = 0$, so $\epsilon_i - \overline{\epsilon} = \epsilon_i$, and thus $\epsilon'\epsilon$ matches the numerator in the basic formula. The denominator in all cases is the degrees of freedom; for example, with k regressors and n data points, there are $n - k$ degrees of freedom.

Given the estimated variance $\hat{\sigma}^2$ for β_i and any constant c, we could write down a test statistic $|\beta_i - c|/\hat{\sigma}$, and then check that statistic against the t_{n-k} distribution to test the claim that $\beta_i = c$. This test bears a close resemblance to the test for the mean of a data set (also a t-distributed scalar statistic) presented on page 307. If you have a joint hypotheses about contrasts among the elements of $\boldsymbol{\beta}$, you can directly apply the above discussion of F tests: just use the estimated mean of $\boldsymbol{\beta}$, its covariance matrix Σ, and $n - k$ degrees of freedom.

Comparison with ANOVA   If a regression consists of nothing but dummy variables, then it can be shown that the process is equivalent to the ANOVA-style categorization methods above.

$\mathbb{Q}_{9.6}$ Alaska is famously low on females, due to its many features that distinguish it from the lower 48 states. Create a dummy variable where 1=Alaska, 0=other state, and regress males per female against both the Alaska dummy and population density (and a constant **1**, of course). Are one or both of the independent variables significant?

$\mathbb{Q}_{9.7}$ Run an independence test on the two-by-two table whose row categories are Alaska and not-Alaska, and whose column categories are males and females. (*Hint*: you will need to combine the population and males per females columns to produce a count for each region, then sum over all regions.)
How does the test differ if you compare the percent male/female or the total count of males and females in each region? What changes in the story underlying the test, and which version better represents the hypothesis?

OLS (along with its friends) has two real advantages over testing via crosstab approaches like ANOVA. First, it readily handles continuous variables, which ANOVA can handle only via approximation by rough categories.

Second, it allows the comparison of a vast number of variables. ANOVAs typically top out at comparing two independent variables against one dependent, but an OLS regression could project the dependent variable into a space of literally hundreds of independent variables. In fact, if you run such a regression, you are basically guaranteed that some number of those variables will be significant.

The multiple testing problem   Freedman (1983) showed the dangers of *data snooping* by randomly generating a set of 51 columns of 100 random numbers each.[9] He set one column to be the dependent variable to be ex-

---

[9]Data snooping used to also be called *data mining*, but that term has lost this meaning, and is now used to refer to categorization techniques such as classification trees.

plained, and the other fifty to be potential regressors. Using a simple exploratory technique, he culled the fifty potential explanatory variables down to 15 variables. He then ran a 15-variable regression, and found that 14 variables were significant with a p-value better than 25%, and six were significant with p better than 5%. Other tests of the regression also indicated a good fit. But the data was pure noise by construction.

Recall from the first paragraph of this book that there are two goals of statistical analysis, and they directly conflict. If a researcher spends too much time looking for descriptive statistics about the data, then he is committing informal data snooping, akin to Freedman's initial exploratory regression, and thus biases the chances of rejecting a null in her favor. But it would be folly for the researcher to not check the data for outliers or other quirks before running the regression, or to embark upon producing an entirely new data set for every regression.

What is the correct balance? Statistics has no answer to this, though most statisticians do. Those in the descriptive-oriented camp are very serious about the importance of good graphical displays and viewing the data every way possible, while those in the testing-oriented camp believe that so much pre-test searching is simply asking for apophenia.

Here is another spin on the issue: people who are testing exactly one hypothesis tend to develop an affection for it, and become reluctant to reject their pet hypothesis. Thus, research as conducted by humans may improve if there are multiple hypotheses simultaneously competing. Chamberlin (1890, p 93) explains:

> Love was long since represented as blind, and what is true in the personal realm is measurably true in the intellectual realm.... The moment one has offered an original explanation for a phenomenon which seems satisfactory, that moment affection for his intellectual child springs into existence; and as the explanation grows into a definite theory, his parental affections cluster about his intellectual offspring, and it grows more and more dear to him, so that, while he holds it seemingly tentative, it is still lovingly tentative.... The mind lingers with pleasure upon the facts that fall happily into the embrace of the theory, and feels a natural coldness toward those that seem refractory.... There springs up, also, an unconscious...pressing of the facts to make them fit the theory.... The search for facts, the observation of phenomena and their interpretation, are all dominated by affection for the favored theory until it appears to its author or its advocate to have been overwhelmingly established. The theory then rapidly rises to the ruling position, and investigation, observation, and interpretation are controlled and directed by it. From an unduly favored child, it readily becomes master, and leads its author whithersoever it will.

His solution, as above, is to test multiple hypotheses simultaneously. "The investigator thus becomes the parent of a family of hypotheses; and, by his parental relation to all, he is forbidden to fasten his affections unduly upon any one." He also points out that maintaining multiple hypotheses allows for complex explanations about how an outcome was partly caused by one factor, partly by another, partly by another. After all, Nature is not compelled to conform to exactly one hypothesis.

Apophenia's model-as-object makes it very easy to test or mix diverse hypotheses, as per Chamberlin's suggestion, and you will see more methods of comparing models in later chapters. But as the number of models grows, the odds of failing to reject at least one model purely by chance grows as well. There is no hard-and fast rule for determining the "correct" number of models to test; just bear in mind that there is a tension among multiple goals and a balance to be struck between them.

Correcting for multiple testing   Moving on from informally poking at the data, consider the case when the experiment's basic design involves a systematic, fixed series of tests, like running a separate test for every genetic marker among a list of a million. This is known as the *multiple testing problem*, and there is a simple means of correcting for it.

Say that a number is drawn from $[0, 1]$, and the draw is less than p with probability p. Then the probability that a draw is greater than p is $1 - p$, and the probability that n independent draws are all greater than p is $(1 - p)^n$, which can be reversed to say that the probability that at least one of n independent draws is less than p is $1 - (1 - p)^n$.

Thus, the probability that, out of n tests with a fixed p-value, at least one will indicate a positive result is $\rho = 1 - (1 - p)^n$. For example, with $p = 0.05$ and $n = 100$, the likelihood of rejecting the null at least once is $1 - (1 - 0.05)^{100} \approx 99.4\%$.

We can instead fix ρ at a value like 0.05 or 0.01, and reverse the above equation to find the p-value for the individual tests that would lead to rejection of all nulls with 5% or 1% likelihood. A line or two of algebra will show that $p = 1 - (1 - \rho)^{1/n}$. For $n = 100$ and $\rho = 0.05$, you would need to set the p-value for the individual tests to 0.0005128. In the example of testing $n = 1,000,000$ genetic markers, if the desired overall $\rho = 0.05$, then the p-value for the individual tests would be 5.129e−8.

There is a wonderfully simple approximation for the above expression: just let $p = \rho/n$. For the first example above, this approximation is $0.05/100 = 0.0005$; for the second it is $0.05/1,000,000 = 5e{-8}$. Both of these approximations are within about $\pm 2.5\%$ of the true value.

Thus, we have a simple rule, known as the *Bonferroni correction*, for multiple tests: just divide the desired overall p-value by the number of tests to get the appropriate individual p-values. The correction is standard in biometrics but virtually unknown in the social sciences. When reading papers with pages of regressions and no corrections for multiple testing, you can easily apply this equation in your head, by multiplying the reported individual p-value by the number of tests and comparing that larger figure to the usual significance levels of 0.05 or 0.01.

> ➤ Because we know their expected mean and covariances, the regression parameters for OLS, IV, WLS, and other such models can be tested individually using the standard t test, or tested as a group via an F test.

> ➤ When running a battery of several tests (based on a regression or otherwise), use the Bonferroni correction to create a more stringent significance level. The common form of calculating the more stringent p-value is to simply divide the one-test p-value by the number of tests.

9.6 GOODNESS OF FIT

This section will present various ways to test claims of the form *the empirical distribution of the data matches a certain theoretical distribution*. For example, we often want to check that the errors from a regression are reasonably close to a Normal distribution.

The visually appealing way to compare two distributions is the Q–Q plot, which stands for quantile–quantile plot. The first (x, y) coordinate plotted is $x_1 =$ the first percentile of your data and $y_1 =$ the first percentile of the distribution you are testing, the second is $x_2 =$ the second percentile of your data and $y_2 =$ the second percentile of the ideal distribution, et cetera. To the extent that the data fits the ideal distribution, the points will draw out the $x = y$ line, while digressions from the line will stand out.

The first half of Listing 9.9 presents an example, displaying a plot to check whether precipitation is Normally distributed. It gathers the data in the usual `apop_-query_to_data` manner, estimates the closest-fitting Normal distribution, and plots the percentiles of the data against the percentiles of the just-estimated distribution. As Figure 9.8 shows, the data set closely fits the Normal distribution (though the extremes of the bottom tail is a bit more elongated and the extreme of the top tail a bit less so).

Modify Listing 9.9 to test whether temperature or log of temperature is Normally distributed. Would any other distribution fit better?

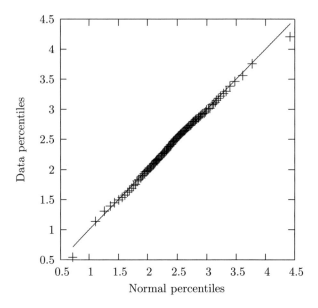

Figure 9.8 Percentiles of precipitation on the y axis plotted against percentiles of the Normal distribution along the x axis.

```
#include <apop.h>

int main(){
    apop_db_open("data−climate.db");
    apop_data *precip = apop_query_to_data("select PCP from precip");
    apop_model *est = apop_estimate(precip, apop_normal);
    Apop_col_t(precip, "PCP", v);
    apop_plot_qq(v, *est, "outfile.gnuplot");

    double var = apop_vector_var(v);
    double skew = apop_vector_skew(v)/pow(var, 3/2);
    double kurt = apop_vector_kurtosis(v)/gsl_pow_2(var) − 3;
    double statistic = v−>size * (gsl_pow_2(skew)/6. + gsl_pow_2(kurt)/24.);
    printf("The skew is %g, the normalized kurosis is %g, "
            "and we reject the null that your data is Normal with %g confidence.\n",
        skew, kurt, gsl_cdf_chisq_P(statistic, 2));
}
```

Listing 9.9 Pull data; estimate the Normal that best fits the data; plot the data against the ideal distribution. The output is presented in Figure 9.8. Online source: `qqplot.c`.

HIGHER MOMENTS A slightly more rigorous alternative means of testing for Normality is to check the higher moments of the Normal distribution (Bowman & Shenton, 1975; Kmenta, 1986, pp 266–267). There is a more general chi-squared goodness-of-fit test for any distribution below.

A Normal distribution has only two parameters, the mean and the standard deviation, and everything else about the Normal distribution is defined therefrom. Notably, the third moment is zero, and the fourth moment is $3\sigma^4$.

We already have everything we need to calculate the distribution of these statistics. The skew and kurtosis are both the mean of an iid process (recall their definitions on page 230: a sum divided by n), so their square is $\sim \chi^2$. Let s be the third moment of the data divided by σ^3 and let κ be the fourth moment divided by σ^4. Then

$$L_s = n \left[\frac{s^2}{6} \right]$$

has a χ^2 distribution with one degree of freedom, as does

$$L_k = n \left[\frac{(\kappa - 3)^2}{24} \right].$$

Some prefer to test both simultaneously using

$$L_{sk} = n \left[\frac{s^2}{6} + \frac{(\kappa - 3)^2}{24} \right],$$

which has a χ^2 distribution with two degrees of freedom.

The second half of Listing 9.9 translates this formula into code. Given the Q–Q plot, it is no surprise that the test soundly fails to reject the null hypothesis that the data is Normally distributed.

Another alternative, keeping with the theme of this book, would be to bootstrap the variance of the kurtosis, which would let you find a confidence interval around $3\sigma^4$ and state with some percentage of certainty that the kurtosis is or is not where it should be; this suggestion is put into practice on page 365.

CHI-SQUARED GOODNESS-OF-FIT TEST

Say that we have a histogram and a vector of points that we claim was drawn from that histogram. It would be nice to test the confidence with which our claim holds; this is a goodness-of-fit test.

Mathematically, it is simple. We have k bins, and two histograms: h0 holds the histogram from which the draws were allegedly made, and h1 holds the data.[10] Then

---

[10]Recall from page 314 that scaling matters for a χ^2 test: the histograms representing two PDFs will each sum to one (by definition), while a histogram representing the density of a population of size n will have bins summing to n (by definition). That means that the χ^2 statistics for a test of the PDFs and a test of the distribution of counts will be different, with the null more likely to be rejected for the distribution of counts. So when investigating a histogram, be careful that you are testing the right hypothesis; claims about the distribution of a population are typically best represented by a test of the PDFs ($\Sigma = 1$) rather than the counts ($\Sigma = n$).

$$\sum_{i=0}^{k} \frac{(\text{h0} \rightarrow \text{bins}[\text{i}] - \text{h1} \rightarrow \text{bins}[\text{i}])^2}{\text{h0} \rightarrow \text{bins}[\text{i}]} \sim \chi^2_{k-1}. \qquad (9.6.1)$$

You will recognize this form as matching the *(observed - expected)^2/expected* form from the ANOVA tests earlier in this chapter.

On page 173, you plotted the leading digit of an arbitrary data set, and saw that it sloped sharply down. Now use a chi-squared goodness of fit test to formally check that your data set fits Equation 5.4.1.

- Write a function to produce a vector of nine elements, with the count of elements in each slot equal to the number of data points with the given leading digit. Don't forget that vectors count from zero but you want the first slot to represent leading digit one, and to rescale your final vector so that it is a PMF (i.e., its elements sum to one).

- Equation 5.4.1 isn't quite a PMF: the sum of its values from one to nine isn't one. Thus, you will need to get the total mass, and rescale your calculations from Benford's equation accordingly.

- Once you have two nine-element vectors of equal mass, you can directly apply Expression 9.6.1 to find the χ^2 statistic and run the χ^2 test.

```
1   #include <apop.h>
2
3   int main(){
4       apop_db_open("data−climate.db");
5       apop_data *precip = apop_query_to_data("select PCP from precip");
6       apop_model *est = apop_estimate(precip, apop_normal);
7       gsl_rng *r = apop_rng_alloc(0);
8       apop_model *datahist = apop_estimate(precip, apop_histogram);
9       apop_model *modelhist = apop_histogram_model_reset(datahist, est, 1e6, r);
10      apop_data_show(apop_histograms_test_goodness_of_fit(datahist, modelhist));
11  }
```

Listing 9.10 The same precipitation data, another test. Online source: `goodfit.c`.

Listing 9.10 tests whether the precipitation data is Normally distributed using the χ^2 goodness-of-fit test.

- Lines 1–6 are a repeat of the query and estimation from `qqplot.c` (page 320).

- Line eight turns the input data into a histogram. Notice that it uses the same `apop_-estimate` form as other models, because a histogram is just another means of expressing a model.

- You can't do a goodness-of-fit test on just any two histograms: the bins have to match, in the sense that the range of each bin in the first histogram exactly matches the range in the corresponding bin of the second. The easiest way to ensure that two histograms match is to generate the second histogram using the first as a template, which is what `apop_histogram_model_reset` does. If we wanted to compare two vectors via this test, this line would use `apop_histogram_vector_reset`.

- The new histogram gets filled via random draws from the model, which means that we need to give `apop_histogram_model_reset` the number of draws to make (here, 1e6), and a `gsl_rng` to provide random numbers. The use of the `gsl_rng` is covered in full on page 357.

- By line ten, we have two histograms representing the data and the model, and they are in sync. Thus, it is a simple matter to send the histograms to `apop_histograms_test_goodness_of_fit` to calculate the statistic in Expression 9.6.1.

KOLMOGOROV'S METHOD   Kolmogorov (1933) suggested considering the steps in a histogram to be a Poisson process, and developed a test based upon this parametrization [see also Conover (1980)]. Given two histograms produced using one of the above-mentioned methods, `apop_test_kolmogorov` finds the maximum difference in the CMFs and find the probability that such a CMF would occur if both histograms were from the same base data set. Because this test uses the ordering of the slices of the PMF, while the Chi-squared test does not, this test generally has higher power.

Kolmogorov's test serves as yet another test for Normality, because it can compare a data set's CDF to that of a Normal distribution.

Q9.10

Is GDP per capita log-normally distributed?

- Pull the log of GDP per capita data from the `data-wb.db` data set.

- Create a histogram (i.e., estimate an `apop_histogram` model).

- Fit a Normal distribution and use it to create a matching histogram using `apop_histogram_model_reset`.

- Send the two histograms to `apop_test_kolmogorov`.

- How do the test results compare with those produced by `apop_histograms_test_goodness_of_fit`?

Q9.11 How about precipitation? Figure 9.8 gave the strong suggestion that the data is Normally distributed; modify the code from the last example to formally test the hypothesis that the data set is drawn from a Normal distribution using the Kolmogorov–Smirnov method.

$\sum$

➤ The Q–Q (quantile-to-quantile) plot gives a quick visual impression of whether the distribution of the data matches that of a theoretical distribution.

➤ We can test the claim that a data set is Normally distributed using the fact that the skew and kurtosis of a Normal distribution are fixed (given the mean and variance).

➤ More generally, we can compare any two distributions by dividing them into bins and comparing the square of the deviation of one distribution from another via a χ^2 test.

➤ The Kolmogorov–Smirnov test offers another method for comparing two distributions, which typically has more power than the Chi-squared method.

MAXIMUM LIKELIHOOD ESTIMATION

> Since the fabric of the universe is most perfect and the work of a most wise Creator,
> nothing at all takes place in the universe in which some rule of maximum or minimum
> does not appear.
>
> —Leonhard Euler[1]

Whether by divine design or human preference, problems involving the search for
optima are everywhere. To this point, most models have had closed-form solutions
for the optimal parameters, but if there is not a nice computational shortcut to
finding them, you will have to hunt for them directly. There are a variety of routines
to find optima, and Apophenia provides a consistent front-end to many of them via
its `apop_maximum_likelihood` function.

Given a distribution $p(\cdot)$, the value at one input, $p(x)$, is *local information*: we
need to evaluate the function at only one point to write down the value. However,
the optimum of a function is *global information*, meaning that you need to know
the value of the distribution at every possible input from $x = -\infty$ up to $x = \infty$ in
order to know where the optimum is located.

This chapter will begin with the simple mathematical theory behind maximum
likelihood estimation (*MLE*), and then confront the engineering question of how
we can find the global optimum of a function by gathering local information about
a small number of points. Once the prerequisites are in place, the remainder of the
chapter covers MLEs in practice, both for description and testing.

---

[1]Letter to Pierre-Louis Moreau de Maupertuis, c. 1740–1744, cited in Kline (1980, p 66).

10.1 LOG LIKELIHOOD AND FRIENDS   To this point, we have met many probability distributions, whose PDFs were listed in Sections 7.2 and 9.2. This section takes such a probability distribution $P(\mathbf{X}, \boldsymbol{\beta})$ as given, and from that produces a few variant and derivative functions that will prove to be easier to work with. Also, a reminder: there is a list of notation for the book on page 12.

Let x_1 and x_2 be two *independent, identical draws* (iid) from a distribution. The independence assumption means that the joint probability is the product of the individual probabilities; that is, $P(\{x_1 \text{ and } x_2\}, \boldsymbol{\beta}) = P(x_1, \boldsymbol{\beta}) \cdot P(x_2, \boldsymbol{\beta})$. The assumption of identical distributions (i.e., that both are draws from the same distribution $P(\cdot, \boldsymbol{\beta})$) allows us to write this more neatly, as

$$P(\{x_1 \text{ and } x_2\}, \boldsymbol{\beta}) = \prod_{i=\{1,2\}} P(x_i, \boldsymbol{\beta}).$$

A probability function gives the probability that we'd see the data that we have given some parameters; a *likelihood function* is the probability that we'd see the specified set of parameters given some observed data. The philosophical implications of this distinction will be discussed further below.

There are three basic transformations of the likelihood function that will appear repeatedly, and are worth getting to know.

Define the *log likelihood function* as $LL \equiv \ln P(\mathbf{x}, \boldsymbol{\beta})|_x$, the *score* as its derivative with respect to β:

$$\mathbf{S} \equiv \begin{bmatrix} \frac{\partial \ln P}{\partial \beta_1} \\ \vdots \\ \frac{\partial \ln P}{\partial \beta_n} \end{bmatrix} = \begin{bmatrix} \frac{\partial LL}{\partial \beta_1} \\ \vdots \\ \frac{\partial LL}{\partial \beta_n} \end{bmatrix},$$

and the *information matrix* as the negation of derivative of the score with respect to β.

$$\mathbb{I} = -\frac{\partial \mathbf{S}}{\partial \boldsymbol{\beta}}$$

$$= - \begin{bmatrix} \frac{\partial^2 LL}{\partial \beta_1^2} & \cdots & \frac{\partial^2 LL}{\partial \beta_n \beta_1} \\ \vdots & \ddots & \vdots \\ \frac{\partial^2 LL}{\partial \beta_1 \beta_n} & \cdots & \frac{\partial^2 LL}{\partial \beta_n^2} \end{bmatrix}.$$

An example: Bernoulli   Say that we have nine draws from a Bernoulli distribution, which you will recall from page 237 means that each draw is one with probability p and is zero with probability $1 - p$, and say that in our case five draws were ones and four were zeros. The likelihood of drawing five ones is p^5; the likelihood of drawing four zeros is $(1 - p)^4$; putting them together via the independence assumption, the likelihood of an arbitrary value of p given this data set $\mathbf{x}$ is

$$P(\mathbf{x}, p)|_x = p^5 \cdot (1 - p)^4.$$

The log likelihood is thus

$$LL(\mathbf{x}, p) = 5 \ln(p) + 4 \ln(1 - p).$$

The score, in this case a one-dimensional vector, is

$$S(\mathbf{x}, p) = \frac{5}{p} - \frac{4}{(1 - p)}, \qquad (10.1.1)$$

and the information value (a 1×1 matrix) is

$$\mathbb{I}(\mathbf{x}, p) = \frac{5}{p^2} + \frac{4}{(1 - p)^2}.$$

Both intuitively and for a number of reasons discussed below, it makes sense to focus on the most likely value of p—that is, the value that maximizes $P(\mathbf{x}, p)$ given our observed data $\mathbf{x}$. Since the log is a monotonic transformation, p maximizes $P(\mathbf{x}, p)$ if and only if it maximizes $LL(\mathbf{x}, p)$. Recall from basic calculus that a necessary condition for a smooth function $f(x)$ to be at a maximum is that $\frac{df}{dx} = 0$—in the case of $LL(\mathbf{x}, p)$, that $S(\mathbf{x}, p) = 0$. Setting Equation 10.1.1 to zero and solving for p gives $p = \frac{5}{9}$.

But a zero derivative can indicate either a maximum or a minimum; to tell which, look at the second derivative. If the second derivative is negative when the first derivative is zero, then the first derivative is about to become negative—the function is beginning to dip downward. At this point, the function must be at a maximum and not a minimum.

Since the information matrix is defined as the negation of the score's derivative, we can check that we are at a maximum and not a minimum by verifying that the information value is positive—and indeed it is: for $p = 5/9$, $\mathbb{I} \approx 4.05$. In more dimensions, the analog is that the information matrix must be *positive definite*; see the exercise on page 269.

To summarize, given $p = \frac{5}{9}$, $P \approx 0.0020649$, $LL \approx -6.182$, $S = 0$, and $\mathbb{I} \approx 4.05$. It is easy to check other values of p to verify that they produce lower probabilities of observing the given data.

Verify the above statements.

- Generate a data set whose matrix element includes five ones and four zeros (in any order). Put the data set in a global variable.

- Produce an output model via apop_estimate(*your_data*, apop_-bernoulli).

- Display the output model via apop_model_show; check that the probability is as it should be (i.e., $5/9 = .\overline{555}$).

- Write a function that takes in an apop_data struct holding a vector with one item, and returns the log likelihood of that parameter given the global data, using apop_log_likelihood.

- Send that function to your plotafunction routine from page 191 (with the range $[0, 1]$) and verify that the log likelihood is at its highest where the parameter is $5/9$.

An example: Probit   Recall the Probit model from page 283. It specified that an agent would act iff its utility from doing so is greater than a Normally-distributed error term. That is, act with probability

$$P(\mathbf{x}, \boldsymbol{\beta}) = \int_{-\infty}^{\mathbf{x}\boldsymbol{\beta}} \mathcal{N}(y|0, 1)dy,$$

where $\mathcal{N}(y|0, 1)$ indicates the standard Normal PDF at y (and so the integral is the CDF up to $\mathbf{x}\boldsymbol{\beta}$).

Reversing this, let $\mathbf{x}^A$ be the set of $\mathbf{x}$'s that led to action and $\mathbf{x}^N$ be the set that led to non-action. Then the likelihood of $\boldsymbol{\beta}$ given the data set is

$$P(\mathbf{x}, \boldsymbol{\beta})|_{\mathbf{x}} = \prod_{i \in A} P(\mathbf{x}_i^A, \boldsymbol{\beta}) \cdot \prod_{i \in N} \left(1 - P(\mathbf{x}_i^N, \boldsymbol{\beta})\right),$$

so

$$LL(\mathbf{x}, \boldsymbol{\beta})|_{\mathbf{x}} = \sum_{i \in A} \ln P(\mathbf{x}_i^A, \boldsymbol{\beta}) + \sum_{i \in N} \ln \left(1 - P(\mathbf{x}_i^N, \boldsymbol{\beta})\right).$$

By the way, the logit (Equation 8.3.1, page 284) tells the same story, but it simplifies significantly, to

$$LL(\mathbf{x}, \boldsymbol{\beta})|_{\mathbf{x}} = \sum_{i \in A} \mathbf{x}_i^A \boldsymbol{\beta} - \sum_{\forall i} \ln\left(1 + e^{\mathbf{x}_i \boldsymbol{\beta}}\right),$$

where the first term counts only those who act, while the second includes everybody. [ℚ: Verify the log likelihood using Equation 8.3.1.]

Unlike the binomial example above, we can not find the optimum of the log likelihood function for either the probit or logit using just a derivative and some quick algebra. We will instead need to use a search method such as those discussed later in this chapter.

⁂ A DIGRESSION: THE PHILOSOPHY OF NOTATION   The probability function has a *frequentist* interpretation: if you give me a fixed distribution, the story behind it, and a fixed parameter β, then after a few million draws, x will occur $P(x, \beta)|_{\beta} \cdot 100$ percent of the time. The likelihood function has no such interpretation, because we assume that the data was produced using one model that had a fixed β, that we happen to be ignorant of. There is no mysterious pool of β's from which ours was drawn with some probability.

Thus, the probability of x given β (and a model) is in many cases an objectively verifiable fact; the likelihood of β given x (and a model) is a subjective construct that facilitates various types of comparison among β's. The integral over all x is always one (i.e., for any fixed β, $\int_{\forall x} P(x, \beta)dx = 1$). The integral over all β of the likelihood function, however, could be anything.

Ronald Aylmer Fisher, the famed eugenicist and statistician whose techniques appear throughout this book, was vehement about keeping a clear distinction: "... [I]n 1922, I proposed the term 'likelihood,' in view of the fact that, with respect to [the parameter], it is not a probability, and does not obey the laws of probability, while at the same time it bears to the problem of rational choice among the possible values of [the parameter] a relation similar to that which probability bears to the problem of predicting events in games of chance.... Whereas, however, in relation to psychological judgment, likelihood has some resemblance to probability, the two concepts are wholly distinct...." (Fisher, 1934, p 287) See Pawitan (2001) for more on the interpretation of likelihood functions.

But as a simple practical matter, the probability of x given fixed parameter β is $P(x, \beta)$, and the likelihood of β given fixed data x is the very same $P(x, \beta)$. At the computer, there is no point writing down separate functions p(x, beta) and likelihood(beta, x)—a single function will serve both purposes. Just fix x to produce a likelihood function over β, and fix β to get a probability distribution of values of x.

We have two choices for notation, both of which can lead to confusion. The first is to use two distinct function names for probability and likelihood—$P(x)$ and $L(x)$ are typical—which clarifies the philosophical differences and leaves it to the reader to recognize that they are numerically identical, and that both are functions of x and β. The second option is to use a single function for both, which clarifies the computational aspects, but leaves the reader to ponder the philosophical implications of a single function that produces an objective probability distribution when viewed one way and a subjective likelihood function when viewed the other way.[2] Because this book is centrally focused on computation, it takes the second approach of listing both probability and likelihood using the same $P(x, \beta)$ form.

MORE ON $LL, S,$ AND $\mathbb{I}$ The log of the likelihood function has a number of divine properties, which add up to making the log likelihood preferable to the plain likelihood in most cases—and wait until you see what the score can do.

First, due to all that exponentiation in the distributions of Sections 7.2 and 9.2, $\ln P$ is often much easier to deal with, yet is equivalent to $P(\cdot)$ for most of our purposes—notably, if we have found a maximum for one, the we have found a maximum for the other.

Also, consider calculating an iid likelihood function given a thousand data points. The probability of observing the data set will have the form $\prod_{i=1}^{1000} P(x_i)$. Since each $P(x_i) \in (0, 1]$, this product is typically on the order of 1×10^{-1000}. As discussed on page 137, such a number is too delicate for a computer to readily deal with. Taking logs, each value of p_i is now a negative number (e.g., $\ln(0.5) \approx -0.69$ and $\ln(0.1) \approx -2.3$), and the product above is now a sum:

$$\ln\left[\prod_{i=1}^{1000} P(x_i)\right] = \sum_{i=1}^{1000} \ln\left(P(x_i)\right).$$

Thus, the log likelihood of our typical thousand-point data set is on the order of -1000 instead of 1×10^{-1000}—much more robust and manageable. You saw an example of these different scales with the nine-point sample in the Bernoulli example, which had $p \approx 0.002$ but $LL \approx -6$.

Analytically, the maximum of the log likelihood function is useful for two reasons with four names: the Cramér–Rao lower bound and the Neyman–Pearson lemma. It all begins with this useful property of the score:[3]

---

[2]There are consistent means of describing subjective probability that accommodate both ways of slicing $P(x, \beta)$. The *subjectivist* approach (closely related to the *Bayesian* approach) takes all views of $P(x, \beta)$ as existing only in our minds—no matter how you slice it, there need not be a physical interpretation. The axiomatic approach, led by Ramsey, Savage, and de Finetti, posits a few rules that lead to 'consistent' subjective beliefs when followed, but places no further constraints on either probability or likelihood. Again, once both probability and likelihood are accepted as subjective beliefs, there is less reason to distinguish them notationally.

[3]All proofs here will be in the case where β is a scalar. Proofs for the multidimensional case are analogous but

Theorem 10.1.1. If $P(\mathbf{x}, \beta)$ satisfies certain regularity conditions as described in the footnote,[4] then for any statistic $f(x)$,

$$E_x(S \cdot f) = \frac{\partial E_x(f)}{\partial \beta}.$$

That is, the score is a sort of derivative machine: the expected value of the score times a statistic is equivalent to the derivative of the expected value of the statistic. Finding an optimum requires finding the point where the derivative is zero and the second derivative is negative, and this theorem gives us an easy trick for finding those derivatives. The next few pages will show how this trick is used.

When reading this theorem, it is worth recalling the sleight-of-notation from page 257: $f(x)$ is a function only of the data, but $E_x(f(x))$ (where x is produced using a certain model and parameters) is a function only of the parameters.

✳**Proof:** The expected value of the score times the statistic is

$$E_x(S \cdot f) = \int_{\forall x} S(\beta) f(x) P(x, \beta) dx$$

$$= \int_{\forall x} \frac{\partial \ln P(x, \beta)}{\partial \beta} f(x) P(x, \beta) dx \qquad (10.1.2)$$

$$= \int_{\forall x} \frac{\frac{\partial P(x,\beta)}{\partial \beta}}{P(x, \beta)} f(x) P(\beta, x) dx \qquad (10.1.3)$$

$$= \int_{\forall x} f(x) \frac{\partial P(x, \beta)}{\partial \beta} dx$$

$$= \frac{\partial \left(\int_{\forall x} f(x) P(x, \beta) dx \right)}{\partial \beta} \qquad (10.1.4)$$

$$= \frac{\partial E_x(f(x))}{\partial \beta} \qquad (10.1.5)$$

---

require more involved notation.

[4]Equation 10.1.4 of the proof uses the claim that $\int f \cdot \frac{dP}{d\beta} dx = \frac{d}{d\beta} \int f \cdot P dx$. If we can't reverse the integral and derivative like this, none of this applies.

The common explanation for when the switch is valid is in the case of any exponential family; the definition of an exponential family will not be covered in this book, but rest assured that it applies to the Normal, Gamma, Beta, Binomial, Poisson, et cetera—just about every distribution but the Uniform.

But it also applies more generally: we need only uniform convergence of the PDF as its parameters go to any given limit (Casella & Berger, 1990, Section 2.4). Roughly, this is satisfied for any PDF whose value and derivative are always finite. For those who prefer the exponential family story, note that any PDF can be approximated arbitrarily closely by a sum of exponential-family distributions (Barron & Sheu, 1991), so for any distribution that fails, there is an arbitrarily close distribution that works. For example, the Uniform$[\beta_1, \beta_2]$ distribution fails because of the infinite slope at either end, but a distribution with a steep slope up between $\beta_1 - 1e{-}10$ and β_1 and a steep slope down between β_2 and $\beta_2 + 1e{-}10$ works fine.

The sequence of events consisted of substituting in the definition of the score (Equation 10.1.2), then substituting the familiar form for the derivative of the log (Equation 10.1.3), and canceling out a pair of $P(x, \beta)$'s. At this point, the simple weighting $P(x)$ (first introduced on page 221) has been replaced with the weighting $dP(x)$. Before, if x_1 was twice as likely as x_2 (i.e., $P(x_1) = 2P(x_2)$), then $f(x_1)$ would get double weighting. Now, if the slope of $P(x)$ at x_1 is twice as steep as the slope at x_2, then $f(x_1)$ gets double weighting.

The final steps state that, under the right conditions, the integral of $f(x)$ using a measure based on $dP(x)$ is equivalent to the derivative of the integral of $f(x)$ using a measure based on $P(x)$. Equation 10.1.4 switched the integral and derivative, using the assumptions in the theorem's footnote, and Equation 10.1.5 recognized the integral as an expectation under the given probability density.   ♦

Corollary 10.1.2.
$$E(S) = 0.$$

Proof: Let the statistic $f(x)$ be the trivial function $f(x) = 1$. Then Theorem 10.1.1 tells us that $E(S \cdot 1) = \partial E(1)/\partial \beta = 0$.

$Q_{10.2}$ | Verify that the expected value of the score is zero for a few of the distributions given in Chapter 7, such as the Exponential on page 248. (*Hint*: you will need to calculate the integral of the expected value of the score over the range from zero to infinity; integration by parts will help.)

Lemma 10.1.3. The *information equality*
$$\text{var}(S) = E(S \cdot S) = E(\mathbb{I}).$$

※**Proof:** The first half comes from the fact that $\text{var}(S) = E(S \cdot S) - E(S) \cdot E(S)$, but we just saw that $E(S) = 0$.

For the second half, write out $E(\mathbb{I})$, using the expansion of $S = \left(\dfrac{\frac{\partial P(x,\beta)}{\partial \beta}}{P(x,\beta)} \right)$ and the usual rules for taking the derivative of a ratio.

$$E\left[\frac{\partial^2 \ln P(x,\beta)}{\partial\beta}\right] = E\left[\frac{\partial\left(\frac{\frac{\partial P(x,\beta)}{\partial\beta}}{P(x,\beta)}\right)}{\partial\beta}\right]$$

$$= E\left[\frac{P(x,\beta)\frac{\partial^2 P(x,\beta)}{\partial\beta^2} - \left(\frac{\partial P(x,\beta)}{\partial\beta}\right)^2}{(P(x,\beta))^2}\right]$$

$$= E\left[\frac{\frac{\partial^2 P(x,\beta)}{\partial\beta^2}}{P(x,\beta)} - S\cdot S\right]$$

$$= -E[S\cdot S]$$

$\mathbb{Q}$: Prove the final step, showing that $E\left[\frac{\frac{\partial^2 P(x,\beta)}{\partial\beta^2}}{P(x,\beta)}\right] = 0$. (*Hint*: use the lessons from the proof of Theorem 10.1.1 to write the expectation as an integral and switch the integral and one of the derivatives.) ♦

The information equality will be computationally convenient because we can re-place a variance, which can be hard to directly compute, with the square of a derivative that we probably had to calculate anyway.

For the culmination of the sequence, we need the Cauchy–Schwarz inequality, which first appeared on page 229. It said that the correlation between any two variables ranges between -1 and 1. That is,

$$-1 \leq \rho(f,g) = \frac{\text{cov}(f,g)}{\sqrt{\text{var}(g)\,\text{var}(f)}} \leq 1$$

$$\frac{\text{cov}(f,g)^2}{\text{var}(g)\,\text{var}(f)} \leq 1$$

$$\frac{\text{cov}(f,g)^2}{\text{var}(g)} \leq \text{var}(f). \tag{10.1.6}$$

Lemma 10.1.4. *The Cramér–Rao lower bound*

Let $f(\mathbf{x},\beta)$ be any statistic, and assume a distribution $P(\mathbf{x},\beta)$ that meets the cri-teria from the prior results. Then

$$-\frac{(\partial E_x\left(f(\mathbf{x},\beta)\right)/\partial\beta)^2}{E(\mathbb{I})} \leq \text{var}(f(\mathbf{x},\beta)). \tag{10.1.7}$$

The proof consists of simply transforming the Cauchy–Schwarz inequality using the above lemmas. Let g in Equation 10.1.6 be the score; then the equation expands to

$$\frac{(E_x(f(\mathbf{x}) \cdot S) - E_x(f(\mathbf{x}))E_x(S))^2}{\text{var}(S)} \leq \text{var}(f(\mathbf{x})) \qquad (10.1.8)$$

The left-hand side has three components, each of which can be simplified using one of the above results:

- $E_x(f(\mathbf{x}, \beta) \cdot S) = \partial f(\mathbf{x}, \beta)/\partial\beta$, by Theorem 10.1.1.
- Corollary 10.1.2 said $E(S)$ is zero, so the second half of the numerator disappears.
- The information equality states that that $\text{var}(S) = E(\mathbb{I})$.

Applying all these at once gives the Cramér–Rao lower bound.

Further, MLEs have a number of properties that let us further tailor the CRLB to say still more.[5] Let the statistic $f(\mathbf{x})$ be the maximum likelihood estimate of the parameter, $MLE(\mathbf{x}, \beta)$.

- MLEs can be biased for finite data sets, but can be shown to be asymptotically unbiased. This means that for n sufficiently large, $E(MLE(\mathbf{x}, \beta)) = \beta$. Therefore, $\partial MLE(\mathbf{x}, \beta)/\partial\beta = 1$, so the numerator on the left-hand side of Equation 10.1.7 is one.
- It can be proven that maximum likelihood estimators actually achieve the CRLB, meaning that in this case the inequality in Equation 10.1.7 is an equality.
- The information matrix is additive: If one data set gives us $\mathbb{I}_1$ and another gives us $\mathbb{I}_2$, then the two together produce $\mathbb{I}_{\text{total}} = \mathbb{I}_1 + \mathbb{I}_2$. For a set of iid draws, each draw has the same amount of information (i.e., the expected information matrix, which is a property of the distribution, not any one draw of data), so the total information from n data points is $n\mathbb{I}$.

The end result is the following form, which we can use to easily calculate the covariance matrix of the MLE parameter estimates.

$$\text{var}(MLE(\mathbf{x}, \beta)) = \frac{1}{nE_x(\mathbb{I})}. \qquad (10.1.9)$$

Equation 10.1.9 makes MLEs the cornerstone of statistics that they are. Given that MLEs achieve the CRLB for large n, they are asymptotically efficient, and if we want to test the parameter estimates we find via a t or F test, there is a relatively

---

[5]See Casella & Berger (1990, pp 310–311) for formal proofs of the statements in this section.

easy computation for finding the variances we would need to run such a test.[6] For many models (simulations especially), we want to know whether the outcome is sensitive to the exact value of a parameter, and the information matrix gives us a sensitivity measure for each parameter.

HOW TO EVALUATE A TEST   A hypothesis test can be fooled two ways: the hypothesis could be true but the test rejects it, or the hypothesis could be false but the test fails to reject it.

There is a balance to be struck between the two errors: as one rises, the other falls. But not all tests are born equal. If a hypothesis has a 50–50 chance of being true, then the coin-flip test, 'heads, accept; tails, reject' gives us a 50% chance of a Type I error and a 50% chance of a Type II error, but in most situations there are tests where both errors are significantly lower than 50%. By any measure we would call those better tests than the coin-flipping test.

> **Evaluation vocab**
>
> Here are some vocabulary terms; if you are in a stats class right now, you will be tested on this:
> Likelihood of a *Type I error* $\equiv \alpha$: rejecting the null when it is true.
> Likelihood of a *Type II error* $\equiv \beta$: accepting the null when it is false.
> *Power* $\equiv 1 - \beta$: the likelihood of rejecting a false null.
> *Unbiased*: $(1 - \beta) \geq \alpha$ for all values of the parameter. I.e., you are less likely to accept the null when it is false than when it is true.
> *Consistent*: the power $\to 1$ as $n \to \infty$.

A big help in distinguishing Type I from Type II error is that one minus the Type II error rate has the surprisingly descriptive name of *power*. To a high-power telescope, every star is slightly different—some things that seem like stars are even galaxies or planets. But to a low-power lens like the human eye, everything just looks like a little dot. Similarly, a high-power test can detect distinctions where a low-power test fails to reject the null hypothesis of no difference. Or, for the more cynical: since most journals have limited interest in publishing null results, a high-power test increases the odds that you will be able to publish results. As you can imagine, researchers are very concerned with maximizing power.

THE NEYMAN–PEARSON LEMMA   The Neyman–Pearson lemma (Neyman & Pearson, 1928a,b) states that a *likelihood ratio test* will have the minimum possible Type II error—the maximum power—of any test with a given level of α. After establishing this fact, we can select a Type I error level and be confident that we did the best we could with the Type II errors by using a likelihood ratio test.

---

[6]Of course, if we run one of the tests from Chapter 9 derived from the CLT, then we need to make sure the CLT holds for the maximum likelihood estimate of the statistic in question. This could be a problem for small data sets; for large data sets or simulations based on millions of runs, it is less of a concern.

Likelihood ratios   Say the cost of a test that correctly accepts or rejects the hypothesis is zero, the cost to a Type I error is C_I, and the cost to a Type II error is C_{II}. Then it is sensible to reject H_0 iff the expected cost of rejecting is less than the expected cost of not rejecting. That is, reject H_0 iff $C_I P(H_0|\mathbf{x}) < C_{II} P(H_1|\mathbf{x})$. We can translate this cost-minimization rule into the ratio of two likelihoods.

Recall Bayes's rule from page 258:

$$P(A|B) = \frac{P(B|A)P(A)}{P(B)}.$$

To apply Bayes's rule to the rejection test, set $A = H_0$ and $B = \mathbf{x}$, so $P(A|B) = P(H_0|\mathbf{x})$ and $P(B|A) = P(\mathbf{x}|H_0)$ (and similarly for H_1). Then:

$$C_I P(H_0|\mathbf{x}) < C_{II} P(H_1|\mathbf{x}) \tag{10.1.10}$$

$$C_I \frac{P(\mathbf{x}|H_0)P(H_0)}{P(\mathbf{x})} < C_{II} \frac{P(\mathbf{x}|H_1)P(H_1)}{P(\mathbf{x})} \tag{10.1.11}$$

$$c < \frac{P(\mathbf{x}|H_1)}{P(\mathbf{x}|H_0)} \tag{10.1.12}$$

Inequality 10.1.10 is the rejection rule from above; Inequality 10.1.11 uses Bayes's rule to insert the likelihood functions; Inequality 10.1.12 does some cross-division, canceling out the $P(\mathbf{x})$'s and defining the *critical value* $c \equiv C_I P(H_1)/C_{II} P(H_0)$, i.e., everything that doesn't depend on $\mathbf{x}$. If you tell me the shape of $P(\cdot|H_1)$ and $P(\cdot|H_0)$ and some number $\alpha \in (0, 1)$, then I can give you a value of c such that Inequality 10.1.12 is true with probability α.[7] The test will then be: gather the data, calculate the likelihood ratio on the right-hand side of Inequality 10.1.12, and reject H_0 iff the inequality is true.

The Neyman–Pearson lemma states that this test is the 'best' in the sense that for a Type I error fixed at α, the LR test minimizes the probability of a Type II error.[8] So we can design any test we like by just fixing α at a value with which we are comfortable (custom says to use 95 or 99%) and calculating a few likelihood functions, and we are assured that we did the best we could regarding Type II errors. Most standard tests can be expressed in a likelihood ratio form, and so Type II errors pretty much never get mentioned, since they're considered taken care of.[9]

---

[7]Alternatively, you could give me a ratio of costs C_I/C_{II} and I could again give you a value of c. Thus, one could draw a relation between the relative costs and the choice of α.

[8]For a proof, see e.g. Amemiya (1994, pp 189–191).

[9]Every test has a Type I and Type II error, but thanks to the Neyman–Pearson Lemma, we just describe a test using the Type I level, with phrases like *a test with 5% p-value*. The introductory chapter of Hunter & Schmidt (2004) is an excellent essay on how such description can be severely misleading. The extreme-case test *always fail to reject the null* has a 0% Type I error rate, but if the null hypothesis is false, then it is wrong 100% of the time.

> ➤ For any sufficiently well-specified model, you can find the probability that a given data set was produced via the model.

> ➤ If the data set consists of independent and identically distributed elements, then the likelihood is a product with one term for each data point. For both computational and analytic reasons, the log likelihood is easier to work with; the product then becomes a sum.

> ➤ The parameters that maximize the likelihood function for a model (or identically, maximize the log likelihood) will have the minimum variance among all unbiased estimators. The variance is a known quantity, given by the Cramér–Rao lower bound.

> ➤ Type I and Type II errors are complementary: as one goes up, the other generally goes down. However, given a Type I error level, different tests will have different Type II error levels. The Neyman–Pearson lemma guarantees that a likelihood ratio test has the minimum probability of Type II error for a fixed Type I error level.

10.2 DESCRIPTION: MAXIMUM LIKELIHOOD ESTIMATORS

Apophenia provides one function to find a model's optimum, `apop_maximum_likelihood`—but what a function it is. It provides a standardized interface to several types of optimization routines that take very different approaches toward finding optima. You will have to provide a log likelihood function, but if you are unable to provide the derivatives, the maximization routines will find them for you. Since the Cramér–Rao lower bound tells us the variance of a most-likely parameter, `apop_maximum_likelihood` will return a parametrized `apop_model` with the variances, along with other useful information. This section gives an overview of some standard optimization methods, and how to choose among them to raise the odds that they will find an optimum for your functions.

You may be wondering why you need to know these details. Isn't finding the optimum of a likelihood function a solved problem?

The answer is decidedly no. Most standard optimization algorithms are built to work well with a smooth, closed-form, globally concave function, such as finding the value of x that maximizes $f(x) = -x^2$. If your function more-or-less meets such conditions, the odds are good that the default optimization routine in any stats package of your choosing will work fine. But *anything* that produces a likelihood value could be a model: a simulation could be a model, where the likelihood is a function of how well the model matches a real-world data set. A dynamic programming problem could be a model. The consumer choosing among goods at the end of Chapter 4.7 was a model. If your model has a stochastic element, has multiple equilibria, or otherwise fails to fulfill the expectation of being a simple globally concave function, then you will need to tailor a method and settings around the problem at hand.[10]

```
1   #include <apop.h>
2
3   double sin_square(apop_data *data, apop_model *m){
4       double x = apop_data_get(m−>parameters, 0, −1);
5           return −sin(x)*gsl_pow_2(x);
6   }
7
8   apop_model sin_sq_model ={"−sin(x) times x^2",1, .p = sin_square};
```

Listing 10.1 A model to be optimized. Online source: `sinsq.c`.

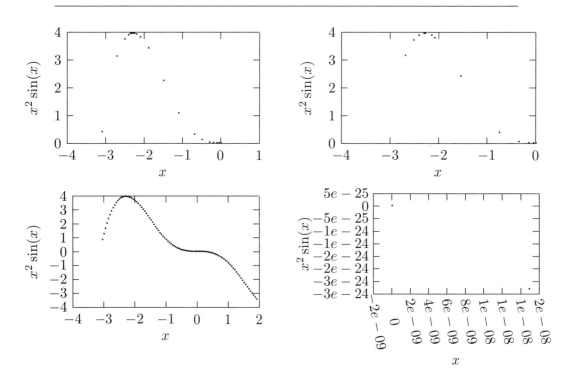

Figure 10.2 Top row: The simplex method and conjugate gradients; bottom row: simulated annealing and root search. [Online source: `localmax_print.c`]

[10]The problem of finding an optimum is so broad that there are large sections of optimization research that this book does not mention. For example, there is the broad and active field of *combinatorial optimization*, which covers questions like the optimal ordering of n elements given an optimization function (which is a problem with $n!$ options). See, e.g., Papadimitriou & Steiglitz (1998) for more on combinatorial optimization methods.

```
1    #include <apop.h>
2
3    apop_model sin_sq_model;
4
5    void do_search(int m, char *name){
6        double p[] = {0};
7        double result;
8            Apop_settings_add(&sin_sq_model, apop_mle, starting_pt, p);
9            Apop_settings_add(&sin_sq_model, apop_mle, method, m);
10           apop_model *out = apop_maximum_likelihood(NULL, sin_sq_model);
11           result = gsl_vector_get(out->parameters->vector, 0);
12           printf("The %s algorithm found %g.\n", name, result);
13   }
14
15   int main(){
16           apop_opts.verbose ++;
17           Apop_settings_add_group(&sin_sq_model, apop_mle, &sin_sq_model);
18           do_search(APOP_SIMPLEX_NM, "N−M Simplex");
19           do_search(APOP_CG_FR, "F−R Conjugate gradient");
20           do_search(APOP_SIMAN, "Simulated annealing");
21           do_search(APOP_RF_NEWTON, "Root−finding");
22   }
```

Listing 10.3 Using `apop_maximum_likelihood` and four types of method to solve for a maximum. Compile with `sinsq.c`. Online source: `localmax.c`.

Listing 10.1 presents a simple model, consisting of the equation $-x^2 \sin(x)$. This is a very simple equation, but it has an infinite number of local modes. As $x \to \infty$, the value of the function at these modes rises in proportion to x^2, so there is no global maximum. Figure 10.2 shows various attempts to search the function, one of which gives a very good picture of the shape of the curve.

- On lines 3–6, the function is defined. Because the system is oriented toward data analysis, the function takes in a data set and an `apop_model` holding the parameters. In this case, the data set is simply ignored.
- Line eight declares a new model with some of the information the MLE function will need, like the name, the number of parameters (one) and the probability function.

Listing 10.3 does the optimization four ways.

- The `apop_model` struct can hold an array of groups of settings. Line 17 adds to the model a group of MLE-appropriate settings. That group includes things like the starting point, method, tolerance, and many other details that you will see below.

- Lines eight and nine change the `starting_pt` and `method` elements of the model's MLE settings group to p and m, respectively.
- All the real work is done by line ten, which calls the maximization. The subsequent two lines interrogate the output.
- The `main` routine does the optimization with four different methods.

By inserting `Apop_settings_add(&sin_sq_model, apop_mle, trace_path, "`*`outfile`*`")` somewhere around line eight or nine, the system writes down the points it tests during its search for an optimum; you can then produce plots like those in Figure 10.2. You can already see disparate styles among the four methods: the simplex and conjugate gradient methods are somewhat similar in this case, but simulated annealing is much more exhaustive—you can clearly see the shape of the curve—while the root search barely leaves zero before deciding that it is close enough.

Chapter 11 will demonstrate another method of producing a picture of a function via random walk, but in the meantime, simulated annealing and `trace_path` provide a pretty effective way to get a graph of a complex and unfamiliar function.

When you run the program, you will see that asking four different methods gives you three different answers, which leads to an important lesson: *never trust a single optimization search*. By redoing the optimization at different points with different methods, you can get a better idea of whether the optimum found is just a local peak or the optimum for the entire function. See below for tips on restarting optimizations.

METHODS OF FINDING OPTIMA   Here is the problem: you want to find the maximum of $f(\mathbf{x})$, but only have time to evaluate the function and its derivative at a limited number of points. For example, $f(\mathbf{x})$ could be a complex simulation that takes in a few parameters and runs for an hour before spitting out a value, and you would like to have an answer this week.

Here are the methods that are currently supported by Apophenia. They are basically a standardization of the various routines provided by the GSL. In turn, the GSL's choice of optimization routines bears a close resemblance to those recommended by Press *et al.* (1988), so see that reference for a very thorough discussion of how these algorithms work, or see below for some broad advice on picking an algorithm. Also, Avriel (2003) provides a thorough, mathematician-oriented overview of optimization, and Gill *et al.* (1981) provides a more practical, modeler-oriented overview of the same topics.

Simplex method—Nelder–Mead   For a d dimensional search, this method draws
[APOP_SIMPLEX_NM]   a polygon with $d + 1$ corners and, at each step,
shifts the corner of the polygon with the small-
est function value to a new location, which may move the polygon or may contract
it to a smaller size. Eventually, the polygon should move to and shrink around the
maximum. When the average distance from the polygon midpoint to the $d + 1$
corners is less than tolerance, the algorithm returns the midpoint.

This method doesn't require derivatives at all.

Conjugate gradient   Including:
Polak–Ribiere [APOP_CG_PR]
Fletcher–Reeves [APOP_CG_FR]
Broyden–Fletcher–Goldfarb–Shanno [APOP_CG_BFGS]

Begin by picking a starting point and a direction, then find the minimum along that
single dimension, using a relatively simple one-dimensional minimization procedure like Newton's Method. Now you have a new point from which to start, and
the conjugate gradient method picks a new single line along which to search.

Given a direction vector $\mathbf{d}_1$, vector $\mathbf{d}_2$ is *orthogonal* iff $\mathbf{d}_1'\mathbf{d}_2 = 0$. Colloquially,
two orthogonal vectors are at right angles to each other. After doing an optimiza-
tion search along $\mathbf{d}_1$, it makes intuitive sense to do the next one-dimensional search
along an orthogonal vector. However, there are many situations where this search
strategy does not do very well—the optimal second direction is typically not at
right angles to the first.

Instead, a *conjugate gradient* satisfies $\mathbf{d}_1'\mathbf{A}\mathbf{d}_2 = 0$, for some matrix $\mathbf{A}$. Orthog-
onality is the special case where $\mathbf{A} = \mathbf{1}$. For quadratic functions of the form
$f(\mathbf{x}) = \frac{1}{2}\mathbf{x}'\mathbf{A}\mathbf{x} - \mathbf{b}\mathbf{x}$, a search along conjugate gradients will find the optimum
in as many steps as there are dimensions in $\mathbf{A}$. However, your function probably
only approximates a quadratic form—and a different quadratic form at every point,
meaning that for approximations at points one and two, $\mathbf{A}_1 \neq \mathbf{A}_2$. It is not necessary to actually calculate $\mathbf{A}$ at any point, but the quality of the search depends on
how close to a quadratic form your function is; the further from quadratic, the less
effective the method.

Polak–Ribiere and Fletcher–Reeves differ only in their choice of method to build
the next gradient from the prior; Press *et al.* (1988) recommend Polak–Ribiere.

The BFGS method is slightly different, in that it maintains a running best guess
about the *Hessian*, and uses that for updating the gradients.[11] However, the same

---

[11]Formally, one could argue that this means that it is not a conjugate gradient method, but I class it with the

general rules about its underlying assumptions hold: the closer your function is to a fixed function with smooth derivatives, the better BFGS will do.

Here is the pseudocode for the three algorithms. The variables in `teletype` are settings that can be tuned before calling the routine, as on lines 8 and 9 of Listing 10.3. For all routines, set `verbose` to 1 to see the progress of the search on screen.

Start at $p_0 = $ `starting_pt`, and gradient vector g_0.
While ($p_i \cdot g_i <$ `tolerance`$|p_i||g_i|$)
    Pick a new candidate point, $p_c = p_i + g_i \cdot$ `step_size`$/|g_i|$.
    If $p_c > p_i$
        $p_{i+1} \leftarrow p_c$
    Else
        $p_{i+1} \leftarrow$ the minimum point on the line between p_i and p_c (to within `tolerance`[12]).
    Find g_{i+1} using a method-specific equation.

We are guaranteed that $p_{i+1} \neq p_i$ because we followed the gradient uphill. If the function continues uphill in that direction, then p_c will be the next point, but if the function along that path goes uphill and then downhill, the next point will be in between p_i and p_c.

The step to calculate g_{i+1} is the only step that differs between the three methods. In all three cases, it requires knowing derivatives of the function at the current point. If your model has no `dlog_likelihood` function, then the system will use a numerical approximation.

Root finding   Including:
                Newton's method [APOP_RF_NEWTON]
Hybrid method [APOP_RF_HYBRID]
Hybrid method; no internal scaling [APOP_RF_HYBRID_NOSCALE]

A root search finds the optimum of the function by searching for the point where the first derivative of the function to be maximized is zero. Notice that this can't distinguish between maxima and minima, though for most likelihood functions, the minima are at $\beta = \pm\infty$.[13]

FR and PR methods because all three routines bear a very close resemblance, and all share the same pseudocode.

[12]Do not confuse the tolerance in these algorithms with a promise (for example, the promise in a Taylor expansion) that if the tolerance is τ and the true value of the parameter is β, then the MLE estimate will be such that $|\hat{\beta}_{MLE} - \beta| < \tau$. There is no way to guarantee such a thing. Instead, the tolerance indicates when the internal measure of change (for most algorithms, $\Delta f(\beta)$) is small enough to indicate convergence.

[13]In fact, it is worth making sure that your models do not have a divergent likelihood, where the likelihood increases as $\beta \to \infty$ or $-\infty$. Divergent likelihoods are probably a sign of a misspecified model.

Let $f(\mathbf{x})$ be the function whose root we seek (e.g., the score), and let ∇f be the matrix of its derivatives (e.g., the information matrix). Then one step in Newton's method follows

$$x_{i+1} \leftarrow x_i - (\nabla f(x_i))^{-1} f(x_i).$$

You may be familiar with the one-dimensional version, which begins at a point x_1, and follows the tangent at coordinate $(x_1, f(x_1))$ down to the $x-$axis; the equation for this is $x_2 = x_1 - f(x_1)/f'(x_1)$, which generalizes to the multidimensional version above.

The hybrid edition of the algorithm imposes a region around the current point beyond which the algorithm does not venture. If the tangent sends the search toward infinity, the algorithm follows the tangent only as far as the edge of the trust region. The basic hybrid algorithm uses a trust region with a different scale for each dimension, which takes some extra gradient-calculating evaluations and could make the trust region useless for ill-defined functions; the no-scaling version simply uses the standard Euclidian distance between the current and proposed point.

The algorithm repeats until the function whose zero is sought (e.g., the score) has a value less than `tolerance`.

Simulated annealing [APOP_SIMAN]   A controlled random walk. As with the other methods, the system tries a new point, and if it is better, switches. Initially, the system is allowed to make large jumps, and then with each iteration, the jumps get smaller, eventually converging. Also, there is some decreasing probability that if the new point is *less* likely, it will still be chosen. One reason for allowing jumps to less likely parameters is for situations where there may be multiple local optima. Early in the random walk, the system can readily jump from the neighborhood of one optimum to another; later it will fine-tune its way toward the optimum. Other motivations for the transition rules will be elucidated in the chapter on Monte Carlo methods.

Here is the algorithm in greater detail, with setting names in appropriate places:

Start with temp = `t_initial`
Let $x_0 \leftarrow$ `starting_point`
While temp $\geq$ `t_min`
    Repeat the following `iters_fixed_T` times:
        Draw a new point x_t, at most `step_size` units away from x_{t-1}.
        Draw a random number $r \in [0, 1]$.
        If $f(x_t) > f(x_{t-1})$
            Jump to the new point: $x_{t+1} \leftarrow x_t$.
        Else if $r < exp(-(x_{t-1} - x_t)/(\mathrm{k} \cdot \mathrm{temp}))$
            Jump to the new point: $x_{t+1} \leftarrow x_t$.
        Else remain at the old point: $x_{t+1} \leftarrow x_{t-1}$.
    Cool: Let temp $\leftarrow$ temp/`mu_t`

Unlike with the other methods, the number of points tested in a simulated anneal-ing run is not dependent on the function: if you give it a specification that reaches `t_min` in 4,000 steps, then that is exactly how many steps it will take. If you know your model is globally convex (as are most standard probability functions), then this method is overkill; if your model is a complex interaction, simulated annealing may be your only bet. It does not use derivatives, so if the derivatives do not exist or are ill-behaved, this is appropriate, but if they are available either analytically or via computation, the methods that use derivative information will converge faster.

If your model is stochastic, then methods that build up a picture of the space (no-tably conjugate gradient methods) could be fooled by a few early bad draws. Sim-ulated annealing is memoryless, in the sense that the only inputs to the next deci-sion to jump are the current point and the candidate. A few unrepresentative draws could send the search in the wrong direction for a while, but with enough draws the search could eventually meander back to the direction in which it should have gone.

Global v local optima   As you saw in the case of $-x^2 \sin(x)$, none of the methods guarantee that the optimum found is the global optimum, since there is no way for a computer to have global knowledge of a function $f(x)$ for all $x \in (-\infty, \infty)$. One option is to restart the search from a variety of starting points, in the hope that if there are multiple peaks, then different starting points will rise to different peaks.

The simulated annealing algorithm deals well with multiple peaks, because its search can easily jump from the neighborhood of one peak to that of another. In fact, as the number of steps in the simulated annealing algorithm $\to \infty$, the algo-rithm can be shown to converge to the global optimum with probability one. How-ever, calculating an infinite number of steps tends to take an unreasonable amount of time, so you will need to select a time-to-confidence trade-off appropriate to your situation.

$\mathbb{Q}_{10.3}$ | Lines 13 and 14 of Listing 10.3 set the key parameters of the method and the starting point. Try various values of both. Which do a better job of jumping out toward the larger modes? [*Bonus*: rewrite the program to take command-line switches using `getopt` so you can do this exercise from a batch file.]

$\mathbb{Q}_{10.4}$ | The `econ101` models from Chapter 4 provide the relatively rare situation where we have an optimization and the analytic values. [Hopefully your own simulations have at least one special case where this is also true.] This is therefore a fine opportunity to try various methods, values of delta, step size, tolerance, method, et cetera Do extreme prices or preferences create problems, and under which optimization settings?

RESTARTING   To reiterate a recommendation from above: *never trust a single optimization search*. But there are a number of ways by which you can order your multiple searches.

- You could start with a large step size and wide tolerance, so the search jumps around the space quickly, then restart with smaller step size and tolerance to hone in on a result.
- Along a similar vein, different search methods have different stopping criteria, so switching between a simplex algorithm and a conjugate gradient algorithm, for example, may lead to a better optimum.
- If you suspect there could be multiple local optima, then try different starting points—either different extremes of the space, or randomly-generated points.
- If you are running a constrained optimization, and one of the constraints binds, then there may be odd interactions between the penalty and the function being optimized. Try a series of optimizations with the penalty declining toward zero, to see if the optimum gets closer to the boundary.
- The `apop_estimate_restart` function will help you to run sequences of optimizations; see the online reference for details.

> ➤ Given an appropriate `apop_data` set and `apop_model`, the `apop_maximum_likelihood` function will apply any of a number of maximum-searching techniques to find the optimal parameters.

> ➤ No computational method can guarantee a global optimum, because the computer can only gather local information about the function. Restarting the search in different locations may help to establish a unique optimum or find multiple optima.

> ➤ You can try various methods in sequence using `apop_estimate_restart`. You can also use the restarting technique to do a coarse search for the neighborhood of an optimum, and then a finer search beginning where the coarse search ended.

10.3 MISSING DATA   Say that your data set is mostly complete, but has an NaN in observation fifty, column three. When you run a regression, the NaN's propagate, and you wind up with NaN's all over the parameter estimates. How can you fill in the missing data?

We could turn this into an MLE problem: we seek the most likely values to fill in for the NaN's, based on some model of how the data was generated.

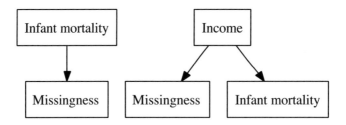

Figure 10.4 Two different flows of causation. At left is MAR: low income causes high infant mortality, and causes infant mortality data to be missing. At right is non-ignorable missingness:
the value of the infant mortality statistic determines whether infant mortality data will
be missing. [online source: `mar.dot`]

But first, we need to distinguish among three types of missingness. Data are *missing completely at random* (MCAR) when the incidence of missing data is uncorrelated to every variable in the data set. This is truly haphazard error: somebody
tripped over the meter's power cord, or one of the surveyors was drunk. The cause
of the missing data is nowhere in the data set.

Data are *missing at random* (MAR) when the incidence of missing data in column i is uncorrelated to the existing data in column i once we condition on the
observed data in all other columns. For example, poor countries tend to have bad
demographic data, so the incidence of a missing infant mortality rate is correlated
to low GNP per capita. Once we have conditioned on GNP per capita, there is no
reason to expect that missingness is correlated to infant mortality. The cause of the
missing values in column i is something in the data set, but not column i. As in
the right-hand diagram in Figure 10.4, there is no flow of causation from infant
mortality to missingness.

Conversely, say that governments are embarrassed by high infant mortality rates.
Statistics bureaux are under orders to measure the rate, but release the measure
only if it falls below a certain threshold. In this case, the incidence of missing data
is directly related to the value of the missing data. The cause of the missing data
is the value of the data. This is known as *missing not at random* (MNAR) or *non-
ignorable missingness*, and is a serious problem because it implies bias almost by
definition.

There are many methods for dealing with censored or otherwise non-ignorable
missingness discussed in many sources, such as Greene (1990). For a full discussion of the many types of missing data, see Allison (2002).

Listwise deletion   One option for dealing with NaN's that are MCAR is listwise deletion. The idea here is supremely simple: if a row is missing data for any variable, then throw out the entire row. This is conceptually simple, does not impose any additional structure or model on the data, and can be executed in one line of code:

```
apop_data *nan_free_data = apop_data_listwise_delete(dirty_data);
```

Alternatively, see page 105 for the syntax to do listwise deletion on the SQL side.

But listwise deletion isn't always appropriate. As a worst-case situation, say that a survey has a hundred questions, and everybody filled out exactly 99 of them. By listwise deletion, we would throw out the entire data set.

But with listwise deletion, the data set is going to be shorter, meaning that we lose information, and if data is MAR (not MCAR), then throwing out observations with missing data means biasing the information among other variables. In the example above, if we throw out countries with missing infant mortality data, we would mostly be throwing out countries with low GDP per capita.

ML imputation   This is where maximum likelihood estimation comes in (Dempster *et al.*, 1977). Let the missing data be β, and the existing data (with holes) be $\mathbf{X}$, as usual. Then our goal is to find the most likely value of β. The first step is to specify a model from which the data was allegedly generated, so that we can evaluate the likelihood of any given β. The norm is that the completed data set has a Multivariate Normal distribution, meaning that the n columns of the data are distributed as an n-dimensional bell curve with mean vector μ and covariance matrix Σ. However, it may make sense for your data to take on any of a number of other forms. But given a parametrized distribution, one could search for the data points that are most likely to have occurred.

In the `data-corruption.db` database, you will find Transparency International's Corruption Perceptions Index from 2003–2006. Because the index depends on about a dozen hard-to-gather measures, there are many missing data points. Listing 10.5 goes through the entire process of filling in those data points, by pulling the data from the database, reducing it to an estimable subset via listwise deletion, fitting a Multivariate Normal to the subset, and then filling in the NaN's in the full data set via maximum likelihood. It may run slowly: filling in about eighty NaN's means a search in an 80-dimensional space. For more missing data than this, you are probably better off finding a means of dividing the data set or otherwise incrementally filling in the blanks.

You are encouraged to look at the results and decide whether they seem plausible.

```
#include <apop.h>

int main(){
    apop_db_open("data-corruption.db");
    apop_data *corrupt = apop_db_to_crosstab("cpi", "country", "year", "score");
    apop_data *clean = apop_data_listwise_delete(corrupt);
    apop_model *mlv = apop_estimate(clean, apop_multivariate_normal);
    apop_ml_imputation(corrupt, mlv);
    apop_crosstab_to_db(corrupt, "cpi_clean", "country", "year", "score");
}
```

Listing 10.5 Filling in NaN's via a Multivariate Normal. Online source: `corrupt.c`.

For example, would you use the data for Yugoslavia? Is a Multivariate Normal the most appropriate model of how the data was formed?

On lengthy surveys, few if any people successfully fill out the entire form. In the worst case, we may have 100 questions, and all subjects answered 99 of them. Listwise deletion would throw out every subject. In this case, the best bet is *pairwise deletion*: calculate the mean of each vector by just ignoring NaN's, and the covariance of each pair of columns by removing only those observations that have an NaN for one of those two variables. Pairwise deletion can introduce odd biases in the covariance matrix, so it should be used as a last resort.

10.4 TESTING WITH LIKELIHOODS

In order to test a hypothesis regarding a statistic, we need to have a means of describing its theoretical distribution. When the statistic is an MLE, the standard means of doing this is via two interlocking approximations: a Taylor expansion and a Normal approximation. This is convenient because the Normal approximation proves to be innocuous in many settings, and the Taylor expansion involves the same cast of characters with which we have been dealing to this point—LL, S, and $\mathcal{I}$.

USING THE INFORMATION MATRIX

The Cramér–Rao lower bound gives us a value for the variance of the parameters: the inverse of the expected information matrix. Given a variance on each parameter, we can do the same t and F tests as before.

It is even easier if we make one more approximation. The *expected* information matrix is an expectation over all possible parameters, which means that it is a property of the model, not of any one set of parameters. Conversely, the *estimated* information matrix is the derivative of the score around the most likely values of the parameters. We can expect that it is different for different parameter estimates.

Efron & Hinkley (1978) found that for most of the applications they consider, the inverse of the estimated information matrix is preferable as an estimator of the variance as the expected information matrix. For *exponential family* distributions, the two are identical at the optimum. From a computational perspective, it is certainly preferable to use the estimated information, because it is a local property of the optimum, not a global property of the entire parameter space. Simulated annealing does a decent job of sampling the entire space, but the other methods go out of their way to not do so, meaning that we would need to execute a second round of data-gathering to get variances. Apophenia's maximum likelihood estimation returns a covariance matrix constructed via the estimated information matrix.

The covariance matrix provides an estimate of the stability of each individual parameter, and allows us to test hypotheses about individual parameters (rather than tests about the model as a whole, as done by the likelihood ratio methods discussed below).[14] However, there are a number of approximations that had to be made to get to this point. Basically, by applying a t test, we are assuming that a few million draws of a parameter's MLE (generated via a few million draws of new data) would be asymptotically Normally distributed. We already encountered this assumption earlier: when testing parameters of a linear regression we assume that the errors are Normally distributed. So the same caveats apply, and if you have a means of generating several data sets, you could test for Normality; if you do not, you could use the methods that will be discussed in Chapter 11 to bootstrap a distribution; or if you are working at a consulting firm, you could just assume that Normality always holds.

There is no sample code for this section because you already know how to run a t test given a statistic's mean and its estimated variance.

USING LIKELIHOOD RATIOS   We can use likelihood ratio (LR) tests to compare models. For example, we could claim that one model is just like another, but with the constraint that $\beta_{12} = 0$, and then test whether the constraint is actually binding via the ratio of the likelihood with the constraint and the likelihood without. Or, say that we can't decide between using an OLS model or a probit model; then the ratio of the likelihood of the two models can tell us the confidence with which one is more likely than the other.

---

[14]In Klemens (2007), I discuss at length the utility of the variance of the MLE as a gauge of which of a simulation's parameters have a volatile effect on the outcome and which have little effect.

A loose derivation   As intimated by the Neyman–Pearson lemma, the ratio of two
                          likelihoods is a good way to test a hypothesis. Given the ratio
of two likelihoods P_1/P_2, the log is the difference $\ln(P_1/P_2) = LL_1 - LL_2$.

Now consider the Taylor expansion of a log likelihood function around $\hat{\beta}$. The
Taylor expansion is a common means of approximating a function $f(x)$ via a series
of derivatives evaluated at a certain point. For example, the second-degree Taylor
expansion around seven would be $f(x) \approx f(7)+(x-7)f'(7)+(x-7)^2 f''(7)/2+\epsilon$,
where ϵ is an error term. The approximation is exactly correct at $x = 7$, and
decreasingly precise (meaning ϵ gets larger) for values further from seven. In the
case of the log likelihood expanded around $\hat{\beta}$, the Taylor expansion is

$$LL(\beta) = LL(\hat{\beta}) + (\beta - \hat{\beta})LL'(\hat{\beta}) + \frac{(\beta - \hat{\beta})^2}{2}LL''(\hat{\beta}) + \epsilon.$$

As per the definitions from the beginning of the chapter, the derivative in the second term is the score, and the second derivative in the third term is $-\mathbb{I}$. When $\hat{\beta}$
is the optimum, the score is zero. Also, as is the norm with Taylor expansions, we
will assume $\epsilon = 0$. Then the expansion simplifies to

$$LL(\beta) = LL(\hat{\beta}) - \frac{(\beta - \hat{\beta})^2}{2}\mathbb{I}(\hat{\beta}). \tag{10.4.1}$$

Typically, the likelihood ratio test involves the ratio of an unrestricted model and
the same model with a constraint imposed. Let LL_c be the constrained log likelihood; then we can repeat Equation 10.4.1 with the constrained log likelihood:

$$LL_c(\beta) = LL_c(\hat{\beta}) - \frac{(\beta - \hat{\beta})^2}{2}\mathbb{I}_c(\hat{\beta}).$$

Now the hypothesis: the constraint is not binding, and therefore both constrained
and unconstrained optimizations find the same value of $\hat{\beta}$. Then

$$-2(LL(\beta) - LL_c(\beta)) = 2LL_c(\hat{\beta}) - 2LL(\hat{\beta}) + (\beta - \hat{\beta})^2\mathbb{I}(\hat{\beta}) - (\beta - \hat{\beta})^2\mathbb{I}_c(\hat{\beta})$$
$$= (\beta - \hat{\beta})^2 \left(\mathbb{I}(\hat{\beta}) - \mathbb{I}_c(\hat{\beta})\right) \tag{10.4.2}$$

The second equation follows from the first because having the same value for $\hat{\beta}$
for constrained and unconstrained optimizations means that $LL(\hat{\beta}) = LL_c(\hat{\beta})$.

But we still haven't said anything about the distribution of $-2(LL(\beta) - LL_c(\beta))$.
Consider the case of the Normal distribution with fixed σ (so the only free param-

eter is the mean μ); there, the Score is

$$S(\mathbf{x}, \mu) = \sum_i (x_i - \mu)/\sigma^2. \qquad (10.4.3)$$

$\mathbb{Q}$: Verify this by finding $\partial LL(\mathbf{x}, \mu)/\partial \mu$ using the probability distribution on page 241.

The right-hand side of Equation 10.4.3 takes the familiar mean-like form upon which the CLT is based, and so is is Normally distributed. Since $E(\mathbb{I}) = E(S \cdot S)$, and a Normally-distributed statistic squared has a χ^2 distribution, Expression 10.4.2 has a χ^2 distribution.

And in fact, this holds for much more than a Normal likelihood function (Pawitan, 2001, p 29). Say that there exists a transformation function $t(x, \beta)$ such that $P(x, \beta) \cdot t(x, \beta)$ is Normally distributed. Then

$$\frac{P(x, \beta) \cdot t(x, \beta)}{P_c(x, \beta) \cdot t(x, \beta)} = \frac{P(x, \beta)}{P_c(x, \beta)}.$$

Instead of canceling out the transformation here, we could also cancel it out in the log likelihood step:

$$LL(x, \beta) + t(x, \beta) - LL_c(x, \beta) - t(x, \beta) = LL(x, \beta) - LL_c(x, \beta).$$

Either way, Expression 10.4.2 is the same with or without the transformation—which means the untransformed version is also $\sim \chi^2$. So provided the likelihood function is sufficiently well-behaved that $t(x, \beta)$ could exist, we don't have to worry about deriving it. This is a specific case of the *invariance principle* of likelihood functions, that broadly says that transformations of the likelihood function do not change the information embedded within the function.

This is what we can use to do the likelihood ratio tests that the Neyman–Pearson lemma recommended. We find the log likelihood of the model in its unconstrained and constrained forms, take two times the difference, and look up the result in the χ^2 tables.

The LR test, constraint case   As above, the typical likelihood ratio test involves the ratio of an unrestricted and a restricted model, and a null hypothesis that the constraint is not binding. Let P be the (not-log, plain) likelihood of the overall model, and P_c be the likelihood of a model with K restrictions, such as K parameters fixed as zero.

In this context, the above discussion becomes

$$-2 \ln \frac{P}{P_c} = -2[\ln P - \ln P_c] \sim \chi_K^2. \qquad (10.4.4)$$

In modeling terms, the unrestricted model could be any of the models discussed earlier, such as the `apop_probit`, `apop_normal`, or even `apop_ols`, because the

OLS parameters ($\beta_{\text{OLS}} = (\mathbf{X'X})^{-1}\mathbf{Xy}$) can be shown to be identical to the maximum likelihood estimate of β.

```
1   #include "eigenbox.h"
2
3   double linear_constraint(apop_data * d, apop_model *m){
4       apop_data *constr = apop_line_to_data((double[]) {0, 0, 0, 1}, 1, 1, 3);
5       return apop_linear_constraint(m->parameters->vector, constr, 0);
6   }
7
8   void show_chi_squared_test(apop_model *unconstrained, apop_model *constrained, int
        constraints){
9       double statistic = 2 * (unconstrained->llikelihood - constrained->llikelihood);
10      double confidence = gsl_cdf_chisq_P(statistic, constraints);
11      printf("The Chi squared statistic is: %g, so reject the null of non-binding constraint "
12          "with %g%% confidence.\n", statistic, confidence*100);
13  }
14
15  int main(){
16      apop_data *d = query_data();
17      apop_model *unconstr = apop_estimate(d, apop_ols);
18      apop_model_show(unconstr);
19
20      Apop_settings_add_group(&apop_ols, apop_mle, &apop_ols);
21      Apop_settings_add(&apop_ols, apop_mle, starting_pt, unconstr->parameters->vector->
            data);
22      Apop_settings_add(&apop_ols, apop_mle, use_score, 0);
23      Apop_settings_add(&apop_ols, apop_mle, step_size, 1e-3);
24      apop_ols.estimate = NULL;
25      apop_ols.constraint = linear_constraint;
26      apop_model *constr = apop_estimate(d, apop_ols);
27      printf("New parameters:\n");
28       apop_vector_show(constr->parameters->vector);
29      show_chi_squared_test(unconstr, constr, 1);
30  }
```

Listing 10.6 Comparing three different models using likelihood ratios: an OLS model, an OLS model with constraint, and a logit model. Online source: `lrtest.c`.

Listing 10.6 presents an unconstrained and a constrained optimization. It uses the query from page 267 that produces a data set whose outcome variable is males per 100 females, and whose independent variables are population and median age. The question *is the coefficient on median age significant?* can be rephrased to: *if we constrain the median age coefficient to zero, does that have a significant effect on the log likelihood?*

• The unconstrained optimization, on line 17, is the ordinary least squares model (which, as above, finds the MLE).

- Lines 20–24 mangle the base OLS model into a constrained model estimated via maximum likelihood. By setting the `estimate` element to `NULL` the estimation on line 25 uses the default method, which is maximum likelihood estimation.

- The constraint function is on line 3, and it uses the `apop_linear_constraint` function to test that an input vector satisfies a constraint expressed as an `apop_-data` set of the type first introduced on page 152. In this case, the constraint is $0 < \beta_3$; since the unconstrained OLS estimation finds that $\beta_3 < 0$, this is a binding constraint.

- Line four uses a new syntactic trick: anonymous structures. The type-in-parens form, (`double []`), looks like a type cast, and it basically acts that way, declaring that the data in braces is a nameless array of `doubles`. The line is thus equivalent to two lines of code, as at the top of the `main` routine in the `ftest.c` program on page 311:

```
double tempvar[] = {0, 0, 0, 1};
apop_line_to_data(tempvar, 1, 1, 3);
```

But we can get away with packing it all onto one line and not bothering with the temp variable. When used in conjunction with designated initializers, anonymous structs can either convey a lot of information onto one line or make the code an unreadable mess, depending on your æsthetic preferences.

- By commenting out the constraint-setting on line 24, you will have an unconstrained model estimated via maximum likelihood, and can thus verify that the OLS parameters and the MLE parameters are identical.

- You will recognize the function on line nine as a direct translation of Expression 10.4.4. It is thus a test of the claim that the constraint is not binding, and it rejects the null with about the same confidence with which the t test associated with the linear regression rejected the null that the third parameter is zero.

- The statistic on line nine, $LL - LL_c$, is always positive, because whatever optimum the constrained optimization found could also be used by the unconstrained model, and the unconstrained model could potentially find something with even higher likelihood. If this term is negative, then it is a sign that the unconstrained optimization is still far from the true optimum, so restart it with a new method, new starting point, tighter tolerances, or other tweaks.

Be sure to compare the results of the test here with the results of the F test on page 311.

The LR test, non-nested case   The above form is a test of two *nested* models, where one is a restricted form of the other, so under the hypothesis of the nonbinding constraint, both can find the same estimate $\hat{\beta}$ and so both can conceivably arrive at the same log likelihood. If this is not the case, then the cancellation of the first part of the Taylor expansion in Equation 10.4.2 does not happen.

In this case (Cox, 1962; Vuong, 1989), the statistic and its distribution is

$$\frac{\ln \frac{P_1}{P_2} - E\left(\ln \frac{P_1}{P_2}\right)}{\sqrt{n}} \sim \mathcal{N}(0,1). \tag{10.4.5}$$

The denominator is simply the square root of the sample size. The first part of the numerator is just $LL_1 - LL_2$, with which we are very familiar at this point. The expected value is more problematic, because it is a global value of the log likelihoods, which we would conceivably arrive at by a probability-weighted integral of $LL_1(\beta) - LL_2(\beta)$ over the entire space of βs.

Alternatively, we could just assume that it is zero. That is, the easiest test to run with Expression 10.4.5 is the null hypothesis of no difference between the expected value of the two logs.

```
#define TESTING
#include "dummies.c"

void show_normal_test(apop_model *unconstrained, apop_model *constrained, int n){
    double statistic = (unconstrained->llikelihood − constrained->llikelihood)/sqrt(n);
    double confidence = gsl_cdf_gaussian_P(fabs(statistic), 1); //one−tailed.
    printf("The Normal statistic is: %g, so reject the null of no difference between models "
        "with %g%% confidence.\n", statistic, confidence*100);
}

int main(){
    apop_db_open("data−metro.db");
    apop_model *m0 = dummies(0);
    apop_model *m1 = dummies(1);
    show_normal_test(m0, m1, m0->data->matrix->size1);
}
```

Listing 10.7 Compare the two Metro ridership models from page 282 Online source: `lrnonnest.c`.

Listing 10.7 reads in the code for the two OLS estimations of Washington Metro ridersiop from page 282, one with a zero-one dummy and one with a dummy for the year's slope.

- If you flip back to the `dummies.c` file, you will see that the `main` function is wrapped by a preprocessor if-then statement: `#ifndef TESTING`. Because `TESTING` is defined here, the `main` function in that file will be passed over.
- Therefore, the next line can read in `dummies.c` directly, without ending up with two `main`s.

- The `main` function here simply estimates two models, and then calls the `show_-normal_test` function, which is a translation of Expression 10.4.5 under the null hypothesis that $E(LL_{\mathrm{OLS}} - LL_{\mathrm{logit}}) = 0$.

Remember, a number of approximations underly both the nested and non-nested LR tests. In the nested case, they are generally considered to be innocuous and are rarely verified or even mentioned. For the non-nested probit and logit models, their log likelihoods behave in a somewhat similar manner (as $n \rightarrow \infty$), so it is reasonable to apply the non-nested statistic above. But for two radically different models, like an OLS model versus an agent-based model, the approximations may start to strain. You can directly compare the two log-likelihoods, and the test statistic will give you a sense of the scale of the difference, but from there it is up to you to decide what these statistics tell you about the two disparate models.

MONTE CARLO

Monte Carlo (Italian and Spanish for Mount Carl) is a city in Monaco famous for its casinos, and has more glamorous associations with its name than Reno or Atlantic City.

Monte Carlo methods are thus about randomization: taking existing data and making random transformations to learn more about it. But although the process involves randomness, its outcome is not just the mere whim of the fates. At the roulette table, a single player may come out ahead, but with millions of suckers testing their luck, casinos find that even a 49–51 bet in their favor is a reliable method of making money. Similarly, a single random transformation of the data will no doubt produce a somehow distorted impression, but reapplying it thousands or millions of times will present an increasingly accurate picture of the underlying data.

This chapter will first look at the basics of random number generation. It will then discuss the general process of describing parameters of a distribution, parameters of a data set, or a distribution as a whole via Monte Carlo techniques. As a special case, bootstrapping is a method for getting a variance out of data that, by all rights, should not be able to give you a variance. Nonparametric methods also make a return in this chapter, because shuffling and redrawing from the data can give you a feel for the odds that some hypothesized event would have occurred; that is, we can write hypothesis tests based on resampling from the data.

```
gsl_rng *apop_rng_alloc(int seed){
    static int first_use = 1;
    if (first_use){
        first_use −−;
        gsl_rng_env_setup();
    }
    gsl_rng *setme = gsl_rng_alloc(gsl_rng_taus2);
    gsl_rng_set(setme, seed);
    return setme;
}
```

Listing 11.1 Allocating and initializing a random number generator.

11.1 RANDOM NUMBER GENERATION

We need a stream of random numbers, but to get any programming done, we need a *replicable* stream of random numbers.[1]

There are two places where you will need replication. The first is with debugging, since you don't want the segfault you are trying to track down to appear and disappear every other run. The second is in reporting your results, because when a colleague asks to see how you arrived at your numbers, you should be able to reproduce them exactly.

Of course, using the same stream of numbers every time creates the possibility of getting a lucky draw, where *lucky* can mean any of a number of things. The compromise is to use a collection of deterministic streams of numbers that have no apparent pattern, where each stream of numbers is indexed by its first value, the *seed*.[2] The GSL implements such a process.

Listing 11.1 shows the innards of the `apop_rng_alloc` function from the Apophenia library to initialize a `gsl_rng`. In all cases, the function takes in an integer, and then sets up the random number generation (RNG) environment to produce new numbers via the Tausworth routine. Fans of other RNG methods can check the GSL documentation for setting up a `gsl_rng` with alternative algorithms. On the first call, the function calls the `gsl_rng_env_setup` function to work some internal magic in the GSL. Listings 11.6 and 11.7 below show an example using this function.

---

[1] Is it valid to call a replicable stream of seemingly random numbers *random*? Because such RNGs are arguably not random, some prefer the term *pseudorandom number generator* (PRNG) to describe them. This question is rooted in a philosophical question into which this book will not delve: what is the difference between perceived randomness given some level of information and true randomness? See, e.g., Good (1972, pp 127–8).

[2] Formally, the RNG produces only one stream, that eventually cycles around to the beginning. The seed simply specifies where in the cycle to begin. But because the cycle is so long, there is little loss in thinking about each seed producing a separate stream.

If you initialize one RNG stream with the value of another, then they are both at the same point in the cycle, and they will follow in lock-step from then on. This is to be avoided; if you need a sequence of streams, you are better off just using a simple list of seeds like 0, 1, 2, ...

RANDOM NUMBER DISTRIBUTIONS   Now that you have a random number generator, here are some functions that use it to
draw from all of your favorite distributions. Input an RNG as allocated above plus
the appropriate parameters, and the GSL will transform the RNG as necessary.

```
double gsl_ran_bernoulli (gsl_rng *r, double p);
double gsl_ran_beta (gsl_rng *r, double a, double b);
double gsl_ran_binomial (gsl_rng *r, double p, int n);
double gsl_ran_chisq (gsl_rng *r, double df);
double gsl_ran_fdist (gsl_rng *r, double df1, double df2);
double gsl_ran_gaussian (gsl_rng *r, double sigma);
double gsl_ran_tdist (gsl_rng *r, double df);
double gsl_ran_flat (gsl_rng *r, double a, double b);
double gsl_rng_uniform (gsl_rng *r);
```

- The *flat* distribution is a Uniform[A,B) distribution. The Uniform[0,1) distribution gets its own no-options function, `gsl_rng_uniform(r)`.

- The Gaussian draw assumes a mean of zero, so if you intend to draw from, e.g., a $\mathcal{N}(7, 2)$, then use `gsl_ran_gaussian(r, 2) + 7`.

- The `apop_model` struct includes a `draw` method that works like the above functions to make random draws, and allows standardization with more exotic models like the histogram below; see the example in Listing 11.5, page 361.

An example: the Beta distribution   The Beta distribution is wonderful for all sorts
of modeling, because it can describe such a
wide range of probability functions for a variable $\in [0, 1]$. For example, you saw
it used as a prior on page 259. But its α and β parameters may be difficult to
interpret; we are more used to the mean and variance. Thus, Apophenia provides
a convenience function, `apop_beta_from_mean_var`, that takes in μ and σ^2 and
returns an appropriate Beta distribution, with the corresponding values of α and β.

As you know, the variance of a Uniform$[0, 1]$ is exactly $\frac{1}{12}$, which means that
the Beta distribution will never have a variance greater than $\frac{1}{12}$ (and close to $\frac{1}{12}$,
perverse things may happen computationally for $\mu \not\approx \frac{1}{2}$).[3] The mean of a function
that has positive density iff $x \in [0, 1]$ must be $\in (0, 1)$. If you send `apop_beta_-`
`from_mean_var` values of μ and σ^2 that are outside of these bounds, the function
will return `GSL_NAN`.

What does a Beta distribution with, say, $\mu = \frac{3}{8}, \sigma^2 = \frac{1}{24} = .041\overline{66}$ look like?
Listing 11.2 sets up an RNG, makes a million draws from a Beta distribution, and
plots the result.

---

[3]More trivia: the Uniform$[0, 1]$ is symmetric, so its skew is zero. Its kurtosis is $\frac{1}{80}$.

```
1   #include <apop.h>
2
3   int main(){
4     int  draws = 1e7;
5     int  bins = 100;
6     double mu  = 0.492; //also try 3./8.
7     double sigmasq = 0.093; //also try 1./24.
8     gsl_rng *r  = apop_rng_alloc(0);
9     apop_data *d  = apop_data_alloc(0, draws, 1);
10      apop_model *m = apop_beta_from_mean_var(mu, sigmasq);
11      for (int i =0; i < draws; i++)
12          apop_draw(apop_data_ptr(d, i, 0), r, m);
13      Apop_settings_add_group(&apop_histogram, apop_histogram, d, bins)
14      apop_histogram_normalize(&apop_histogram);
15      apop_plot_histogram(&apop_histogram, NULL);
16  }
```

Listing 11.2 Building a picture of a distribution via random draws. Online source: `drawbeta.c`.

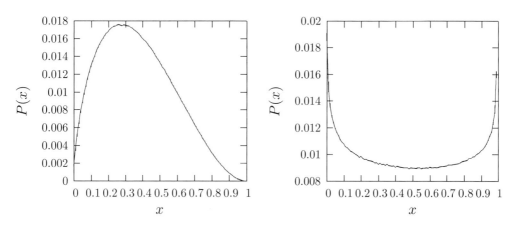

Figure 11.3 The flexible Beta distribution. Run via `drawbeta | gnuplot`.

- Most of the code simply names constants; the first real action occurs on line 11, where `apop_beta_from_mean_var` takes in μ and σ^2 and returns an `apop_model` representing a Beta distribution with the appropriate parameters.

- In line 14, the data set `d` is filled with random draws from the model. The `apop_draw` function takes in a pointer-to-`double` as the first argument, and puts a value in that location based on the RNG and model sent as the second and third arguments. Thus, you will need to use `apop_data_ptr`, `gsl_vector_ptr`, or `gsl_matrix_ptr` with `apop_draw` to get a pointer to the right location. The slightly awkward pointer syntax means that no copying or reallocation is necessary, so there is less to slow down drawing millions of numbers.

- The final few lines of code take the generic `apop_histogram` model, set it to use

the data d and the given number of bins, normalize the histogram thus produced to
have a total density of one, and plot the result.

The output of this example (using 1e7 draws) is at left in Figure 11.3; by contrast,
the case where $\mu = 0.492$, $\sigma^2 = 0.093$ is pictured at right.

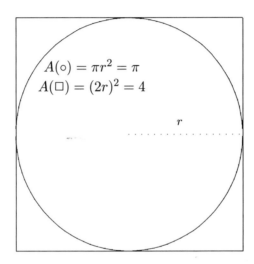

Figure 11.4 A circle inscribed in a square. The ratio of the area of the circle to the area of the square
is $\pi/4$. Online source: `squarecircle.gnuplot`.

$Q_{11.1}$

As per Figure 11.4, when a circle is inscribed inside a square, the ratio of
the area of the circle to the square is $\pi/4$. Thus, if we randomly draw 100
points from the square, we expect $100\pi/4$ to fall within the circle.
Estimate π via random draws from a square. For i in zero to about 1e8:

- Draw x_i from a Uniform$[-1, 1]$ distribution; draw y_i from a
  Uniform$[-1, 1]$ distribution.

- Determine whether (x_i, y_i) falls within the unit circle, meaning that
  $\sqrt{x_i^2 + y_i^2} \leq 1$ (which is equivalent to $x_i^2 + y_i^2 \leq 1$).

- Every 10,000 draws, display the proportion of draws inside the circle
  times four. How close to π does the estimate come? (*Hint*: it may be
  clearer to display `fabs(M_PI - pi_estimate)` instead of the estimate itself.)

DRAWING FROM YOUR OWN DATA   Another possibility, beyond drawing from famous distributions that your data theoretically approximates, would be to draw from your actual data.

If your data are in a vector, then just draw a random index and return the value at that index. Let r be an appropriately initialized gsl_rng, and let *your_data* be a gsl_vector from which you would like to make draws. Then the following one-liner would make a single draw:

gsl_vector_get(your_data, gsl_rng_uniform_int(r, your_data−>size));

⁂ *Drawing from histograms*   There are a few reasons for your data to be in a histogram form—a rough probability mass function—like the ones used for plotting data in Chapter 5, for describing data in Chapter 7, for goodness of fit tests in Chapter 9, and as the output from the apop_update function. Here, we will draw from them to produce artificial data sets.

```
1   #include <apop.h>
2   gsl_rng *r;
3
4   void plot_draws(apop_model *m, char *outfile){
5       int draws = 2e3;
6       apop_data *d = apop_data_alloc(0, draws,1);
7           for(size_t i=0; i < draws; i++)
8               apop_draw(apop_data_ptr(d, i, 0), r, m);
9           apop_model *h3 = apop_estimate(d, apop_histogram);
10          apop_histogram_print(h3, outfile);
11  }
12
13  int main(){
14      r = apop_rng_alloc(1);
15      apop_db_open("data−wb.db");
16      apop_data *pops = apop_query_to_data("select population+0.0 p from pop where p>500");
17      apop_model *h = apop_estimate(pops, apop_histogram);
18      apop_histogram_print(h, "out.hist");
19      plot_draws(apop_estimate(pops, apop_lognormal), "lognormal_fit");
20      printf("set xrange [0:2e5]; set yrange [0:0.12]; \n \
21              plot 'out.hist' with boxes, 'lognormal_fit' with lines\n");
22  }
```

Listing 11.5 Draw from the actual histogram of country populations, and from the exponential distribution that most closely fits. Online source: drawfrompop.c.

For example, say that we would like to generate communities whose populations are distributed the way countries of the world are distributed. Listing 11.5 does this two ways. The simplest approach is to simply generate a histogram of world populations and make draws from that histogram.

- Line 17 creates a filled histogram by filling the un-parametrized base `apop_-histogram` model with a list of populations.

In some cases, there are simply not enough data for the job. The World Bank data set lists 208 countries; if your simulation produces millions of communities, the repetition of 208 numbers could produce odd effects. The solution presented here is to estimate a Lognormal distribution, and then draw from that ideal distribution. Line 21 does the model fit and then sends the output of the `apop_estimate` function to the `plot_draws` function, which makes a multitude of draws from the ideal distribution, and then plots those. You can see that the result is smoother, without the zero-entry bins that the real-world data has.

- Lines 7–8 fill column zero of `d` with data, then line 9 turns that data into a histogram.
- The easiest way to put two data sets on one plot is to write both of them to separate files (lines 18 and 10), and then call those files from a Gnuplot script (lines 20–21).

※ *Seeding with the time*   There are situations where a fixed seed is really not what you want—you want different numbers every time you run your program. The easiest solution is to seed using the `time` function. The standard library function `time(NULL)` will return the number of seconds that have elapsed since the beginning of 1970, which is roughly when UNIX and C were first composed. As I write this, `time` returns 1,147,182,523—not a very good way to tell the time. There are a number of functions that will turn this into hours, minutes, or months; see any reference on C's standard library (such as the GNU C Library documentation) for details. But this illegible form provides the perfect seed for an RNG, and you get a new one every second.

Listing 11.6, `time.c`, shows a sample program that produces ten draws from an RNG seeded with the time. This is not industrial-strength random number generation, because patterns could conceivably slip in. For an example of the extreme case, try compiling `time.c`, and run it continuously from the shell. In a Bourne-family shell (what you are probably using on a POSIX system), try

```
while true; do ./time; done
```

You should see the same numbers a few dozen times until the clock ticks over, and then another stream repeated several times. [You can get your command prompt

```
#include <time.h>
#include <apop.h>

int main(){
    gsl_rng *r = apop_rng_alloc(time(NULL));
    for (int i =0; i< 10; i++)
        printf("%.3g\t", gsl_rng_uniform(r));
    printf("\n");
}
```

Listing 11.6 Seeding an RNG using the time. Online source: `time.c`.

back using <ctrl-c>.] If you have multiple processors and run one simulation on each, then runs that start in the same second will be replicating each other. Finally, if you ever hope to debug this program, then you will need to write down the time started so that you can replicate the runs that break:

```
//I assume the database is already open and has a one-column
//table named runs. The time is a long integer
//in the GNU standard library, so its printf tag is %li.
long int right_now = time(NULL);
apop_query("insert into runs (%li);", right_now);
gsl_rng *r = apop_rng_alloc(right_now);
```

Caveats aside, if you just want to see some variety every time the program runs, then seeding with the time works fine.

※ *The standard C RNG*   If the GSL is not available, the standard C library includes a `rand` function to make random draws and an `srand` function to set its seed. E.g.:

```
#include <stdlib.h>
srand(27);
printf("One draw from a U[0,1]: %g", rand()/(RAND_MAX +0.0));
```

The GSL's RNGs are preferable for a few reasons. First, `gsl_ran_max(r)` is typically greater than `RAND_MAX`, giving you greater variation and precision. Second, the C language standard specifies that there must be a `rand` function, but not how it works, meaning two machines may give you different streams of random numbers for the same seed.

Finally, `rand` gives your entire program exactly one stream of numbers, while you can initialize many `gsl_rng`s that will be independent of each other. For example, if you give every agent in a simulation its own RNG, you can re-run the simulation with one agent added or removed (probably at a breakpoint in GDB) and are guaranteed that the variation is due to the agent, not RNG shifts. Here is some sample code to clarify how such a setup would be initialized:

```
typedef struct agent{
    long int agent_number;
    gsl_rng *r;
    ...
} agent;

void init_agent(agent *initme, int agent_no){
    initme->agent_number = agent_no;
    initme->r = apop_rng_init(agent_no);
    ...
}
```

➤ Random number generators produce a deterministic sequence of values. This is a good thing, because we could never debug a program or replicate results without it. Change the stream by initializing it with a different seed.

➤ Given a random number generator, you can use it to draw from any common distribution, from a histogram, or from a data set.

11.2 DESCRIPTION: FINDING STATISTICS FOR A DISTRIBUTION

For many statistic-distribution pairs, there exists a closed-form solution for the statistic: the kurtosis of a $\mathcal{N}(\mu, \sigma)$ is $3\sigma^4$, the variance of a Binomial distribution is $np(1 - p)$, et cetera. You can also take recourse in the Slutsky theorem, that says that given an estimate r for some statistic ρ and a continuous function $f(\cdot)$, then $f(r)$ is a valid estimate of $f(\rho)$. Thus, sums or products of means, variances, and so on are easy to calculate as well.

However, we often find situations where we need a global value like a mean or variance, but have no closed-form means of calculating that value. Even when there is a closed-form theorem that begins *in the limit as $n \to \infty$, it holds that...*, there is often evidence of the theorem falling flat for the few dozen data points before us.

One way to calculate the expected value of a statistic $f(\cdot)$ given probability distribution $p(\cdot)$ would be a numeric integral over the entire domain of the distribution. For a resolution of 100,000 slices, write a loop to sum

$$E[f(\cdot)|p(\cdot)] = \frac{f(-500.00) \cdot p(-500.00)}{100,000} + \frac{f(-499.99) \cdot p(-499.99)}{100,000}$$
$$+ \cdots + \frac{f(499.99) \cdot p(499.99)}{100,000} + \frac{f(500.00) \cdot p(500.00)}{100,000}.$$

This can be effective, but there are some details to be hammered out: if your distribution has a domain over $(-\infty, \infty)$, should you integrate over $[-3, 3]$ or $[-30, 30]$? You must decide up-front how fine the resolution will be, because (barring some tricks) each resolution is a new calculation rather than a modification of prior calculations. If you would like to take this approach, the GSL includes a set of numerical integration functions.

Another approach is to evaluate $f(\cdot)$ at values randomly drawn from the distribution. Just as Listing 11.2 produced a nice picture of the Beta distribution by taking enough random draws, a decent number of random draws can produce a good estimate of any desired statistic of the overall distribution. Values will, by definition, appear in proportion to their likelihood, so the $p(\cdot)$ part takes care of itself. There is no cutoff such that the tails of the distribution are assumed away. You can incrementally monitor $E[f(\cdot)]$ at 1,000 random draws, at 10,000, and so on, to see how much more you are getting with the extra time.

An example: the kurtosis of a t distribution   You probably know that a t distribution is much like a Normal distribution but with fatter tails, but probably not how much fatter those tails are. The kurtosis of a vector is easy to calculate—just call `apop_vector_kurtosis`. By taking a million or so draws from a t distribution, we can produce a vector whose values cover the distribution rather well, and then find the kurtosis of that vector.

Listing 11.7 shows a program to execute this procedure.

- The `main` function just sets up the header of the output table (see Table 11.8) and calls `one_df` for each df.
- The `for` loop on lines 6–7 does the draws, storing them in the vector v.
- Once the vector is filled, line eight calculates the partially normalized kurtosis. That is, it calculates raw kurtosis over variance squared; see the box on page 230 on the endless debate over how best to express kurtosis.

The closed-form formula for the partially-normalized kurtosis of a t distribution with $df > 4$ degrees of freedom is $(3df - 6)/(df - 4)$. For $df \leq 4$, the kurtosis is undefined, just as the variance is undefined for a Cauchy distribution (i.e., a t distribution with $df = 1$). At $df = 5$, it is finite, and it monotonically decreases as df continues upwards.

```
1   #include <apop.h>
2
3   void one_df(int df, gsl_rng *r){
4       long int i, runct = 1e6;
5       gsl_vector *v = gsl_vector_alloc(runct);
6           for (i=0; i< runct; i++)
7               gsl_vector_set(v, i, gsl_ran_tdist(r, df));
8           printf("%i\t %g", df, apop_vector_kurtosis(v)/gsl_pow_2(apop_vector_var(v)));
9           if (df > 4)
10              printf("\t%g", (3.*df − 6.)/(df−4.));
11          printf("\n");
12          gsl_vector_free(v);
13  }
14
15  int main(){
16      int df, df_max = 31;
17      gsl_rng *r = apop_rng_alloc(0);
18          printf("df\t k (est)\t k (actual)\n");
19          for (df=1; df< df_max; df++)
20              one_df(df, r);
21  }
```

Listing 11.7 Monte Carlo calculation of kurtoses for the t distribution family. Online source: tdistkurtosis.c.

Table 11.8 shows an excerpt from the simulation output, along with the true kurtosis.

The exact format for the variance of the estimate of kurtosis will not be given here (ℚ: use the methods here to find it), but it falls with df: with $df < 4$, we may as well take the variance of the kurtosis estimate as infinite, and it shrinks as df grows. Correspondingly, the bootstrap estimates of the kurtosis are unreliable for $df = 5$ or 6, but are more consistent for df over a few dozen. You can check this by re-running the program with different seeds (e.g., replacing the zero seed on line 17 of the code with time(NULL)).

➤ By making random draws from a model, we can make statements about global properties of the model that improve in accuracy as the number of draws → ∞.

➤ The variance of the Monte Carlo estimation of a parameter tends to mirror the variance of the underlying parameter. The maximum likelihood estimator for a parameter achieves the Cramér–Rao lower bound, so the variance of the Monte Carlo estimate will be larger (perhaps significantly so).

df	k (est)	k (analytic)
1	183640	
2	5426.32	
3	62.8055	
4	19.2416	
5	8.5952	9
6	6.00039	6
7	4.92161	5
8	4.52638	4.5
9	4.17413	4.2
10	4.00678	4
15	3.55957	3.54545
20	3.37705	3.375
25	3.281	3.28571
30	3.23014	3.23077

Table 11.8 The fatness of the tails of t distributions at various df.

11.3 INFERENCE: FINDING STATISTICS FOR A PARAMETER

The t distribution example made random draws from a closed-form distribution, in order to produce an estimate of a function of the distribution parameters. Conversely, say that we want to estimate a function of the data, such as the variance of the mean of a data set, $\widehat{\text{var}}(\hat{\mu}(\mathbf{X}))$. We have only one data set before us, but we can make random draws from $\mathbf{X}$ to produce several values of the statistic $\hat{\mu}(\mathbf{X}_r)$, where $\mathbf{X}_r$ represents a random draw from $\mathbf{X}$, and then estimate the variance of those draws. This is known as the *bootstrap* method of estimating a variance.

BOOTSTRAPPING THE STANDARD ERROR The core of the bootstrap is a simple algorithm:

Repeat the following m times:
    Let $\tilde{\mathbf{X}}$ be n elements randomly drawn from the data, with replacement.
    Write down the statistic(s) $\beta(\tilde{\mathbf{X}})$.
Find the *standard error* of the m values of $\beta(\tilde{\mathbf{X}})$.

This algorithm bears some resemblance to the steps demonstrated by the CLT demo in Listing 9.1 (page 298): draw m iid samples, find a statistic like the mean of each, and then look at the distribution of the several statistics (rather than the underlying data itself). So if $\beta(\tilde{\mathbf{X}})$ is a mean-like statistic (involving a sum over

n), then the CLT applies directly, and the artificial statistic approaches a Normal distribution. Thus, it makes sense to apply the usual Normal distribution-based test to test hypotheses about the true value of β.

```
1   #include "oneboot.h"
2
3   int main(){
4      int rep_ct = 10000;
5      gsl_rng *r = apop_rng_alloc(0);
6      apop_db_open("data−census.db");
7      gsl_vector *base_data = apop_query_to_vector("select in_per_capita from income where
            sumlevel+0.0 =40");
8      double RI = apop_query_to_float("select in_per_capita from income where sumlevel+0.0
            =40 and geo_id2+0.0=44");
9      gsl_vector *boot_sample = gsl_vector_alloc(base_data−>size);
10     gsl_vector *replications = gsl_vector_alloc(rep_ct);
11     for (int i=0; i< rep_ct; i++){
12         one_boot(base_data, r, boot_sample);
13         gsl_vector_set(replications, i, apop_mean(boot_sample));
14     }
15     double stderror = sqrt(apop_var(replications));
16     double mean = apop_mean(replications);
17     printf("mean: %g; standard error: %g; (RI−mean)/stderr: %g; p value: %g\n",
18         mean, stderror, (RI−mean)/stderror, 2*gsl_cdf_gaussian_Q(fabs(RI−mean), stderror));
19  }
```

Listing 11.9 Bootstrapping the standard error of the variance in state incomes per capita. Online source: `databoot.c`.

```
#include "oneboot.h"

void one_boot(gsl_vector *base_data, gsl_rng *r, gsl_vector* boot_sample){
    for (int i =0; i< boot_sample−>size; i++)
        gsl_vector_set(boot_sample, i,
                gsl_vector_get(base_data, gsl_rng_uniform_int(r, base_data−>size)));
}
```

Listing 11.10 A function to produce a bootstrap draw. Online source: `oneboot.c`.

Listings 11.10 and 11.9 shows how this algorithm is executed in code. It tests the hypothesis that Rhode Island's income per capita is different from the mean.

• Lines 4–8 of Listing 11.9 are introductory material and the queries to pull the requisite data. For Rhode Island, this is just a scalar, used in the test below, but for the rest of the country, this is a vector of 52 numbers (one for each state, commonwealth, district, and territory in the data).

- Lines 11–14 show the main loop repeated m times in the pseudocode above, which makes the draws and then finds the mean of the draws.

- Listing 11.10 is a function to make a single bootstrap draw, which will be used in a few other scripts below. The `one_boot` function draws with replacement, which simply requires repeatedly calling `gsl_rng_uniform_int` to get a random index and then writing down the value at that index in the data vector.

- Lines 15 and 16 of Listing 11.10 find the mean and standard error of the returned data, and then lines 17–18 run the standard hypothesis test comparing the mean of a Normal distribution to a scalar.

- Recall the discussion on page 300 about the standard deviation of a data set, which is the square root of its variance, σ; and the standard deviation of the mean of a data set, which is $\sigma/\sqrt{n}$. In this case, we are interested in the distribution of $\beta(\tilde{\mathbf{X}})$ itself, not the distribution of $E(\beta(\tilde{\mathbf{X}}))$s, so we use σ instead of $\sigma/\sqrt{n}$.

$\mathbb{Q}_{11.2}$

Rewrite `databoot.c` to calculate the standard error of `base_data` directly (without bootstrapping), and test the same hypothesis using that standard error estimate rather than the bootstrapped version. Do you need to test with your calculated value of $\hat{\sigma}$ or with $\hat{\sigma}/\sqrt{n}$?

```
1    #include <apop.h>
2
3    int main(){
4        int draws = 5000, boots = 1000;
5        double mu = 1., sigma = 3.;
6        gsl_rng *r = apop_rng_alloc(2);
7        gsl_vector *d = gsl_vector_alloc(draws);
8        apop_model *m = apop_model_set_parameters(apop_normal, mu, sigma);
9        apop_data *boot_stats = apop_data_alloc(0, boots, 2);
10       apop_name_add(boot_stats−>names, "mu", 'c');
11       apop_name_add(boot_stats−>names, "sigma", 'c');
12       for (int i =0; i< boots; i++){
13           for (int j =0; j< draws; j++)
14               apop_draw(gsl_vector_ptr(d, j), r, m);
15           apop_data_set(boot_stats, i, 0, apop_vector_mean(d));
16           apop_data_set(boot_stats, i, 1, sqrt(apop_vector_var(d)));
17       }
18       apop_data_show(apop_data_covariance(boot_stats));
19       printf("Actual:\n var(mu) %g\n", gsl_pow_2(sigma)/draws);
20       printf("var(sigma): %g\n", gsl_pow_2(sigma)/(2*draws));
21   }
```

Listing 11.11 Estimate the covariance matrix for the Normal distribution Online source: `normalboot.c`.

A Normal example   Say that we want to know the covariance matrix for the estimates of the Normal distribution parameters, $(\hat{\mu}, \hat{\sigma})$. We could look it up and find that it is

$$\begin{bmatrix} \frac{\sigma^2}{n} & 0 \\ 0 & \frac{\sigma^2}{2n} \end{bmatrix},$$

but that would require effort and looking in books.[4] So instead we can write a program like Listing 11.11 to generate the covariance for us. Running the program will show that the computational method comes reasonably close to the analytic value.

- The introductory material up to line 11 allocates a vector for the bootstrap samples and a data set for holding the mean and standard deviation of each bootstrap sample.
- Instead of drawing permutations from a fixed data set, this version of the program produces each artificial data vector via draws from the Normal distribution itself, on lines 13–14, and then lines 15–16 write down statistics for each data set.
- In this case, we are producing two statistics, μ and σ, so line 18 finds the covariance matrix rather than just a scalar variance.

 Q₁₁.₃   On page 321, I mentioned that you could test the claim that the kurtosis of a Normal distribution equals $3\sigma^4$ via a bootstrap. Modify `normalboot.c` to test this hypothesis. Where the program currently finds means and variances of the samples, find each sample's kurtosis. Then run a t test on the claim that the mean of the vector of kurtoses is $3\sigma^4$.

Σ
- ➤ To test a hypothesis about a model parameter, we need to have an estimate of the parameter's variance.

- ➤ If there is no analytic way to find this variance, we can make multiple draws from the data itself, calculate the parameter, and then find the variance of that artificial set of parameters. The Central Limit Theorem tells us that the artificial parameter set will approach a well-behaved Normal distribution.

---

[4]E.g., Kmenta (1986, p 182). That text provides the covariance matrix for the parameters $(\hat{\mu}, \hat{\sigma}^2)$. The variance of $\hat{\mu}$ is the same, but the variance of $\hat{\sigma}^2 = \frac{2\sigma^4}{n}$.

11.4 DRAWING A DISTRIBUTION To this point, we have been searching for global parameters of either a distribution or a data set, but the output has been a single number or a small matrix. What if we want to find the entire distribution?

We have at our disposal standard RNGs to draw from Normal or Uniform distributions, and this section will present are a few techniques to transform those draws into draws from the distribution at hand. For the same reasons for which random draws were preferable to brute-force numeric integration, it can be much more efficient to produce a picture of a distribution via random draws than via *grid search*: the draws focus on the most likely points, the tails are not cut off at an arbitrary limit, and a desired level of precision can be reached with less computation.

And remember, *anything* that produces a nonnegative univariate measure can be read as a subjective likelihood function. For example, say that we have real-world data about the distribution of firm sizes. The model on page 253 also gives us an artificial distribution of firm sizes given input parameters such as the standard error (currently hard-coded into the model's `growth` function). Given a run, we could find the distance $d(\sigma)$ between the actual distribution and the artificial. Notice that it is a function of σ, because every σ produces a new output distribution and therefore a new distance to the actual. The most likely value of σ is that which produces the smallest distance, so we could write down $L(\sigma) = 1/d(\sigma)$, for example.[5] Then we could use the methods in this chapter and the last to find the most likely parameters, draw a picture of the likelihood function given the input parameters, calculate the variance given the subjective likelihood function, or test hypotheses given the subjective likelihood.

IMPORTANCE SAMPLING There are many means of making draws from one distribution to inform draws from another; Train (2003, Chapter 9) catalogs many of them.

For example, *importance sampling* is a means of producing draws from a new function using a well-known function for reference. Let $f(\cdot)$ be the function of interest, and let $g(\cdot)$ be a well-known distribution like the Normal or Uniform. We want to use the stock Normal or Uniform distribution RNGs to produce an RNG for an arbitrary function.

For $i = 1$ to a few million:
>   Draw a new point, x_i, from the reference distribution $g(\cdot)$.
>   Give the point weighting $f(x_i)/g(x_i)$.
Bin the data points into a histogram (weighting each point appropriately).

---

[5]Recall the invariance principle from page 351, that basic transformations of the likelihood function don't change the data therein, so we don't have to fret over the exact form of the subjective likelihood function, and would get the same results using $1/d$ as using $1/d^2$ or $1/(d + 0.1)$, for example.

At the end of this, we have a histogram that is a valid representation of $f(\cdot)$, which can be used for making draws, graphing the function, or calculating global information.

What reference distribution should you use? Theoretically, any will do, but engineering considerations advise that you pick a function that is reasonably close to the one you are trying to draw from—you want the high-likelihood parts of $g(\cdot)$ to match the high-likelihood parts of $f(\cdot)$. So if $f(\cdot)$ is generally a bell curve, use a Normal distribution; if it is focused near zero, use a Lognormal or Exponential; and if you truly have no information, fall back to the Uniform (perhaps for a rough exploratory search that will let you make a better choice).

MARKOV CHAIN MONTE CARLO   *Markov Chain Monte Carlo* is an iterative algorithm that starts at one state and has a rule defining the likelihood of transitioning to any other given state from that initial state—a Markov Chain. In the context here, the state is simply a possible value for the parameter(s) being estimated. By jumping to a parameter value in proportion to its likelihood, we can get a view of the probability density of the parameters. The exact probability of jumping to another point is given a specific form to be presented shortly.

This is primarily used for Bayesian updating, so let us review the setup of that method. The analysis begins with a prior distribution expressing current beliefs about the parameters, $P_{\text{prior}}(\beta)$, and a likelihood function $P_{\text{L}}(\mathbf{X}|\beta)$ expressing the likelihood of an observation given a value of the parameters. The goal is to combine these two to form a posterior; as per page 258, Bayes's Rule tells us that this is

$$P_{\text{post}}(\beta|\mathbf{X}) = \frac{P_{\text{L}}(\mathbf{X}|\beta)P_{\text{prior}}(\beta)}{\int_{\forall B \in \mathbb{B}} P_{\text{L}}(\mathbf{X}|B)P_{\text{prior}}(\beta)dB}$$

The numerator is easy to calculate, but the denominator is a global value, meaning that we can not know it with certainty without evaluating the numerator at an infinite number of points. The Metropolis–Hastings algorithm, a type of MCMC, offers a solution.

The gist is that we start at an arbitrary point, and draw a new candidate point from the prior distribution. If the candidate point is more likely than the current point (according to P_{L}), jump to it, and if the candidate point is less likely, perhaps jump to it anyway. After a burn-in period, record the values. The histogram of recorded values will be the histogram of the posterior distribution. To be more precise, here is a pseudocode description of the algorithm:

Begin by setting β_0 to an arbitrary starting value.
For $i = 1$ to a few million:

    Draw a new proposed point, β_p, from P_{prior}.
    If $P_{\text{L}}(\mathbf{X}|\beta_p) > P_{\text{L}}(\mathbf{X}|\beta_{i-1})$
        $\beta_i \leftarrow \beta_p$
    else
        Draw a value $u \sim \mathcal{U}[0, 1]$
        If $u < P_{\text{L}}(\mathbf{X}|\beta_p)/P_{\text{L}}(\mathbf{X}|\beta_{i-1})$
            $\beta_i \leftarrow \beta_p$
        else
            $\beta_i \leftarrow \beta_{i-1}$
    If $i > 1,000$ or so, record β_i.

Report the histogram of β_is as the posterior distribution of β.

As should be evident from the description of the algorithm, it is primarily used to go from a prior to a posterior distribution. To jump to a new point β_{new}, it must first be chosen by the prior (which happens with probability $P_{\text{prior}}(\beta_{\text{new}})$) and then will be selected with a likelihood roughly proportional to $P_{\text{L}}(\mathbf{x}|\beta_{\text{new}})$, so the recorded draw will be proportional to $P_{\text{prior}}(\beta_{\text{new}})P_{\text{L}}(\mathbf{x}|\beta_{\text{new}})$. This rough intuitive argument can be made rigorous to prove that the points recorded are a valid representation of the posterior; see Gelman *et al.* (1995).

Why not just draw from the prior and multiply by the likelihood directly, rather than going through this business of conditionally jumping? It is a matter of efficiency. The more likely elements of the posterior are more likely to contribute a large amount to the integral in the denominator of the updating equation above, so we can estimate that integral more efficiently by biasing draws toward the most likely posterior values, and the jumping scheme achieves this with fewer draws than the naïve draw-from-the-prior scheme does.

You have already seen one concrete example of MCMC: the simulated annealing algorithm from Chapter 10. It also jumps from point to point using a decision rule like the one above. The proposal distribution is basically an improper uniform distribution—any point is as likely as any other—and the likelihood function is the probability distribution whose optimum the simulated annealing is seeking. The difference is that the simulated annealing algorithm makes smaller and smaller jumps; combined with the fact that MCMC is designed to tend toward more likely points, it will eventually settle on a maximum value, at which point we throw away all the prior values. For the Bayesian updating algorithm here, jumps are from a fixed prior, and tend toward the most likely values of the posterior, but the $1,000$th jump is as likely to be a long one as the first, and we use every point found to produce the output distribution.

At the computer, the function to execute this algorithm is `apop_update`. This function takes in a parametrized prior model, an unparametrized likelihood and data, and outputs a parametrized posterior model.

The output model will be one of two types. As with the Beta/Binomial example on page 259, there are many well-known *conjugate distributions*, where the posterior will be of the same form as the prior distribution, but with updated parameters. For example, if the prior is named *Gamma distribution* (which the `apop_gamma` model is named) and the likelihood is named *Exponential distribution* (and `apop_-exponential` is so named), then the output of `apop_update` will be a closed-form Gamma distribution with appropriately updated parameters. But if the tables of closed form conjugates offer nothing applicable, the system will fall back on MCMC and output an `apop_histogram` model with the results of the jumps.

Q11.4

Check how close the MCMC algorithm comes to the closed-form model.

- Pick an arbitrary parameter for the Exponential distribution, β_I. Generate ten or twenty thousand random data points.

- Pick a few arbitrary values for the Gamma distribution parameters, β_{prior}.

- Estimate the posterior, by sending the two distributions, the randomly-generated data, and β_{prior} to `apop_update`.

- Make a copy of the `apop_exponential` model and change its name to something else (so that `apop_update` won't find it in the conjugate distribution table). Re-send everything to `apop_update`.

- Use a goodness-of-fit test to find how well the two output distributions match.

Q11.5

The output to the updating process is just another distribution, so it can be used as the prior for a new updating step. In theory, distributions repeatedly updated with new data will approach putting all probability mass on the true parameter value. Continuing from the last exercise:

- Regenerate a new set of random data points from the Exponential distribution using the same β_I.

- Send the new data, the closed-form posterior from the prior exercise, and the same Exponential model to the updating routine.

- Plot the output.

- Once you have the generate/update/plot routine working, call it from a `for` loop to generate an animation beginning with the prior and continuing with about a dozen updates.

- Repeat with the MCMC-estimated posteriors.

11.5 NON-PARAMETRIC TESTING   Section 11.3 presented a procedure for testing a claim about a data set that consisted of drawing a bootstrap data set, regenerating a statistic, and then using the many draws of the statistic and the CLT to say something about the data. But we can test some hypotheses by simply making draws from the data and checking their characteristics, without bothering to produce a statistic and use the CLT.

To give a concrete example of testing without parametric assumptions, consider the permutation test. Say that you draw ten red cards and twenty black cards from what may be a crooked deck. The hypothesis is that the mean value of the red cards you drew equals the mean of the black cards (counting face cards however you like). You could put the red cards in one pile to your left, the black cards in a pile to your right, calculate $\bar{x}_{red}$, $\hat{\sigma}_{red}$, $\bar{x}_{black}$, and $\hat{\sigma}_{black}$, assume $\bar{x}_{black} - \bar{x}_{red} \sim t$ distribution, and run a traditional t test to compare the two means.

But if μ_{red} really equals μ_{black}, then the fact that some of the cards you drew are black and some are red is irrelevant to the value of $\bar{x}_{left} - \bar{x}_{right}$. That is, if you shuffled together the stacks of black and red cards, and dealt out another ten cards to your left and twenty to your right, then $\bar{x}_{left} - \bar{x}_{right}$ should not be appreciably different. If you deal out a few thousand such shuffled pairs of piles, then you can draw the distribution of values for $\bar{x}_{left} - \bar{x}_{right}$. If $\bar{x}_{red} - \bar{x}_{black}$ looks as if it is very far from the center of that distribution, then we can reject the claim that the color of the cards is irrelevant.

What is the benefit of all this shuffling and redealing when we could have just run a t test to compare the two groups? On the theoretical level, this method relies on the assumption of the bootstrap principle rather than the assumptions underlying the CLT. Instead of assuming a theoretical distribution, you can shuffle and redraw to produce the distribution of outcome values that matches the true sample data (under the assumption that $\mu_{red} = \mu_{black}$), and then rely on the bootstrap principle to say that what you learned from the sample is representative of the population from which the sample was drawn.

On a practical level, the closer match to the data provides real benefits. Lehmann & Stein (1949) showed that the permutation test is more powerful—it is less likely to fail to reject the equality hypothesis when the two means are actually different.

In pseudocode, here is the test procedure:[6]

---

[6]The test is attributed to Chung & Fraser (1958). From p 733: "The computation took one day for programming and two minutes of machine time."

$\mu \leftarrow |\overline{\mathbf{x}}_{\text{red}} - \overline{\mathbf{x}}_{\text{black}}|$
$\mathbf{d} \leftarrow$ the joined test and control vectors.
Allocate a million-element vector $\mathbf{v}$.
For $i = 1$ to a million:
    Draw (without replacement) from $\mathbf{d}$ a vector the same size as $\overline{\mathbf{x}}_{\text{red}}$, $\mathbf{L}$.
    Put the other elements of $\mathbf{d}$ into a second vector, $\mathbf{R}$.
    $\mathbf{v}_i \leftarrow |\overline{\mathbf{L}} - \overline{\mathbf{R}}|$.
$\alpha \leftarrow$ the percentage of $\mathbf{v}_i$ less than μ.
Reject the claim that $\mu = 0$ with confidence α.
Fail to reject the claim that $\mu = 0$ with confidence $1 - \alpha$.

Q11.6   Baum *et al.* (2008) sampled genetic material from a large number of cases with bipolar disorder and controls who were confirmed to not have bipolar disorder. To save taxpayer money, they pooled the samples to form the groups listed in the file `data-genes` in the code supplement. Each line of that file lists the percent of the pool where the given SNP (single nucleotide polymorphism) has a given marker.
For which of the SNPs can we reject the hypothesis of no difference between the cases and controls? Write a program to read the data into a database, then use the above algorithm to test whether we reject the hypothesis of no difference in marker frequency for a given SNP, and finally write a `main` that runs the test for each SNPs in the database. The data here is cut from 550,000 SNPs, so base the Bonferroni correction on that many tests.

Other nonparametric tests tend to follow a similar theme: given a null hypothesis that some bins are all equiprobable, we can develop the odds that the observed data occurred. See Conover (1980) for a book-length list of such tests, including Kolmogorov's method from page 323 of this book.

TESTING FOR BIMODALITY   As the finale to the book, Listing 11.12 shows the use of kernel densities to test for multimodality. This involves generating a series of kernel density estimates of the data, first with the original data, and then with a few thousand bootstrap estimates, and then doing a nonparametric test of the hypothesis that the distribution has fewer than n modes, for each value of n.

Recall the kernel density estimate from page 262, which was based on the form

$$\hat{f}(t, \mathbf{x}, h) = \frac{\sum_{i=1}^{n} \mathcal{N}((t - x_i)/h)}{n \cdot h},$$

where $\mathbf{x}$ is the vector of n data points observed, $\mathcal{N}(y)$ is a Normal$(0, 1)$ PDF evaluated at y, and $h \in \mathbb{R}^+$ is the bandwidth. Let k be the number of modes.

Figure 7.15 (p 262) showed that as h rises, the spike around each point spreads out and merges with other spikes, so k falls, until eventually the entire data set is subsumed under one single-peaked curve, so $k = 1$. Thus, there is a monotonic relationship between h and the number of modes in the density function induced by h.

Silverman (1981) offers a bootstrap method of testing a null hypothesis of the form *the distribution has more than k modes*. Let h_0 be the smallest value of h such that $\hat{f}(t, X, h_0)$ has more than k modes; then the null hypothesis is that h_0 is consistent with the data.

We then draw a few hundred bootstrap samples from the data, $x'_1, \ldots, x'_{200}$; write down $\hat{f}(t, x'_1, h_0), \ldots, \hat{f}(t, x'_{200}, h_0)$; and count the modes of each of these functions. Silverman shows that, thanks to the monotonic relationship between h and the number of modes, the percentage of bootstrap distributions with more than k modes matches the likelihood that $\hat{f}(t, x, h_0)$ is consistent with the data. If we reject this hypothesis, then we would need a lower value of h, and therefore more modes, to explain the data.

Listing 11.12 shows the code used to produce these figures and test a data set of television program ratings for bimodality. Since it draws a few hundred bootstrap samples, link it with `oneboot.c` from page 368.

- The `countmodes` function makes two scans across the range of the data. The first generates a histogram: at each point on the x-axis, it piles up the value at that point of a Normal distribution centered at every data point. The result will be a histogram like those pictured in Figure 7.15, page 262, or those that are produced in the Gnuplot animation the program will write to the `kernelplot` file. The second pass checks for modes, by simply asking each point whether it is higher than the points to its left and right.
- $h \rightarrow k$: For each index `i`, `ktab[i]` is intended to be the smallest value of `h` that produces fewer than `i` modes. The `fill_kmap` function calls `countmodes` to find the number of modes produced by a range of values of `h`. The function runs from the largest `h` to the smallest, so the last value of `h` written to any slot will be the smallest value that produces that mode count.[7]
- $k \rightarrow p$: Now that we have the smallest bandwidth h that produces a given k using the actual data, `boot` does the bootstrapping: `one_boot` (p 368) uses a straightforward `for` loop to produce a sample, then the same `countmodes` function used above is applied to the artificial sample. If the mode count is larger than the number of modes in the original, it is a success for the hypothesis. The function then

---

[7] Also, the `for` loop in this function demonstrates a pleasant trick for scanning a wide range: instead of stepping by a fixed increment, it steps by percentages, so it quickly scans the hundreds but looks at the lower end of the range in more detail.

returns the percent successes, which we can report as a p value for the hypothesis that the distribution has more than k modes.

- In some parts of the output, such as for $k = 12$, the battery of tests finds low confidence that there are more than k modes and high confidence that there are more than $k + 1$ modes. If both tests were run with the same value of h, this would be inconsistent, but we are using the smallest level of smoothing possible in each case, so the tests for k and $k + 1$ parameters have little to do with each other. For small k, the output is monotonic, and shows that we can be very confident that the data set has more than three modes, and reasonably confident that it has more than four.

```
#include "oneboot.h"

double modect_scale, modect_min, modect_max,
        h_min=25, h_max=500, max_k = 20,
        boot_iterations = 1000,
        pauselength = 0.1;
char outfile[] = "kernelplot";

void plot(apop_data *d, FILE *f){
    fprintf(f, "plot '-' with lines\n");
    apop_data_print(d, NULL);
    fprintf(f, "e\npause %g\n", pauselength);
}

int countmodes(gsl_vector *data, double h, FILE *plothere){
    int len =(modect_max-modect_min)/modect_scale;
    apop_data *ddd = apop_data_calloc(0, len, 2);
    double sum, i=modect_min;
        for (size_t j=0; j < ddd->matrix->size1; j ++){
            sum = 0;
            for (size_t k = 0; k< data->size; k++)
                sum += gsl_ran_gaussian_pdf((i-gsl_vector_get(data,k))/h,1)/(data->size*h);
            apop_data_set(ddd, j, 0, i);
            apop_data_set(ddd, j, 1, sum);
            i+=modect_scale;
        }
    int modect =0;
    for (i = 1; i< len-1; i++)
        if(apop_data_get(ddd,i,1)>=apop_data_get(ddd,i-1,1)
            && apop_data_get(ddd,i,1)>apop_data_get(ddd,i+1,1))
                modect++;
    if (plothere) plot(ddd, plothere);
    apop_data_free(ddd);
    return modect;
}

void fill_kmap(gsl_vector *data, FILE *f, double *ktab){
    for (double h = h_max; h> h_min; h*=0.99){
```

```
            int val = countmodes(data, h, f);
            if (val <max_k)
                ktab[val − 1] = h;
        }
}

double boot(gsl_vector *data, double h0, int modect_target, gsl_rng *r){
    double over_ct = 0;
    gsl_vector *boots = gsl_vector_alloc(data−>size);
        for (int i=0; i < boot_iterations; i++){
            one_boot(data, r, boots);
            if (countmodes(boots, h0, NULL) > modect_target)
                over_ct++;
        }
        gsl_vector_free(boots);
        return over_ct/boot_iterations;
}

apop_data *produce_p_table(gsl_vector *data, double *ktab, gsl_rng *r){
    apop_data *ptab = apop_data_alloc(0, max_k, 2);
        apop_name_add(ptab−>names, "Mode", 'c');
        apop_name_add(ptab−>names, "Likelihood of more", 'c');
        for (int i=0; i< max_k; i++){
            apop_data_set(ptab, i, 0, i);
            apop_data_set(ptab, i, 1, boot(data, ktab[i], i, r));
        }
        return ptab;
}

void setvars(gsl_vector *data){ //rescale based on the data.
        double m1 = gsl_vector_max(data);
        double m2 = gsl_vector_min(data);
        modect_scale = (m1−m2)/200;
        modect_min = m2−(m1−m2)/100;
        modect_max = m1+(m1−m2)/100;
}

int main(){
        APOP_COL(apop_text_to_data("data−tv", 0,0), 0, data);
        setvars(data);
        FILE *f = fopen(outfile, "w");
        apop_opts.output_type = 'p';
        apop_opts.output_pipe = f;
        double *ktab = calloc(max_k, sizeof(double));
        fill_kmap(data, f, ktab);
        fclose(f);
        gsl_rng *r = apop_rng_alloc(3);
        apop_data_show(produce_p_table(data, ktab, r));
}
```

Listing 11.12 Silverman's kernel density test for bimodality. Online source: `bimodality.c`

Another way to speed the process in `bimodality.c` is to clump the data before summing the Normal distributions. If there are three points at $-.1$, 0, and .1, then this will require three calculations of the Normal PDF for every point along the real line. If they are clumped to 0, then we can calculate the Normal PDF for $\mu = 0$ times three, which will run three times as fast.

Q11.7

Write a function to group nearby data points into a single point with a given weight (you can use the `weight` element of the `apop_data` structure to record it). Rewrite the `countmodes` function to use the clumped and weighted data set. How much clumping do you need to do before the results degrade significantly?

Σ

➤ Via resampling, you can test certain hypotheses without assuming parametric forms like the t distribution.

➤ The typical kernel density estimate consists of specifying a distribution for every point in the data set, and then combining them to form a global distribution. The resulting distribution is in many ways much more faithful to the data.

➤ As the bandwidth of the sub-distributions grows, the overall distribution becomes smoother.

➤ One can test hypotheses about multimodality using kernel densities, by finding the smallest level of smoothing necessary to achieve n-modality, and bootstrapping to see the odds that that level of smoothing would produce n-modality.

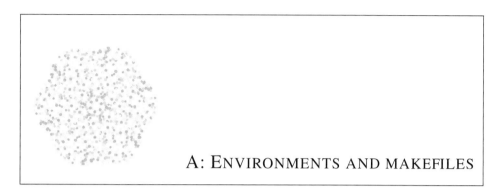

A: ENVIRONMENTS AND MAKEFILES

Since C is a standard, not a product from a single corporation or foundation, different C development environments have minor, potentially maddening differences.

The problem of whether a decent compiler and useful libraries are available on a given system has basically been surmounted: if something is missing, just ask the package manager to install it from the Internet. But then the next question is: where did the package manager put everything? Some systems like to put libraries in /usr/local, some like to put them in /opt/, and neither of these locations even makes sense for the ever-eccentric Windows operating system.

The solution is to set *environment variables* that will specify how your compiler will find the various elements it needs to produce your program. These variables are maintained by the operating system, as specified by the POSIX standard.[1]

A.1 ENVIRONMENT VARIABLES

If you type env at the command prompt, you will get a list of the environment variables set on your system. Here is a sampling from my own prompt:

```
SHELL=/bin/bash
USER=klemens
```

---

[1]Later members of the Windows family of operating systems claim POSIX compliance, meaning that most of this works from the Windows command prompt, although this may require downloading Microsoft's Interix package. However, the Cygwin system has its own prompt which basically keeps its own set of environment variables, and I will assume you are using Cygwin's shell rather than that of Windows.

```
LD_LIBRARY_PATH=/usr/local/lib:
PATH=/home/klemens/tech/bin:/usr/local/bin:/usr/bin:/bin:/usr/games:/sbin
HOME=/home/klemens
```

Every time a program spawns a new subprogram, the environment variables are duplicated and passed to the child program. Notably, when you start a program from the shell prompt, all of the environment variables you saw when you typed env are passed on to the program.

The env command works well with grep (see page 404). For example, to find out which shell you are using, try env | grep SHELL.

Setting   Now that you know which shell you are using, the syntax for setting environment variables differs slightly from shell to shell.

You are probably using bash or another variant of the Bourne shell.[2] In these systems, set environment variables using the shell's export command:

```
export USER=Stephen
```

Some shells are picky about spacing, and will complain if there is a space before or after the equals sign. Others do not even accept this syntax, and require a two-line version:

```
USER=Stephen
export USER
```

This form clarifies that the process of setting an environment veriable in bash consists of first setting a local variable (which is not passed on to child programs) and then moving it from the set of local variables to the set of environment variables. If you do not do much shell programming, there is no loss in setting all variables in the environment.

The other family of shells is the C shell, which bears a vague, passing resemblance to the C programming language. In csh, use setenv:

```
setenv USER Stephen
```

---

[2]bash=Bourne-again shell, since it is an overhaul of the shell Stephen Bourne wrote for UNIX in 1977.

Getting   In all shells, and many shell-like programs, you can get the value of a variable by putting a dollar sign before the variable name. For example,

> **echo** $USER

will print the current value of the USER variable to the screen. To give a more useful example,

> **export** PATH=$PATH:$HOME/bin

will extend the PATH, replacing that variable with a copy of itself plus another directory at the end. For my own already lengthy path, listed above, this command would result in the following new path:

> PATH=/home/klemens/tech/bin:/usr/local/bin:/usr/bin:/bin:/usr/games:/sbin:/home/klemens/bin

Setting for good   Every time your shell starts, it reads a number of configuration files; see your shell's manual page (man bash, man csh, ...) for details. But most all of them read a file in your home directory whose name begins with a dot and ends in rc, such as .bashrc or .cshrc.

It is a POSIX custom that if a file begins with a dot, then it is hidden, meaning that the usual directory listing such as ls will not show these files. However, if you explicitly ask for hidden files, via ls -a, ls -d .*, or your GUI file browser's *show hidden* option, you will certainly find a number of .rc files in your home directory.[3]

The shell's .rc files are plain text, and generally look like the sort of thing you would type on a command line—because that is what they are. If you put an export or setenv command in this file, then those commands will be read every time your shell starts, and the variables will thus be set appropriately for all of your work.

• The file does not auto-execute after you edit. To see the effects of your edit, either exit and re-enter the shell, or use source .bashrc or source .cshrc to explicitly ask the shell to read the file.

---

[3]By the way, when making backups of your home directory, it is worth verifying that hidden files are also backed up.

✳ **GETTING AND SETTING FROM C**   It is easy to read and write environment variables in a POSIX system because it is easy to read and write them in C. Listing A.1 shows that the `getenv` function will return a string holding the value of the given environment variable, and `setenv` will set the given environment variable (on line 6, USER) to the given value ("Stephen"). The third argument to `setenv` specifies whether an existing environment variable should be overwritten. There are a few other useful environment-handling functions; see your standard C library reference for details.

```
1   #include <stdlib.h> //environment handling
2   #include <stdio.h>
3
4   int main(){
5       printf("You are: %s\n", getenv("USER"));
6       setenv("USER", "Stephen", 1);
7       printf("But now you are: %s\n", getenv("USER"));
8   }
```

Listing A.1 Getting and setting environment variables. Online source: `env.c`.

Why is it safe to run the program in Listing A.1, which will overwrite an important environment variable? Because a child process (your program) can only change its own copy of the environment variables. It can not overwrite the variables in the environment of a parent process (the shell). Overriding this basic security precaution requires a great deal of cooperation from the parent process. But this does not make `setenv` useless: if your program starts other programs via a function like `system` or `popen`, then they will inherit the current set of environment variables. When the shell opened your program, it passed in its environment in exactly this manner.

➤ The operating system maintains a set of variables describing the current overall environment, such as the user's name or the current working directory.

➤ These environment variables are passed to all child programs.

➤ They can be set or read on the command line or from within a program.

A.2 PATHS On to the problem of getting your C compiler to find your libraries.

When you type a command, such as `gcc`, the shell locates that program on your hard drive and loads it into memory. But given that an executable could be anywhere (maybe /bin, /opt, or c:\Program Files), how does the shell know where to look? It uses an environment variable named PATH that lists a set of directories separated by colons (except on non-Cygwin Windows systems, where they are separated by semicolons). You saw a number of examples of such a variable above, and can run `env | grep PATH` to check your own path.

When you type a command, the shell checks the first directory on the path for the program you requested. If it finds an executable with the right name, then the shell runs that program; else it moves on to the next element of the path and tries again. If it makes its way through the entire path without finding the program, then it gives up and sends back the familiar *command not found* error message.

The current directory, `./`, is typically not in the path, so programs you can see with `ls` will not be found by searching the path.[4] The solution is to either give an explicit path, like `./run_me`, or extend the current directory to your path:

> **export** PATH=$PATH:./

Recall that by adding this line to your `.bashrc` or `.cshrc`, the path will be extended every time you log in.

There are two paths that are relevant to compilation of a C program: the include path and the library path.[5] The procedure is the same as with the shell's executable path: when you `#include` a file (e.g., lines 1 and 2 of Listing A.1), the preprocessor steps through each directory listed in the include path checking for the requested header file. It uses the first file that matches the given name, or gives up with a *header not found* error. When the linker needs to get a function or structure from a library, it searches the library path in the same manner.

But unlike the executable path, there are several ways for a directory to appear on the include or library path.

---

[4]Many consider putting the current directory in the path to be a security risk (e.g., Frisch (1995, p 226)); in security lingo, it allows a *current directory attack*. If the current directory is at the head of the path, a malicious user could put a script named `ls` in a common directory, where the script contains the command `rm -rf $HOME`. When you switch to that location and try to get a directory listing, the shell instead finds the malicious script, and your home directory is erased. Putting the current directory at the end of the path provides slightly more protection, but the malicious script could instead be named after a common typo, such as `c` or `mr`. But given that many of today's POSIX computers are used by one person or a few close associates, adding `./` to the path is not a real risk in most situations.

[5]Java users will notice a close resemblance between these paths and the CLASSPATH environment variable, which is effectively an include and a library path in one.

- There is a default include path and a default library path, built in to the compiler and linker. This typically includes /usr/include and /usr/local/include in the include path, and /usr/lib and /usr/local/lib in the libpath.

- The environment variables INCLUDEPATH and LIBPATH are part of the path. These names are sometimes different, depending on the linker used: gcc uses ld, so it looks for LD_LIBRARY_PATH—except on Mac systems, where it looks for DYLD_- LIBRARY_PATH. On HP-UX systems, try SHLIB_PATH. Or more generally, try man ld or man cc to find the right environment variable for your system.

- You can add to the paths on the compiler's command line. The flag
-I/usr/local/include
would add that directory to the include path, and the flag
-L/usr/local/lib
would add that directory to the library path.

Also, some libraries are *shared*, which in this context means that they are linked not during compilation, but when actually called in execution. Thus, when you execute your newly-complied program, you could still get a missing library error; to fix this you would need to add the directory to your libpath environment variable.

Searching   So, now that you know the syntax of adding a directory to your PATHs, which directory should you add? Say that you know you need to add the header for the TLA library, which will be a file with a name like *tla.h*. The command find *dir* -name *tla.h* will search *dir* and every subdirectory of *dir* for a file with the given name. Searching the entire hierarchy beginning at the root directory, via find / -name *tla.h*, typically takes a human-noticeable amount of time, so try some of the typical candidates for *dir* first: /usr, /opt, /local, or for Mac OS X, /sw.

Compiled libraries have different names on different systems, so your best bet is to search for all files beginning with a given string, e.g., find *dir* -name '*tla.**'.

Assembling the compiler command line   There is nothing in your source code that tells the system which libraries are to be linked in, so the libraries must be listed on the command line. First, you will list the non-standard directories to search, using the capital -L flag, then you will need another lower-case -l flag for each library to be called in.

Order matters when specifying libraries. Begin with the most-dependent library or object file, and continue to the least dependent. The Apophenia library depends on SQLite3 and the GSL library—which depends on a BLAS library like the GSL's companion CBLAS. If your program already has an object file named source.o and you want the output program to be named `run_me`, then you will need to specify something like

> gcc source.o −lapophenia −lgsl −lcblas −lsqlite3
> or
> gcc source.o −lapophenia −lsqlite3 −lgsl −lcblas

The Sqlite3 library and GSL/CBLAS are mutually independent, so either may come first. The rest of the ordering must be as above for the compilation to work.[6] You may also need to specify the location of some of these libraries; the path extension will have to come first, so the command now looks like

> gcc −L/usr/local/lib source.o −lapophenia −lgsl −lcblas −lsqlite3

This is an immense amount of typing, but fortunately, the next section offers a program designed to save you from such labor.

> ➤ Many systems, including the shell, the preprocessor, and the linker, search a specified directory path for requested items.
>
> ➤ The default path for the preprocessor and linker can be extended via both environment variables and instructions on the compiler's command line.

A.3 MAKE   The `make` program provides a convenient system for you to specify flags for warnings and debugging symbols, include paths, libraries to link to, and who knows what other details. You write a makefile that describes what source files are needed for compilation and the flags the compiler needs on your system, and then `make` assembles the elaborate command line required to invoke the compiler. Many programs, such as `gdb`, `vi`, and `EMACS`, will even let you run `make` without leaving the program. Because assembling the command line by hand is so time consuming and error-prone, you are strongly encouraged to have a makefile for each program.

Fortunately, once you have worked out the flags and paths to compile one program, the makefile will probably work for every program on your system. So the effort of writing a makefile pays off as you reuse copies for every new program.

The discussion below is based on Listing A.2, a makefile for a program named

---

[6]With some linkers, order does not matter. If this is the case on your system, consider yourself lucky, but try to preserve the ordering anyway to ease any potential transition to another system.

run_me, which has two .c files and one header file. It is not very far removed from the sample makefile in the online code supplement.

```
 1  OBJECTS = file1.o file2.o
 2  PROGNAME = run_me
 3  CFLAGS = −g −Wall −Werror −std=gnu99
 4  LINKFLAGS = −L/usr/local/lib −lgsl −lgslcblas −lsqlite
 5  COMPILE = gcc $(CFLAGS) −c $< −o $@
 6
 7  executable: $(OBJECTS)
 8          gcc $(CFLAGS) $(OBJECTS) $(LINKFLAGS) −o$(PROGNAME)
 9
10  file1.o: file1.c my_headers.h
11          $(COMPILE)
12
13  file2.o: file2.c my_headers.h
14          $(COMPILE)
15
16  run: executable
17          ./$(PROGNAME) $(PARGS)
```

Listing A.2 A sample makefile for a program with two source files.

The make program does two types of operation: expanding variable names and following dependencies.

Variable expansion   Variable names in a makefile look much like environment variables, and behave much like them as well. They are slightly easier to set—as on lines 1–5, just use $VAR = value$. When referencing them, use a dollar sign and parens, so line five will replace $(CFLAGS) with -g -Wall -Werror -std=gnu99. All of the environment variables are also available this way, so your makefile can include definitions like:

```
INCLUDEPATH = $(INCLUDEPATH):$(HOME)/include
```

As with environment variables, variables in makefiles are customarily in all caps, and as with constants in any program, it is good form to group them at the top of the file for ease of reference and adjustment.

The $@ and $< variables on line five are special variables that indicate the target and the file used to build it, respectively, which brings us to the discussion of dependency handling.

Dependencies   The remainder of the file is a series of target/dependency pairs and actions. A *target* is something to be produced. Here the targets are object files and an executable, and on page 188, there is a makefile whose targets are PDFs and graphics. The *dependency* is the source containing the data used to produce the target. An object file depends on the C source code defining the functions and structures, an executable depends on the object files being linked together, and a PDF document depends on the underlying text. In most of these cases, the dependency is data that you yourself wrote, while the target is always computer generated.

The lines with colons, such as line ten, are target/dependency descriptions: the single item before the colon is the target name, and the one or several items after the colon are the files or other targets upon which the target depends.

After each target line, there are instructions for building the target from the dependencies. For a simple C program, the instructions are one line (typically a call to gcc). The makefile on page 188 shows some more involved target build scripts that span several lines.

An important annoyance: the lines describing the building process are indented by tabs, not spaces. If your text editor replaces tabs with a set of spaces, make will fail.[7]

Having described the format of the makefile, let us go over the process that the system will go through when you type make on the command line. The first target in a makefile is the default target, and is where make begins if you do not specify a target on the command line. So make looks at the target specification on line seven and, using its powers of variable expansion, sees that run_me depends on file1.o and file2.o. Let us assume that this is the first time the program is being built, so neither object file exists. Then make will search for a rule to build them, and thus digresses to the later targets. There, it executes the command specified by $(COMPILE) on lines 11 and 14. Having created the subsidiary targets, it then returns to the original target, and runs the command to link together the two object files.

Let us say that you then modify file1.c (but none of the other files) and then call make again. The program again starts at the top of the tree of targets, and sees that it needs to check on file1.o and file2.o. Checking the dependencies for file1.o, it sees that the file is not up-to-date, because its dependency has a newer time stamp.[8] So file1.o is recompiled, and run_me then recreated. The system knows that file2.o does not need to be recompiled, and so it does not bother to do so. In a larger project, where recompilation of the whole project can take several seconds or even minutes, make will thus save you time as well as typing.

---

[7] The odds are good that <ctrl-V> will let you insert an odd character into your text. E.g., <ctrl-V><tab> will insert a tab without converting it to spaces.

[8] What should you do if the time stamp on a file is broken, because it was copied from a computer in a different time zone or was otherwise mis-stamped? The touch command (e.g., touch *.c) will update the time stamp on a file to the current time. If you want make to recompile file1.o, you can do so with touch file1.c; make.

Compile && run   Almost without fail, the next step after a successful compile is to
                     run the program. The makefile above gives you the option of do-
ing this via the `run` dependency on lines 16–17, which depends on the `executable`
dependency on line 7. You can specify targets on the command line; if you simply
type `make`, then the system assumes the first target in the file (line 7), and if you
type the command `make  run`, then it will use the instructions under the `run` target.

`make` halts on the first error, so if the program fails to compile, the system stops
and lets you make changes; if it compiles correctly, then `make` goes on to run the
program. It considers a compilation with warnings but no errors to be successful,
but you should heed all warnings. The solution is to add `-Werror` to the `CFLAFS`
command line, to tell the compiler to treat warnings as errors.

There is no easy way to pass switches and other command-line information from
`make`'s command line to that of your program. The makefile here uses the hack of
defining an environment variable `PARGS`. On the command line, you can `export`
`PARGS=' -b |gnuplot -persist'`, and `make` will run your program with the `-b`
flag and pipe the output through `gnuplot`.

If you find the `PARGS` hack to be too hackish, you can just run the program from
the command line as usual. Recall that in C syntax, the `b` in `(a && b)` is evaluated only if `a` is true. Similarly on the command line: the command `make &&`
`./run_me` will first run `make`, then either halt if there are errors or continue on to
`./run_me` if `make` was successful. Again, you have one command line that does
both compilation and execution; coupled with most shells' ability to repeat the last
command line (try the up-arrow or `!!`), you have a very fast means of recompiling
and executing.

.

Take one of the sample programs you wrote in Chapter 2 and create a directory for it.

- Create a new directory, and copy over the .c file.

- Copy over the sample makefile from the online sample code.

- Modify the PROGNAME and OBJECTS line to correspond to your project.

- If you get errors that the system can not find certain headers, then you will need to modify the INCLUDEPATH line. If you get errors from the linker about symbols not found, you will need to change the LINKFLAGS line.

- Type make and verify that the commands it prints are what they should be, and that you get the object and executable files you expected.

Now that you have a makefile that works for your machine, you can copy it for use for all your programs with only minimal changes (probably just changing the PROGNAME or OBJECTS).

➤ The makefile summarizes the method of compiling a program on your system.

➤ At the head of the makefile, place variables to describe the flags needed by your compiler and linker.

➤ The makefile also specifies what files depend on what others, so that only the files that need updating are updated.

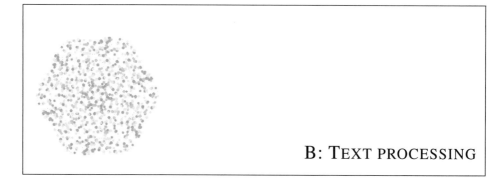

If you are lucky, then your data set is in exactly the format it needs to be in to be read by your stats package, Apophenia, Graphviz, et cetera.

By definition, you will rarely get lucky, so this appendix will explain how to modify a text file to conform to an input standard.

The book so far has included a great deal on writing new text files via the `printf` family, but not much on the process of modifying existing files. That is because modifying files in place is, simply put, frustrating. Something as simple as replacing 10 with 100 requires rewriting the entire file from that point on.

Thus, this appendix will look at some means of modifying text via existing programs that go through the tedium of modifying files for you. Its goal, narrowly defined by the rest of the text, is to show you how to convert a data file either into the input format required by a stats package or into a command file in the language of Gnuplot, a Graphviz-family tool, or SQL.

As with the rest of the analysis, you should be able to execute these procedures in a fully automated manner. This is because you will no doubt receive a revised data set next week with either corrections or more data, and because writing down a script produces an audit trail that you or your colleagues can use to check your work.

The primary technique covered in this appendix is the parsing of *regular expressions*, wherein you specify a certain regularity, such as *numbers followed by letters*, and a program searches for that regular expression in text and modifies it as per your instructions.

A secondary theme is the POSIX water metaphor. Data flows in a stream from one location to another, burbling out of data files and piping through filters like `sed` (the stream editor). Eventually, the stream of data reaches its final destination: a graphic, a database, or another file holding clean, filtered data.

Along the way, you will see how to effectively search your files. As your projects get large, you will desire tools that find every use of, say, `buggy_variable` in your program or every reference to an author in your documents. Regular expression tools will help you with this.

After the discussion of how one assembles a pipeline from small parts, this appendix will cover simple searching, as an introduction to regular expressions. The final section applies these tools to produce several scripts to reformat text files into a format appropriate for your data-handling systems.

B.1 SHELL SCRIPTS   As with Gnuplot or SQLite's command-line utility, you can operate the *shell* (what many just call the command line) by typing in commands or by writing a script and having the shell read the script. Thus, you can try the commands below on the command line, and when they work as desired, cut and paste them into a script as a permanent record.

Your script can do more than just reformat text files: it can begin with some `perl` or `sed` commands to reformat the data file, then call your C program to process the file, and finally call Gnuplot to display a graph.

The quickest way to make a script executable is to use the `source` command. Given a list of commands in the text file *myscript*, you can execute them using `source` *myscript*.[1]

Write a short script to compile and execute `hello_world.c`. On the first line of the script, give the compilation command (as per the exercise on page 18), and on the second line, execute `./a.out`. Then run your script from the command line.

---

[1]If you are going to run the script more than a few times, you may want to make it directly executable. This is not the place for a full POSIX tutorial, but `chmod 755` *myscript* or `chmod +x` *myscript* will do the trick. Having made the script executable, you can call a script in the current directory (aka `./`) from the shell using `./myscript`. [Why do you need the `./`? See the discussion of paths in Appendix A.]

Shell symbol	Meaning	C equivalent
>	send output to a file	`fopen("filename", "w");` `fprintf...;`
>>	append output to a file	`fopen("filename", "a");` `fprintf ...;`
<	read output from a file	`fopen("filename", "r");` `fgets...;`

|	redirect output to an-other program	`popen("progname", "w");`
		`fprintf...;`
		`popen("progname", "r");`
		`fgets...;`

<<	take input from an in-line script	exercise for the reader (use `fgets`)

Table B.1 Redirection via the command prompt and C

REDIRECTION Each program has a standard input stream and a standard output stream, named `stdin` and `stdout`, plus a third stream typically used for error messages, `stderr`. This section will cover the various means of bending and redirecting these streams, and the next section will cover a few tools you join together to filter streams of text.

By default `stdin` and `stdout` are the keyboard and screen, respectively. For example, `sed` reads from `stdin` and writes to `stdout`, so the command

```
sed −n p
```

simply prints to the screen whatever is typed in. If you try it, you will see that this means simply that `sed` repeats whatever you type.

```
shell prompt> sed −n p
Hello.
Hello.
How are you?
How are you?
Stop imitating me.
Stop imitating me.
<ctrl−D>
```

Now for the shifting of the streams, via the shell symbols listed in Table B.1. A clause of the form > *filename* tells the system to write whatever it would have put on the screen to a file. Thus, if you put the command `sed -n p > `*`outfile`* on the command line, then nothing will print to the screen, but whatever you type (until you hit <ctrl-D> to exit) will be written to *`outfile`*.

As a variant, `>> `*`outfile`* will append to a file, rather than overwriting it as `> `*`outfile`* would.

A `<` tells the system that it should take input not from the keyboard but from a file. Thus, `sed -n p < `*`input_file`* will dump the contents of *`input_file`* to the screen.

Q*B.2* | Use `sed -n p` and redirection of both input and output to copy a file.

Your final redirection option is the *pipe*. The pipe connects `stdout` for one program to `stdin` for another.

Question: what exponents of four end in a six? On page 209 (and in the code supplement), you will find a program, getopt, that prints exponents to `stdout`. Below, you will see that `grep` will search its input for a certain pattern; for example, `grep '6$' <`*`myfile`* will search *`myfile`* for lines ending in a six. Thus, we could redirect the output for getopt to a file, then input that file to grep.

> **Shell expansion**
>
> POSIX-compliant shells use both 'single-ticks' and "double-ticks" to delimit strings, but they behave differently. The single-tick form does not expand *$VAR* to the value of the *VAR* environment variable, and does not treat \ or ! as special. Conversely, the "double-tick" form makes all of these changes before the called program sees the string. Single-ticks tend to be easier to use for regular expressions, because of the preponderance of special characters. Notably, it is basically impossible to put a ! inside double-ticks; you will need a form like `"start string--pause string"\!"--and continue."`

```
./getopt 4 > powers_of_four
grep '6$' < powers_of_four
```

The pipe, |, streamlines this by directly linking `stdout` from the first program and `stdin` to the second, bypassing the need for a file in between the two commands.

```
./getopt 4 | grep '6$'
```

This line prints the filtered output of getopt to the screen, at which point the pattern in the exponents is eminently clear.

Now that you have seen >, >>, and <, you may be wondering about <<. Appending to input does not quite make sense; instead, this form allows you to put small scripts that would normally be in a separate file directly on the command line. Here is a sample usage:

```
gnuplot << end−of−script
set term postscript
set out "redirect_test.ps"
plot sin(x)
end−of−script
```

The *end-of-script* marker can be any string (*EOF* is a popular choice), and indicates when the script will end. Everything from the first line until the concluding string (on its own line) is sent to Gnuplot as if it were typed in directly.

※ stderr There is one more stream after stdin and stdout: stderr, which is intended to print error messages or diagonstics to the screen, even when stdout is dumping data to a file. There are reasons to redirect this stream as well, such as when your program is producing so many errors that they scroll faster than you can read them.

- In csh, use >& anywhere you would use > to redirect both stdout and stderr to the given destination, e.g., make >& errors.txt.
- In bash, write stderr to a file is to use 2> as you would > above. For example, make 2>errors.txt will dump all compilation errors to the errors.txt file.

※ *Streams and C* C and UNIX co-evolved, so all of the above methods of redirecting inputs and outputs have an easy C implementation.

As you saw on pages 166*ff*, you can use fprintf to send data to a file or another program. You may have noticed the close similarity between opening/writing to a file and opening/writing to a pipe:

```
FILE *f = fopen("write_to_me", "w");
FILE *g = popen("gnuplot", "w");
fprintf(f, "The waves roll out.\n");
fprintf(g, "set label 'The waves roll out.';set yrange [−1:1]; plot −1");
fclose(f);
pclose(g);
```

If this code were in a program named water, then the fopen and fprintf instructions are analogous to water > write_to_me. The popen and fprintf pair are analogous to water | gnuplot.

This book has spent little time on the subject of reading data from `stdin` (look up `fscanf` in your favorite C reference), but by changing the `"w"`s to `"r"`s in the above code, the flow of data is from the `FILE*` to the program itself, analogous to `water < write_to_me` and `gnuplot | water`.

> Modify `getopt.c` to filter its output through `grep`.
>
> • Popen `grep '6$'` for writing.
>
> • Modify the `printf` statement to an `fprintf` statement.

The three standard pipes are defined by C's standard library, and can be used as you would any other popened stream. That is, `fprintf(stdout, "Data forth-coming:\n")` and `fprintf(stderr, "danger!\n")` will write to `stdout` and `stderr` as expected. What is the difference between `fprintf(stdout, "data")` and `printf("data")`? There is none; use whichever form is most convenient for your situation.

Finally, there is the `system` function, which simply runs its argument as if it were run from the command line. Most POSIX machines come with a small program named `fortune`, that prints fortune-cookie type sayings. Most of us can't eat just one, but consume a dozen or so at a time. Listing B.2 shows a program that automates the process via a `for` loop. The program does not use `printf` and does not `#include <stdio.h>`, because the C program itself is not printing anything—the child programs are.[2]

```
#include <stdlib.h>

int main(){
    for (int i=0; i< 12; i++){
        system("fortune");
        system("echo");
        system("sleep 6");
    }
}
```

Listing B.2 Print fortunes. Online source: `fortunate.c`.

There are two differences between a `system` command and a `popen` command. First, there is the fact that there is no way for the rest of the program to access `system`'s input and output, while `popen` allows input or output from within the

---

[2]This program is longer than it needs to be, because all three `system` lines could be summarized to one: `system("fortune; echo; sleep 6");`. Notice also that the `apop_system` convenience function allows you to give a command with printf-style placeholders; e.g., `apop_system("cp %s %s", from_file, to_file)`.

program. But also, `system` will stop your program and wait for the child program to finish, while `popen` will open the pipe and then move on to the next step in your program. So if the command has to finish before you can continue, use `system`, and if the command is something that should run in the background or should wait for data to become available, use `popen`.

Σ

➤ All programs have standard inputs and outputs, which are by default the keyboard and screen.

➤ Both `stdin` and `stdout` can be redirected. `stdin` can read from a file using the `< `*infile* form; `stdout` can write to a file using the `> `*outfile* form. Use `>` *outfile* to overwrite and `>>` *outfile* to append.

➤ Programs can send their output to other programs using the pipe, |.

➤ You can do redirection from inside C, using `fopen` to effect the command-line's `<` *infile* and `>` *outfile*, and `popen` to open a pipe.

B.2 SOME TOOLS FOR SCRIPTING Depending on your system, you can type somewhere between a few hundred and a few thousand commands at the command prompt. This section points out a few that are especially useful for the purposes of dealing with data in text format; they are summarized in Table B.3. All but a few (`column`, `egrep`, `perl`) are POSIX-standard, and you can expect that scripts based on the standard commands will be portable. The basic file handling functions (`mkdir`, `ls`, `cp`, ...) are listed for reference, but are not discussed here. If you are not familiar with them, you are encouraged to refer to any of an abundance of basic UNIX and POSIX tutorials online or in print (e.g., Peek *et al.* (2002)).

Most of these commands are in some ways redundant with methods elsewhere in the book: you could use `cut` to pull one column of data, or you could read the data into a database and `select` the column; you could use `wc` to count data points, or use `select count(*)` to do the same.[3] Generally, command-line work is good for pure text manipulation and quick questions (e.g., do I have the thousands of data points I was expecting?), while you will need to write a program or query for any sort of numeric analysis or to answer more detailed questions about the data (e.g., how many rows are duplicates?).[4]

---

[3] There is even a `join` command, which will do database-style joins of two files. However, it is hard to use for any but clean numeric codes. For example, try using `join` to merge `data-wb-pop` and `data-wb-gdp`.

[4] `apop_query_to_double("select count(*) from `*data*`")` - `apop_query_to_double("select count(*) from (select distinct * from `*data*`)")`

	Basic file handling
`mkdir/rmdir`	Make or remove a directory
`cd`	Change to a new current directory
`ls`	List directory contents
`cp/mv`	Copy or move/rename a file
`rm`	Remove a file
	Reading
`cat`	List a file, or concatenate multiple files
`head/tail`	List only the first/last few lines of a file
`less/more`	Interactively browse through a file
`column`	Display a file with its columns aligned
	Writing
`sort`	Sort
`cut/paste`	Modify files by columns rather than lines
`nl`	Put a line number at the head of each line
`uniq`	Delete duplicate lines
	Information
`man`	Manual pages for commands
`wc`	Count words, lines, or characters
`diff`	Find differences between two files
	Regular expressions
`sed`	Stream editor: add/delete lines, search/replace
`grep/egrep`	Search a file for a string
`perl`	All of the above; used here for search/replace

Table B.3 Some commands useful for handling text files.

man   What does the 755 in chmod 755 do? Ask man chmod. What did the echo com-
       mand in Listing B.2 do? Try man echo. The manual pages list the basic function
       of a program and its various *switches*. They are not a very good way to learn to use
       a POSIX system, but are the definitive reference for command-line details.

cat   The simplest thing you could do with a text file is print it, and this is what
       cat does. Instead of opening a file in your text editor, you can simply type cat
       file_to_read on the command prompt for a quick look.

       The other use of cat, for which it was named, is concatenation. Given two files a
       and b, cat a b > c writes a and, immediately thereafter, b, into c.

less/more   These are the paging programs, useful for viewing longer files. The POSIX
             standard, more, is so named because for files of more than a screenful of
       text, it displays a page of text with a banner at the bottom of the screen reading
       more. The successor, less, provides many more features (such as paging up as
       well as down). These programs can also read from stdin, so you can try combi-
       nations like sort *data_file* | less.

head/tail   head *myfile* will print the first ten lines of *myfile*, while tail *myfile*
             will print the last ten. These are good for getting a quick idea of what is
       in a large file. Also, tail -f *myfile* will follow *myfile*, first showing the last
       ten lines, and then updating as new lines are added. This can provide reassurance
       that a slow-running program is still working.

sed   The stream editor will be discussed in much more detail below, but for now it
       suffices to know that it can be used to replicate head or tail. The command sed
       -n '1,3p' < a_file will print the first through third lines of a_file. The last
       line is indicated by $, so sed -n '2,$p' < a_file will print the second through
       last lines of the file.

sort   This is as self-descriptive as a command can get. The data-wb-pop file (in the
       code supplement) is sorted by population; sort data-wb-pop would output
       the file in alphabetical order. Sort has many useful switches, that will sort ignoring
       blanks, by the second (or nth) column of data, in reverse order, et cetera. See man
       sort for details.

$\mathbb{Q}_{B.4}$ | Sorting `data-wb-pop` like this sorts the two header lines in with all the countries. Write a command that first calls `sed` to delete the headers, then pipes the output to `sort`.

cut/paste   Almost all of the commands here operate on rows; `cut` and `paste` operate on columns. Specify the delimiter with the `-d` flag, and then the field(s) with the `-f` flag (where the first column is column one, not zero). To get a list of just countries from the World Bank data, try `cut -d"|" -f 1 data-wb-pop`; if you would like to see just the population numbers, use `cut -d"|" -f 2 data-wb-pop`.

`paste` puts its second input to the right of its first input—it is a vertical version of `cat`.

$\mathbb{Q}_{B.5}$ | Use the command string from the last exercise to sort the population and GDP data into two new files. `cut` the second file down to include only the column of numbers. Then use `paste sorted_pop sorted_gdp> combined_data` to produce one file with both types of data. Verify that the lines of data are aligned correctly (i.e., that the Albania column does not list the GDP of Algeria).

wc   This program (short for word count) will count words, lines, and characters. The default is to print all three, but `wc -w`, `wc -1`, and `wc -c` will give just one of the three counts.

When writing a text document, you could use it for a word count as usual: `wc -w report.tex`. If you have a data file, the number of data points is probably `wc -1 data` (if there is a header, you will need to subtract those lines).

This program is especially useful in tandem with `grep` (also discussed at length below). How many lines of the `data-wb-pop` file have missing data (which the WB indicates by `..`)? `grep '\.\.' data-wb-pop` will output those lines where the string `..` appears; `wc -1` will count the lines of the input; piping them together, `grep '\.\.' data-wb-pop | wc -1` will count lines with `..`s.

nl   This command puts line numbers at the beginning of every line. Remember that SQLite puts a `rowid` on every row of a table, which is invisible but appears if explicitly `selected`. But if you want to quickly put a counter column in your data set, this is the way to do it.

column   [This command is not POSIX-standard, but is common for GNU-based sys-
         tems.] Text data will be easy to read for either a computer or a human, but
         rarely both. Say that a data file is tab-delimited, so each field has exactly one tab
         between columns:

> Name<tab>Age
> Rumpelstiltskin<tab>78
> Dopey<tab>35

This will read into a spreadsheet perfectly, but when displayed on the screen, the
output will be messy, since Rumpelstiltskin's tab will not be aligned with Dopey's.
The column command addresses exactly this problem, by splitting columns and
inserting spaces for on-screen human consumption. For example, column -s"|"
-t data-wb-pop will format the data-wb-pop file into columns, splitting at the
| delimiters. The first column will include the header lines, but you already know
how to get sed to pipe headerless data into column.

diff   diff *f1 f2* will print only the lines that differ between two files. This can
       quickly be overwhelming for files that are significantly different, but after a
       long day of modifying files, you may have a few versions of a file that are only
       marginally different, and diff can help to sort them out. By the end of this chap-
       ter, that will be the case, so various opportunities to use diff will appear below.
       EMACS, vim, and some IDEs have modes that run diff for you and simultaneously
       display the differences between versions of a file.

To give another use, say that you have a running program, but want to clean up
the code a little. Save your original to a backup, and then run the backup program, writing the output to *out1*. Then clean up the code as you see fit, and run
the cleaned code, writing to out2. If you did no damage, then diff *out1 out2*
should return nothing.

uniq   This program removes many duplicate lines from a file. uniq *dirty > clean*
       will write a version of the input file with any *successive* duplicate lines deleted.
       For example, given the input

```
1   a;
2   b;
3
4   b;
5   c;
6   c;
7   a;
```

the output will omit line five but leave the remainder of the file intact. If order does not matter, then `sort your_file | uniq` will ensure that identical lines are sequential, so `uniq` will easily be able to clean them all. Alternatively, read the file to a database and use `select distinct * from your_table`.

B.3 REGULAR EXPRESSIONS The remainder of this appendix will cover programs to parse *regular expressions*. Regular expressions comprise a language of their own, which provides a compact means of summarizing a text pattern. Knowledge of regexes comes in handy in a wide variety of contexts, including standard viewers like `less`, a fair number of online search tools, and text editors like `vi` and `EMACS`. Entire books have been written on this one seemingly small subject, the most comprehensive being Friedl (2002).

This chapter will use grep, sed, and Perl because it is trivial to use them for regular expression parsing and substitution from the command line. Other systems, such as C or Python, require a regex compilation step before usage, which is nothing but an extra step in a script or program, but makes one-line commands difficult.

※ *Standard syntax, lack thereof* Unfortunately, there are many variants on the regex syntax, since so many programs process them, and the author of each program felt the need to make some sort of tweak to the standard. Broadly, the basic regular expression (*BRE*) syntax derives from ed, a text editor from back when the only output available to computers was the line printer. Most POSIX utilities, such as `sed` or `grep` or `awk`, use this basic regex syntax.

There is a barely-modified variant known as *extended regular expression* (*ERE*) syntax, available via `egrep` and a switch to many other utilities. New features like the special meaning of characters like (,), +, or | evolved after a host of BRE programs were already in place, so in the BRE syntax, these characters match an actual (,), +, or | in the text, while they can be used as special operators when preceded by a backslash (e.g., \+, \|, ...). EREs used these characters to indicate special operations from the start, so + is a special operator, while \+ indicates a plain plus sign (and so on for most other characters).

The scripting language Perl introduced a somewhat expanded syntax for regular expressions with significantly more extensions and modifications, and many post-Perl programs adopted Perl-compatible regular expressions (*PCRE*s).

This chapter covers Perl and GNU grep and sed as they exist as of this writing, and you are encouraged to check the manual pages or online references for the subtle shifts in backslashes among other regular expression systems.

THE BASIC SEARCH   The simplest search is for a literal string of text. Say that you have a C program, and are searching for every use of a variable, *var*. You can use the command-line program grep to search for *var* among your files.[5] The syntax is simple: grep *var* *.c will search all files in the current directory ending in .c for the string *var*, and print the result on the command line.

This is already enough to do a great deal. In olden times, programmers would keep their phone book in a simple text file, one name/number per line. When they needed Jones's number, they could just run grep Jones my_phone_book to find it. If you need to know Venezuela's GDP, grep will find it in no time: grep Venezuela data-wb-gdp.

$\mathbb{Q}_{B.6}$ | Use grep to search all of your .c files for uses of printf. The -C n option (e.g., grep -C 2) outputs n context lines before and after the match; repeat your search with two context lines before and after each printf.

Bracket expressions   Now we start to add special characters from the regex language. An expression in square brackets always represents a single character, but that single character could be one of a number of choices. For example, the expression [fs] will match either f or s. Thus, the expression [fs]printf will match both fprintf or sprintf (but not printf), so grep [fs]printf *.c will find all such uses in your C files. More bits of bracket expression syntax:

- You can use ranges: [A-Z] searches for English capital letters, [A-Za-z] searches for English capital or lowercase letters, and [0-9] matches any digit.

- You can match any character except those in a given list by putting ^ as the first item in the bracketed list. Thus, [^fs] matches one single character that is not f or s.[6]

- There are a few common sets of characters that are named for the POSIX utilities. For example, [[:digit:]] will search for all numbers, [0-9]. Notice that [:digit:] is the character class for digits, so [[:digit:]] is a bracket expression matching any single digit. See Table B.4 for a partial list. These character sets are locale-dependent. For example, Ü is not in [A-Z], but if it is a common letter in the language your computer speaks, then it will be an element of [:alpha:].

---

[5]The name comes from an old **ed** command (we'll meet **ed** later) for global regular expression printing: g/re/p. Many just take it as an acronym for general regular expression parser.

[6]If you are searching for the ^ character itself, just don't put it first in the bracket. Conversely, if you need a literal dash that does not indicate a range, put it first. To find all lines with carats or dashes, try grep "[-^]" *myfile*.

POSIX	Perl	English
[:digit:]	\d	Numeric characters.
[:alpha:]		Alphabetic characters.
[:upper:]		Uppercase alphabetic characters.
[:lower:]		Lowercase alphabetic characters.
[:alnum:]	\w	Alphanumeric characters.
[:blank:]		Space and tab characters.
[:space:]	\s	Any white space, including space, tab (\t), newline (\n).
[:punct:]		Punctuation characters
[:print:]		Printable characters, including white space.
[:graph:]		Printable characters, excluding white space.
	\D, \W, \S	Not numeric chars, not alphanumeric chars, not white space

Table B.4 Some useful regular expression character sets.

chars	main regex meaning	meaning in brackets
[ ]	Bracket expression	as [: :], a named group
^	Beginning of line	don't match following chars
.	Any character	plain period
-	A plain dash	a range, e.g., 0-9

Table B.5 Some characters have different meanings inside and outside brackets.

All of these can be combined into one expression. For example,

$$[-+eE[:digit:]]$$

will match any digit, a minus or plus, or the letter e. Or, if you named a variable opt but grep opt *.c spits out too many comment lines about options, try grep opt[^i] *.c.

Brackets can be a bit disorienting because there is nothing in the syntax to visually offset a group from the other elements in the bracket, and no matter how long the bracketed expression may be, the result still represents exactly one character in the text. Also, as demonstrated by Table B.5, the syntax inside brackets has nothing to do with the syntax outside brackets.

Regular expressions are vehemently case-sensitive. `grep perl search_me` will not turn up any instances of `Perl`. The first option is to use the `-i` command-line switch to search without case: `grep -i perl search_me`. The second option, occasionally useful for more control, is to use a bracket: `grep [Pp]erl search_me`.

Alternatives   Recall that Apophenia includes various printing functions, such as `apop_data_print` and `apop_matrix_print`. Say that you would like to see all such examples. Alternatives of the form *A or B* are written in basic regex notation as (A|B). Notice how this analogizes nicely with C's (A||B) form.

This is where the various flavors of regular expression diverge, and the backslashes start to creep in. Plain `grep` uses BREs, and so needs backslashes before the parens and pipe: \(A\|B\). When run as `grep -E`, `grep` uses EREs. Most systems have an `egrep` command that is equivalent to `grep -E`.[7]

Also, recall the difference between 'single-ticks' and "double-ticks" on the command line, as discussed in the box on page 395: the single-tick form does not treat backslashes in a special manner, so for example, you can find a single literal period with '\.' instead of "\\.".

All that said, here are a few commands that will search for both `apop_data_print` and `apop_matrix_print`:

```
grep "apop_\\(data\\|matrix\\)_print" *.c
grep 'apop_\(data\|matrix\)_print' *.c
grep −E "apop_(data|matrix)_print" *.c
egrep "apop_(data|matrix)_print" *.c
```

A few special symbols   You sometimes won't care at all about a certain segment. A dot matches any character, and you will see below that a star can repeat the prior atom. Therefore

```
grep 'apop.*_print' *.c
```

will find any line that has `apop`, followed by anything, eventually followed by `_print`.

• Once again, the symbols . and * mean different things in different contexts. In a regex, the dot represents any character, and the star represents repetition; on the

---

[7]Officially, `egrep` is obsolete and no longer part of the POSIX standard.

command line (where wildcard matching is referred to as *globbing*), the dot is just a dot, and the star represents any character.

- The caret and dollar sign (^ and \$) indicate the beginning and end of the line, respectively. grep "^int" *.c will only match lines where int is right at the beginning of the line, and grep "{\$" *.c will only match open-braces at the end of the line.[8]

- A single space or tab are both characters, meaning that a dot will match them. In grep, the atom \W will match any single space or tab, and the atom \w will match anything that is not white space. Since there are frequently an unknown bunch of tabs and spaces at the head of code files, a better way to search for int declarations would be

> grep '^\W*int' *.c

> ➤ The quickest way to search on the command line is via grep. The syntax is grep 'regular expression' files_to_search.

> ➤ Without special characters, grep simply searches for every line that matches the given text. To find every instance of *fixme* in a file: grep 'fixme' filename.

> ➤ A bracketed set indicates a single character.

> > – The set can include single characters: [Pp]erl matches both Perl and perl (though you maybe want a case-insensitive search with grep -i).

> > – The set can include ranges: [A-Za-z] will match any standard English letter.

> > – The set can exclude elements: [^P-S] will find any character except P, Q, R, and S.

> ➤ Express alternatives using the form (A|B).

> ➤ Different systems have different rules about what is a special character. The ERE/PCRE form for alternation used by egrep and perl is (A|B); the BRE form used by grep is \(A\|B\).

> ➤ A single dot (.) represents any character including a single space or tab, and grep understands \W to mean any white space character and \w to mean any non-white space character.

---

[8]It is difficult to search across lines. Perl has the m option (m//). A more reliable and universal option is to simply remove newlines, turning your input stream into a single long line. The easiest way to do this is via a form like tr '\n' '|' < infile | grep ..., where tr translates elements of the first set (in this case, the newline) into the elements of the second (the pipe).

REPLACING   grep is wonderful for searching, but is a purely read-only program. If you want to change what you find, you will need to use a more complex program, such as perl or sed (the stream editor). sed will easily handle 95% of data-filtering needs, but perl adds many very useful features to the traditional regular expression parsing syntax. If you find yourself using regexes frequently, you may want to get to know how they are handled in a full Perl-compatible regex scripting system like Python, Ruby, or Perl itself. These programs complement C nicely, because they provide built-in facilities for string handling, which is decidedly not C's strongest suit.

Perl and sed syntax   sed is certainly installed on your system, since it is part of the POSIX standard; perl is almost certainly installed, since many modern system utilities use it. If it is not installed, you can install it via your package manager.

Both programs generally expect that you will use a command file, but you can also give a command directly on the command line via the -e command.

For example,

```
perl −e 'print "Hello, world.\n"'
```

will run the Perl command print "Hello, world.\n".

sed always operates on an input file, so we might as well start by emulating grep:[9]

```
sed −n −e '/regex/p' < file_to_search
```

Sed likes its regexes between slashes, and the p is short for *print*, so the command /*regex*/p means 'print every line that matches *regex*.' Sed's default is to print every input line, and then the print command will repeat lines that match *regex*, which leads for massive redundancy. The -n command sets sed to print only when asked to, so only matching lines print.

Generally, a sed command has up to three parts: the lines to work on, a single letter indicating the command, and some optional information for the command.

Above, the specification of lines to work on was a regular expression, meaning that the command (p) should operate only on those lines that match /regex/.

---

[9]As you saw in the sed examples in the section on redirection, the -e flag is optional using this form, but it never hurts.

- You can replace this regex with a fixed set of lines, such as 1,10 to mean the first ten lines.

- In this context, $ means the last line of the file, so 11,$p would print everything but the first ten lines.

- You can print everything but a given set of lines using an ! after the range, so 1,10!p would also print everything but the first ten lines.[10]

The box presents the five single-letter sed commands discussed here; sed has a few more but this is already enough for most relevant work. The third part of the command, following the single-letter command, will vary depending on the command. As in the examples above, the third part is blank for the p command.

Sed commands	
p	Print.
s	Substitute.
d	Delete.
i	Insert before.
a	Append after.

Search and replace syntax   The basic syntax for a search and replace is s/replace me/with me/g. The slashes distinguish the command (s), the text to search (replace me), the text for replacement (with me), and the modifiers (the g; see box).[11]

If you have only one file to scan, you can use these programs as filters:

```
perl −p −e "s/replace me/with me/g" <file_to_modify >modified_file
sed −e "s/replace me/with me/g" <file_to_modify >modified_file
```

Perl's -p switch is the opposite of sed's -n switch. As opposed to sed, Perl's default is to never print unless the pattern matches, which means that if you do not give it the -p switch, it will pass over any line that does not match the regex. With -p, non-matching lines appear as-is and lines that match have the appropriate substitutions made, as we want. In this case, sed is doing the right thing by default, so you do not need the -n switch.

Why the /g?

Back when regexes were used for real-time editing of files, it made sense to fix only one instance of an expression on a line. Thus, the default was to modify only the first instance, and to specify replacing only the second instance of a match with s/.../.../2, the third instance with s/.../.../3, and all instances with s/.../.../g (global). This is still valid syntax that is occasionally useful, but when filtering a data file, you will almost always want the /g option.

---

[10]The ! is used by the bash shell for various command history purposes, and as a result has absolutely perverse behavior in "double-ticks", so using 'single-ticks' is essential here. To print all the lines not beginning with a #, for example, use sed -n '/#/!p' < infile.

[11]A convenient trick that merits a footnote is that you can use any delimiter, not just a slash, to separate the parts of a search and replace. For example, s|replace me|with me|g works equally well. This is primarily useful if the search or replace text has many slashes, because using a slash as a delimiter means you will need to escape slashes in the text—s/\/usr\/local/\/usr/g—while using a different delimiter means slashes are no longer special, so s|/usr/local|/usr|g would work fine. The norm is slashes, so that is what the text will use.

Since you probably want to search and replace on every line in the file, there is nothing preceding the single-letter command s. But if you want to replace replace me only on lines beginning with #, you can specify the sed command /#/ s/replace me/with me/g. In this case, sed is taking in two regular expressions: it first searches vertically through the file for lines that match #. When it finds such a line, it searches horizontally through the line for replace me.

$\mathbb{Q}_{B.7}$ | Say that your database manager does not accept pipe-delimited text, but wants commas instead. Write a script to replace every pipe in the data-classroom file with a comma. Then modify the script to replace only the pipes in the data, leaving the pipe in the comments unchanged.

Another alternative is to replace-in-place, using the -i switch:[12]

```
perl −p −i.bak −e 's/replace me/with me/g' files_to_modify
sed −i.bak −e 's/replace me/with me/g' files_to_modify
```

With the -i option, Perl and sed do not write the substitutions to the screen or the > file, but make them directly on the original file. The .bak extension tells Perl and sed to make a backup before modification that will have a .bak extension. You may use any extension that seems nice: e.g., -i˜ will produce a backup of a file named *test* that would be named *test*˜. If you specify no suffix at all (-i), then no backup copy will be made before the edits are done. It is up to you to decide when (if ever) you are sufficiently confident to not make backups.[13]

$\mathbb{Q}_{B.8}$ | Create a file named about_me with one line, reading I am a teapot. Use either perl or sed to transform from a teapot to something else, such as a kettle or a toaster. Verify your change using diff.

You are welcome to include multiple commands on the line, by the way. In perl, separate them with a semicolon, as in C. In sed or perl, you may simply specify additional -e commands, that will be executed in order—or you can just use a pipe.

```
perl −pi.bak −e 's/replace me/with me/g; s/then replace this/with this/g' files_to_modify
perl −pi.bak −e 's/replace me/with me/g' −e's/then replace this/with this/g' files_to_modify
sed −i.bak −e 's/replace me/with me/g' −e's/then replace this/with this/g' files_to_modify
```

---

[12]The -i switch is not standard in sed, but works on GNU sed, which you probably have. Bear this in mind if you are concerned about a script's portability.

[13]If editing a file in place is so difficult, how does sed do it? By writing a new file and then switching it for the original when it is finished. So if you are filtering a 1GB file 'in place' with no backup and you do not have 1GB of free space on your hard drive, you will get a disk full error.

```
perl −p −e 's/replace me/with me/g' < modify_me | perl −p −e's/then replace this/with this/g'
    >modified_version
sed 's/replace me/with me/g' <modify_me | sed 's/then replace this/with this/g' >
    modified_version
```

Replacing with modifications   Parentheses, used above to delimit conditional segments, can also be used to store sections for later
use. For example, say that your file reads There are 4 monkeys at play, but
your copyeditor feels that the sentence as written is imprecise. You forgot the
number of monkeys, so you are tempted to use s/[0-9] monkeys/Callimico
goeldii/g—but this will lose the number of monkeys.

However, you can put a segment of the search term in parentheses, and then use
\1 in sed or $1 in Perl to refer to it in the replacement. Thus, the correct command
lines for replacing an unknown number of monkeys are

```
sed −i~ −e 's/\([0−9]*\) monkeys/\1 Callimico goeldii/g' monkey_file
sed −r −i~ −e 's/([0−9]*) monkeys/\1 Callimico goeldii/g' monkey_file
perl −p −i~ −e 's/([0−9]*) monkeys/$1 Callimico goeldii/g' monkey_file
```

- The \1 or $1 will be replaced by whatever was found in the parentheses ([0-9]*).
- sed normally uses BREs, but the GNU version of sed uses EREs given the -r
  flag.

If there are multiple substitutions to be made, you will need higher numbers.
Say that we would like to formalize the sentence the 7 monkeys are fighting
the 4 lab owners. Do this using multiple parens, such as

```
sed −i~ −e 's/\([0−9]*\) monkeys are fighting the \([0−9]*\) lab owners/\1 Callimico goeldii
    are fighting the \2 Homo sapiens/g' monkey_file
```

Repetition   Say that values under ten in the data are suspect, and should be replaced
with "NaN". The search s/[0-9]/"NaN"/g won't work, because 45
will be replaced by "NaN""NaN", since both 4 and 5 count as separate matches. Or
say that values over ten are suspect. The search s/[1-9][0-9]/"NaN"/g won't
work, because 100 would be replaced by "NaN"0. We thus need a means of specifying repititions more precisely.

Here are some symbols to match repetitions or to make an element optional.

> *   the last atom appears zero or more times
> \+   the last atom appears one or more times (GNU grep/sed: \\+)
> ?   the last atom appears zero or one times (GNU grep/sed: \\?)

To replace all of the integers less than 100 in a file:

```
perl −pi~ −e 's/([^0−9]|^)([0−9][0−9]?)([^0−9]|\$)/$1NaN$3/g' search_me
sed −i~ 's/\([^0−9]\|^\)\([0−9][0−9]\?\)\([^0−9]\|\$\)/\1NaN\3/g' search_me
sed −r −i~ 's/([^0−9]|^)([0−9][0−9]?)([^0−9]|\$)/\1NaN\3/g' search_me
```

This is a mess, but a very precise mess (which will be made more legible below).

- The first part, (`[^0-9]|^`) will match either a non-numeric character or the beginning of the line.
- The second part, (`[0-9][0-9]?`), matches either one or two digits (but never three).
- The third part, (`[^0-9]|$`) will match either a non-numeric character or the end of the line.

Thus, we have precisely communicated that we want something, then an integer under 100, then something else. Since the search string used three sets of parens, the output string can refer to all three. It repeats the first and last verbatim, and replaces the second with NaN, as desired.

Structured regexes   Chapter 2 presented a number of techniques for writing code that is legible and easy to debug:

- Break large chunks of code into subfunctions.
- Debug the subfunctions incrementally.
- Write tests to verify that components act as they should.

All of this applies to regular expressions. You can break expresssions down using variables, which your shell can then substitute into the expression. For example the illegible regexes above can be somewhat clarified using a few named subexpressions. This is the version for bash or csh.[14]

---

[14] You can also replace variables in the shell using a form like `${digits}`. This is useful when the variable name may merge with the following text, such as a search for exponential numbers of a certain form: `${digits}e$digits`.

```
notdigit='\([^0−9]\|^\|$\)'
digits='\([0−9][0−9]\?\)'
replacecmd="s/$notdigit$digits$notdigit/\1NaN\\3/g"
sed −i~ $replacecmd search_me
```

The `echo` command is useful for testing purposes. It dumps its text to `stdout`, so it makes one-line tests relatively easy. For example:

```
echo 123 | sed $replacecmd
echo 12 | sed $replacecmd
echo 00 | sed $replacecmd
```

$\mathbb{Q}_{B.9}$ Modify the search to include an optional decimal of arbitrary length. Write a test file and test that your modification works correctly.

B.4   ADDING AND DELETING   Recall the format for the sed command to print a line from page 408: `/find_me/p`. This consists of a location and a command. With the slashes, the location is: those lines that match `find_me`, but you can also explicitly specify a line number or a range of lines, such as `7p` to print line seven, `$p` to print the last line of the input file, or `1,$p` to print the entire file.

You can use the same line-addressing schemes with `d` to delete a line. For example, `sed "$d" <infile` will print all of `infile` but the last line.

$\mathbb{Q}_{B.10}$ Use `d` to produce a version of `data-classroom` with the comments removed.

You can also use `i` to insert above the given line(s), and `a` to append after the given line. A few examples:

```
#Add a pause after a Gnuplot data block
sed −i~ −e "/^e$/ a pause pauselength" plotfile
#Put the text plot '−' at the head of a file.
sed −i~ −e "1 i plot '−'" plotfile
#Pretend missing data does not exist
sed −i~ −e "/NaN/ d" plotfile
```

By the way, if you really want to get `sed` to print "Hello, world" you can do it by inserting at line one and ignoring the rest of the file:

```
sed −n −e "1 i Hello, world." < any_random_file
```

Perl can do all of these things easily from inside a Perl script, but inserting and deleting lines from the command line is not as pleasant as using sed.[15]

Q_B.11
Refer to the exercise on page 184, which read in a text file and produced a Graphviz-readable output file. That exercise read the text to an apop_-data set and then wrote the output, which is sensible when pulling many classes from a database. But given the text file data-classroom in the code supplement, you can modify it directly into Graphviz's format. Write a sed script to

- Delete the header line.

- Replace each pipe with a ->.

- Replace each number n with node$n$.

- Add a header (as on page 184).

- Add the end-brace on the last line.

Pipe your output through neato to check that your processing produced a correctly neato-readable file and the right graph.

Q_B.12
Turn the data-classroom file into an SQL script to create a table.

- Delete the header line.

- Add a header create table class(ego, nominee). [Bonus points: write the search-and-replace that converts the existing header into this form.]

- Replace each pair of numbers n|m with the SQL command insert into class(n, m);.

- For added speed, put a begin;—commit; wrapper around the entire file.

Pipe the output to sqlite3 to verify that your edits correctly created and populated the table.

---

[15] A reader recommends the following for inserting a line:
```
perl -n -e 'print; print "pause pauselength" if /^e$/'
```
For adding a line at the top of a file:
```
perl -p -e 'print "plot '-'\n" unless $a; $a=1'
```
For deleting a line:
```
perl -n -e 'print unless /NaN/'
```

➤ Your best bet for command-line search-and-replace is `perl` or `sed`. Syntax: `perl -pi "s/replace me/with me/g" data_file` or `sed -i "s/replace me/with me/g" data_file`.

➤ If you have a set of parens in the search portion, you can refer to it in Perl's replace portion by $1, $2, . . . ; and in sed's replace via \1, \2, ...

➤ The `*`, `+`, and `?` let you repeat the previous character or bracketed expression zero or more, one or more, and zero or one times, respectively.

B.5 MORE EXAMPLES

Now that you have the syntax for regexes, the applications come fast and easy.

Quoting and unquoting Although the dot (to match any character) seems convenient, it is almost never what you want. Say that you are looking for expressions in quotes, such as `"anything"`. It may seem that this translates to the regular expression `".*"`, meaning any character, repeated zero or more times, enclosed by quotes. But consider this line: `"first bit", "second bit"`. You meant to have two matches, but instead will get only one: `first bit", "second bit`, since this is everything between the two most extreme quotes. What you meant to say was that you want anything that is not a `"` between two quotes. That is, use `"[^"]+"`, or `"[^"]*"` if a blank like `""` is acceptable.

Say that the program that produced your data put it all in quotes, but you are reading it in to a program that does not like having quotes. Then this will fix the problem:

```
perl -pi~ -e 's/"([^"]*)"/$1/g' data_file
```

Getting rid of commas Some data sources improve human readability by separating data by commas; for example, `data-wb-gdp` reports that the USA's GDP for 2005 was 12,455,068 millions of dollars. Unfortunately, if your program reads commas as field delimiters, this human-readability convenience ruins computer readability. But commas in text are entirely valid, so we want to remove only commas between two numbers. We can do this by searching for a number-comma-number pattern, and replacing it with only the numbers. Here are the `sed` command-line versions of the process:

```
sed −i~ 's/\([0−9]\),\([0−9]\)/\1\2/g' fixme.txt
sed −r −i~ 's/([0−9]),([0−9])/\1\2/g' fixme.txt
```

Suspicious non-printing characters Some sources like to put odd non-printing characters in their data. Since they don't print, they are hard to spot, but they produce odd effects like columns in tables with names like pop□ÖÉ□tion. It is a hard problem to diagnose, but an easy problem to fix. Since [:print:] matches all printing characters, [^[:print:]] matches all non-printing characters, and the following command replaces all such characters with nothing:

```
sed −i~ 's/[^[:print:]]//g' fixme.txt
```

Blocks of delimiters Some programs output multiple spaces to approximate tabs, but programs that expect input with whitespace as a delimiter will read multiple spaces as a long series of blank fields.[16] But it is easy to merge whitespace, by just finding every instance of several blanks (i.e., spaces or tabs) in a row, and replacing them with a single space.

```
sed −i~ 's/[[:blank:]]\+/ /g' datafile
```

You could use the same strategy for reducing any other block of delimiters, where it would be appropriate to do so, such as s/,+/,/.

Alternatively, a sequence of repeated delimiters may not need merging, but may mark missing data: if a data file is pipe-delimited, then 7||3 may be the data-producer's way of saying 7|NaN|3. If your input program has trouble with this, you will need to insert NaN's yourself.

This may seem easy enough: s/||/|NaN|/g. But there is a catch: you will need to run your substitution twice, because regular expression parsers will not overlap their matches. We expect the input ||| to invoke two replacements to form |NaN|NaN|, but the regular expression parser will match the first two pipes, leaving only a single | for later matches; in the end, you will have |NaN|| in the output. By running the replacement twice, you guarantee that every pipe finds its pair.

```
sed −i~ datafile −e 's/||/|NaN|/g' −e 's/||/|NaN|/g'
```

---

[16]The solution at the root of the problem is to avoid using whitespace as a delimiter. I recommend the pipe, |, as a delimiter, since it is virtually never used in human text or other data.

Text to database   The `apop_text_to_db` command line program (and its corresponding C function) can take input from `stdin`. Thus, you can put it at the end of any of the above streams to directly dump data to an SQLite database.

For many POSIX programs that typically take file input, the traditional way to indicate that you are sending data from `stdin` instead of a text file is to use - as the file name. For example, after you did all that work in the exercise above to convert data to SQL commands, here is one way to do it using `apop_text_to_-db`:[17]

> sed ’/#/ d’ data−classroom | apop_text_to_db ’−’ friends classes.db

```
gnuplot
```

Database to plot   Apophenia includes the command-line program `apop_plot_-query`, which takes in a query and outputs a Gnuplottable file. It provides some extra power: the -H option will bin the data into a histogram before plotting, and you can use functions such as `var` that SQLite does not support. But for many instances, this is unnecessary.

SQLite will read a query file either on the command line or from a pipe, and Gnuplot will read in a formatted file via a pipe. As you saw in Chapter 5, turning a column of numbers (or a matrix) into a Gnuplottable file simply requires putting `plot ’-’` above the data. If there is a query in the file `queryfile`, then the sequence is:

> sqlite3 −separator " " data.db < queryfile | sed "1 i set key off\nplot ’−’" | gnuplot −persist

The -`separator` " " clause is necessary because Gnuplot does not like pipes as a delimiter. Of course, if you did not have that option available via sqlite3, you could just add -e "s/|/ /" to the sed command.

    Write a single command line to plot the yearly index of surface temperature anomalies (the `year` and `temp` columns from the `temp` table of the `data-climate.db` database).

---

[17]Even this is unnecessary, because the program knows to read lines matching `^#` as comments. But the example would be a little too boring as just `apop_text_to_db data-classroom friends classes.db`.

UNIX versus Windows: the end of the line   If your file is all one long line with no breaks but a few funny characters interspersed, or has ^L's all over the place, then you have just found yourself in the crossfire of a long-standing war over how lines should end. In the style of manual typewriters, starting a new line actually consists of two operations: moving horizontally to the beginning of the line (a carriage return, CR), and moving vertically down a line (a line feed, LF). The ASCII character for CR is <ctrl-M>, which often appears on-screen as the single character ^M; the ASCII character for LF is ^L.

The designers of AT&T UNIX decided that it is sufficient to end a line with just a LF, ^L, while the designers of Microsoft DOS decided that a line should end with a CR/LF pair, ^M^L. When you open a DOS file on a POSIX system, it will recognize ^L as the end-of-line, but consider the ^M to be garbage, which it leaves in the file. When you open a POSIX file in Windows, it can't find any ^M^L pairs, so none of the lines end.[18] As further frustration, some programs auto-correct the line endings while others do not, meaning that the file may look OK in your text editor but fall apart in your stats package.

Recall from page 61 that \r is a CR and \n is a LF. Going from DOS to UNIX means removing a single CR from each line, going from UNIX to DOS means adding a CR to each line, and both of these are simple sed commands:

```
#Convert a DOS file to UNIX:
sed −i~ 's/\r$//' dos_file

#Convert a UNIX file to DOS:
sed −i~ 's/$/\r/' unix_file
```

Some systems have dos2unix and unix2dos commands that do this for you,[19] but they are often missing, and you can see that these commands basically run a single line of sed.

---

[18]Typing ^ and then M will not produce a CR. ^M is a single special character confusingly represented on-screen with two characters. In most shells, <ctrl-V> means 'insert the next character exactly as I type it,' so the sequence <ctrl-V> <ctrl-M> will insert the single CR character which appears on the screen as ^M; and <ctrl-V> <ctrl-L> will similarly produce a LF.

[19]Perhaps ask your package manager for the dosutils package.

C: GLOSSARY

See also the list of notation and symbols on page 12.

Acronyms

ANSI: American National Standards Institute

ASCII: American Standard Code for Information Interchange

ANOVA: analysis of variance [p 312]

BLAS: Basic Linear Algebra System

BLUE: best linear unbiased estimator [p 221]

BRE: basic regular expression [p 403]

CDF: cumulative density function [p 236]

CMF: cumulative mass function [p 236]

CLT: central limit theorem [p 296]

df: degree(s) of freedom

ERE: extended regular expression [p 403]

erf: error function [p 284]

GCC: GNU Compiler Collection [p 48]

GDP: gross domestic product

GLS: generalized least squares [p 277]

GNU: GNU's Not UNIX

GSL: GNU Scientific Library [p 113]

GUI: graphical user interface

IDE: integrated development environment

IEC: International Electrotechnical Commission

IEEE: Institute of Electrical and Electronics Engineers

IIA: independence of irrelevant alternatives [p 286]

iid: independent and identically distributed [p 326]

ISO: International Standards Organization

IV: instrumental variable [p 275]

LR: likelihood ratio [p 351]

MAR: missing at random [p 345]

MCAR: missing completely at random [p 345]

MCMC: markov chain Monte Carlo [p 372]

ML: maximum likelihood

MLE: maximum likelihood estimation/estimate [p 325]

MNAR: missing not at random [p 345]

MSE: mean squared error [p 223]

OLS: ordinary least squares [p 270]

PCA: principal component analysis [p 265]

PCRE: Perl-compatible regular expression [p 403]

PDF: probability density function [p 236]

PDF: portable document format

PRNG: pseudorandom number generator [p 357]

PMF: probability mass function [p 236]

RNG: random number generator [p 357]

SSE: sum of squared errors [p 227]

SSR: sum of squared residuals [p 227]

SST: total sum of squares [p 227]

SQL: Structured Query Language [p 74]

SVD: singular value decomposition [p 265]

TLA: three-letter acronym

UNIX: not an acronym; see main glossary

WLS: weighted least squares [p 277]

Terms

affine projection: A linear projection can always be expressed as a matrix $\mathbf{T}$ such that $\mathbf{x}$ transformed is $\mathbf{xT}$. But any such projection maps $\mathbf{0}$ to $\mathbf{0}$. An affine projection adds a constant, transforming $\mathbf{x}$ to $\mathbf{xT} + \mathbf{k}$, so $\mathbf{0}$ now transforms to a nonzero value. [p 280]

ANOVA: "The analysis of variance is a body of statistical methods of analyzing measurements assumed to be of the structure $[y_i = x_{1i}\beta_1 + x_{2i}\beta_2 + \cdots + x_{pi}\beta_p + e_i, i = 1, 2, \ldots, n]$, where the coefficients $\{x_{ji}\}$ are integers usually 0 or 1" (Scheffé, 1959) [p 312]

apophenia: The human tendency to see patterns in static. [p 1]

array: A sequence of elements, all of the same type. An array of text characters is called a *string*. [p 30]

arguments: The inputs to a function. [p 36]

assertion: A statement in a program used to test for errors. The statement does nothing, but should always evaluate to being true; if it does not, the program halts with a message about the failed assertion. [p 71]

bandwidth: Most distribution-smoothing techniques, including some *kernel density estimates*, gather data points from a fixed range around the point being evaluated. The span over which data points are gathered is the bandwidth. For cases like the Normal kernel density estimator, whose tails always span $(-\infty, \infty)$, the term is still used to indicate that as bandwidth gets larger, more far-reaching data will have more of an effect. [p 261]

Bernoulli draw: A single draw from a fixed process that produces a one with probability p and a zero with probability $1 - p$. [p 237]

bias: The distance between the expected value of an estimator of β and β's true value, $|E(\hat{\beta}) - \beta|$. See *unbiased statistic*. [p 220]

binary tree: A set of structures, similar to a *linked list*, where each structure consists of data and two pointers, one to a next-left structure and one to a next-right structure. You can typically go from the head of the tree to an arbitrary element much more quickly than if the same data were organized as a *linked list*. [p 200]

BLUE: The Estimator $\hat{\beta}$ is a Linear function, Unbiased, and Best in the sense that $\text{var}(\hat{\beta}) \leq \text{var}(\tilde{\beta})$ for all linear unbiased estimators $\tilde{\beta}$. [p 221]

bootstrap: Repeated sampling with replacement from a population produces a sequence of artificial samples, which can be used to produce a sequence of *iid* statistics. The *Central Limit Theorem* then applies, and you can find the expected value and variance of the statistic for the entire data set using the set of iid draws of the statistic. The name implies that using samples from the data to learn about the data is a bit like pulling oneself up by the bootstraps. See also *jackknife* and the *bootstrap principle*. [p 367]

bootstrap principle: The claim that samples from your data sample will have properties matching samples from the population. [p 296]

call-by-address: When calling a function, sending to the function a copy of an input variable's location (as opposed to its value). [p 54]

call-by-value: When calling a function, sending to the function a copy of an input variable's value. [p 39]

Cauchy–Schwarz inequality: Given the *correlation coefficient* between any two vectors **x** and **y**, ρ_{xy}, it holds that $0 \le \rho_{xy}^2 \le 1$. [p 229]

Central Limit Theorem: Given a set of means, each being the mean of a set of n *iid* draws from a data set, the set of means will approach a Normal distribution as $n \to \infty$. [p 297]

central moments: Given a data vector **x** and mean $\overline{x}$, the kth central moment of $f(\cdot)$ is $\frac{1}{n} \sum_{x \in \mathbf{x}} \left(f(x) - \overline{f(x)} \right)^k$. In the continuous case, if x has distribution $p(x)$, then the kth central moment of $f(x)$ is $\int_{-\infty}^{\infty} \left(f(x) - \overline{f(x)} \right)^k p(x) dx$. In both cases, the first central moment is always zero (but see *noncentral moment*). The second is known as the *variance*, the third as *skew*, and the fourth as *kurtosis*. [p 230]

closed-form expression: An expression, say $x^2 + y^3$, that can be written down using only a line or two of notation and can be manipulated using the usual algebraic rules. This is in contrast to a function algorithm or an empirical distribution that can be described only via a long code listing, a histogram, or a data set.

compiler: A non-interactive program (e.g., `gcc`) to translate code from a human-readable *source file* to a computer-readable *object file*. The compiler is often closely intertwined with the *preprocessor* and *linker*, to the point that the preprocessor/-compiler/linker amalgam is usually just called the compiler. Compare with *interpreter*. [p 18]

conjugate distributions: A prior/likelihood pair such that if the prior is updated using the likelihood, the posterior has the same form as the prior (but updated parameters). For example, given a Beta distribution prior and a Binomial likelihood function, the posterior will be a Beta distribution. Unrelated to conjugate gradient methods. [p 374]

consistent estimator: An estimator $\hat{\beta}(\mathbf{x})$ is consistent if, for some constant c, $\lim_{n \to \infty} P(|\hat{\beta}(\mathbf{x}) - c| > \epsilon) = 0$, for any $\epsilon > 0$. That is, as the sample size grows, the value of $\hat{\beta}(\mathbf{x})$ converges in probability to the single value c. [p 221]

consistent test: A test is consistent if the *power* $\to 1$ as $n \to \infty$. [p 335]

contrast: A hypothesis about a linear combination of coefficients, like $3\beta_1 - 2\beta_2 = 0$. [p 309]

correlation coefficient: Given the square roots of the *covariance* and *variances*, σ_{xy}, σ_x, and σ_y, the correlation coefficient $\rho_{xy} \equiv \frac{\sigma_{xy}}{\sigma_x \sigma_x}$. [p 229]

covariance: For two data vectors $\mathbf{x}$ and $\mathbf{y}$, $\sigma^2_{\mathbf{xy}} \equiv \frac{1}{n}\sum_i (x_i - \overline{\mathbf{x}})(y_i - \overline{\mathbf{y}})$. [p 228]

Cramér–Rao lower bound: The elements of the covariance matrix of the estimate of a parameter vector must be equal to or greater than a limit that is constant for a given PDF, as in Equation 10.1.7 (page 333). For an MLE, the CRLB reduces to $1/(n\mathbb{I})$, where $\mathbb{I}$ is the *information matrix*. [p 333]

crosstab: A two-dimensional table, where each row represents values of one variable (y), each column represents values of another variable (x), and each (row, column) entry provides some summary statistic of the subset of data where y has the given row value and x has the given column value. See page 101 for an example. [p 101]

cumulative density function: The integral of a *PDF*. Its value at any given point indicates the likelihood that a draw from the distribution will be equal to or less than the given point. Since the PDF is always non-negative, the CDF is monotonically nondecreasing. At $-\infty$, the CDF is always zero, and at ∞ the CDF is always one. [p 236]

cumulative mass function: The integral of a *PMF*. That is, a CDF when the distribution is over discrete values. [p 236]

data mining: Formerly a synonym for *data snooping*, but in current usage, methods of categorizing observations into a small set of (typically nested) bins, such as generating trees or separating hyperplanes.

data snooping: Before formally testing a hypothesis, trying a series of preliminary tests to select the form of the final test. Such behavior can taint *inferential statistics* because the statistic *parameter from one test* has a very different distribution from the statistic *most favorable parameter from fifty tests*. [p 316]

debugger: A standalone program that runs a program, allowing the user to halt the program at any point, view the *stack* of *frames*, and query the program for the value of a variable at that point in the program's execution. [p 43]

declaration: A line of code that indicates the *type* of a variable or function. [p 28]

degrees of freedom: The number of dimensions along which a data set varies. If all n data points are independent, then $df = n$, but if there are restrictions that reduce the data's dimensionality, $df < n$. You can often think of the df as the number of independent pieces of information. [p 222]

dependency: A statement in a makefile indicating that one file depends on another, such as an *object file* that depends on a *source file*. When the depended-on file changes, the dependent file will need to be re-produced. [p 388]

descriptive statistics: The half of probability and statistics aimed at filtering useful patterns out of a world of overwhelming information. The other half is *inferential statistics*. [p 1]

dummy variable: A variable that takes on discrete values (usually just zero or one) to indicate the category in which an observation lies. [p 281]

efficiency: A parameter estimate that comes as close as possible to achieving the *Cramér–Rao lower bound*, and thus has as small a variance as possible, is dubbed an efficient estimate. [p 220]

error function: The *CDF* of the Normal(0, 1) distribution. [p 284]

environment variable: A set of variables maintained by the system and passed to all child programs. They are typically set from the *shell*'s export or setenv command. [p 381]

expected value: The first *noncentral moment*, aka the mean or average. [p 221]

frame: A collection of a function and all of the variables that are in *scope* for the function. [p 37]

GCC: The GNU Compiler Collection, which reads *source files* in a variety of languages and produces *object files* accordingly. This book uses only its ability to read and compile C code. [p 48]

Generalized Least Squares: The *Ordinary Least Squares* model assumes that the covariance matrix among the observations is $\Sigma = \sigma^2 \mathbf{I}$ (i.e., a scalar times the *identity matrix*). A GLS model is any model that otherwise conforms to the OLS assumptions, but allows Σ to take on a different form. [p 277]

globbing: The limited *regular expression* parsing provided by a *shell*, such as expanding *.c to the full list of file names ending in .c. Uses an entirely different syntax from standard regular expression parsers. [p 407]

graph: A set of nodes, connected by edges. The edges may be directional, thus forming a directional graph. Not to be confused with a *plot*. [p 182]

grid search: Divide the space of inputs to a function into a grid, and write down the value of the function at every point on the grid. Such an exhaustive walk through the space can be used to get a picture of the function (this is what most graphing packages do), or to find the optimum of the function. However, it is a last resort for most purposes; the search and random draw methods of Chapters 10 and 11 are much more efficient and precise. [p 371]

hat matrix: Please see *projection matrix*. [p 272]

header file: A C file consisting entirely of declarations and type definitions. By `#include`-ing it in multiple C files, the variables, functions, and types declared in the header file can be defined in one file and used in many. [p 49]

Hessian: The matrix of second derivatives of a function evaluated at a given point. Given a log likelihood function $LL(\boldsymbol{\theta})$, the negation of its Hessian is the *information matrix*. [p 341]

heteroskedasticity: When the errors associated with different observations have different variances, such as observations on the consumption rates of the poor and the wealthy. This violates an assumption of OLS, and can therefore produce inefficient estimates; *weighted least squares* solves this problem. [p 277]

identically distributed: A situation where the process used to produce all of the elements of a data set is considered to be identical. For example, all data points may be drawn from a Poisson(0.4) distribution, or may be individuals randomly sampled from one population. [p 326]

identity matrix: A square matrix where every non-diagonal element is zero, and every diagonal element is one. Its size is typically determined by context, and it is typically notated as **I**. There are really an infinite number of identity matrices (a 1×1 matrix, a 2×2 matrix, a 3×3 matrix, ...), but the custom is to refer to any one of them as *the identity matrix*.

iff: If and only if. The following statements are equivalent: $A \iff B$; *A iff B*; $A \equiv B$; *A is defined to be B*; *B is defined to be A*.

iid: Independent and identically distributed. These are the conditions for the *Central Limit Theorem*. See *independent draws* and *identically distributed*. [p 326]

importance sampling: A means of making draws from an easy-to-draw-from distribution to make draws from a more difficult distribution. [p 371]

independence of irrelevant alternatives: The ratio of (likelihood of choice A being selected)/(likelihood of choice B being selected) does not depend on what other options are available—adding or deleting choices C, D, and E will not change the ratio. [p 286]

independent draws: Two events x_1 and x_2 (such as draws from a data set) are independent if $P(x_1 \cap x_2)$—that is, the probability of (x_1 and x_2)—is equal to $P(x_1) \cdot P(x_2)$. [p 326]

inferential statistics: The half of probability and statistics aimed at fighting against *apophenia*. The other half is *descriptive statistics*. [p 1]

information matrix: The negation of the derivative of the *Score*. Put differently, given a log likelihood function $LL(\theta)$, the information matrix is the negation of its *Hessian* matrix. See also the *Cramér–Rao lower bound*. [p 326]

instrumental variable: If a variable is measured with error, then the OLS parameter estimate based on that variable will be *biased*. An instrumental variable is a replacement variable that is highly correlated with the measured-with-error variable. A variant of OLS using the instrumental variable will produce unbiased parameter estimates. [p 275]

interaction: An element of a model that contends that it is not x_1 or x_2 that causes an outcome, but the combination of both x_1 and x_2 simultaneously (or x_1 and not x_2, or not x_1 but x_2). This is typically represented in OLS regressions by simply multiplying the two together to form a new variable $x_3 \equiv x_1 \cdot x_2$. [p 281]

interpreter: A program to translate code from a human-readable language to a computer's *object code* or some other binary format. The user inputs individual commands, typically one by one, and then the interpreter produces and executes the appropriate machine code for each line. Gnuplot and the sqlite3 command-line program are interpreters. Compare with *compiler*.

jackknife: A relative of the *bootstrap*. A subsample is formed by removing the first element, then estimating $\hat{\beta}_{j1}$; the next subsample is formed by replacing the first element and removing the second, then re-estimating $\hat{\beta}_{j2}$, et cetera. The multitude of $\hat{\beta}_{jn}$'s thus formed can be used to estimate the variance of the overall parameter estimate $\hat{\beta}$. [p 131]

join: Combining two database tables to form a third, typically including some columns from the first and some from the second. There is usually a column on which the join is made; e.g., a first table of names and heights and a second table of names and weights would be joined by matching the names in both tables. [p 87]

kernel density estimate: A method of smoothing a data set by centering a standard PDF (like the Normal) around every point. Summing together all the sub-PDFs produces a smooth overall PDF. [p 262]

kurtosis: The fourth *central moment*. [p 230]

lexicographic order: Words in the dictionary are first sorted using only the first letter, completely ignoring all the others. Then, words beginning with *A* are sorted by their second letter. Those beginning with the same first two letters (aandblom, aard-wolf, aasvogel, ...) are sorted using their third letter. Thus, a lexicographic ordering sorts using only the first characteristic, then breaks ties with a second characteristic, then breaks further ties with a third, and so on. [p 91]

library: A set of functions and variables that perform tasks related to some specific task, such as numeric methods or *linked list* handling. The library is basically an *object file* in a slightly different format, and is typically kept somewhere on the library *path*. [p 52]

likelihood function: The likelihood $P(\mathbf{X}, \beta)|_\mathbf{X}$ is the probability that we'd have the parameters β given some observed data $\mathbf{X}$. This is in contrast to the probability of a data set given fixed parameters, $P(\mathbf{X}, \beta)|_\beta$. See page 329 for discussion. [p 326]

likelihood ratio test: A test based on a statistic of the form P_1/P_2. This is sometimes logged to $LL_1 - LL_2$. Many tests that on their surface seem to not fit this form can be shown to be equivalent to an LR test. [p 335]

linked list: A set of structures, where each structure holds data and a pointer to the next structure in the list. One could traverse the list by following the pointer from the head element to the next element, then following that element's pointer to the next element, et cetera. [p 198]

linker: A program that takes in a set of *libraries* and *object files* and outputs an executable program. [p 51]

Manhattan metric: Given distances in several dimensions, say $d_x = |x_1 - x_2|$ and $d_y = |y_1 - y_2|$, the standard Euclidian metric combines them to find a straight-line distance via $\sqrt{d_x^2 + d_y^2}$. The Manhattan metric simply adds the distance on each dimension, $d_x + d_y$. This is the distance one would travel by first going only along East–West streets, then only along North–South streets. [p 150]

make: A program that keeps track of *dependencies*, and runs commands (specified in a makefile) as needed to keep all files up-to-date as their dependencies change. Usually used to produce executables when their *source files* change. [p 387]

macro: A rule to transform strings of text with a fixed pattern. For example, a *preprocessor* may replace every occurrence of `GSL_MIN(a,b)` with `((a) < (b) ? (a) : (b))`. [p 212]

metadata: Data about data. For example, a *pointer* is data about the location of base data, and a *statistic* is data summarizing or transforming base data. [p 128]

mean squared error: Given an estimate of β named $\hat{\beta}$, the MSE is $E(\hat{\beta} - \beta)^2$. This can be shown to be equivalent to $\text{var}(\hat{\beta}) + \text{bias}^2(\hat{\beta})$. [p 220]

memory leak: If you lose the address of space that you had allocated, then the space remains reserved even though it is impossible for you to use. Thus, the system's usable memory is effectively smaller. [p 62]

missing at random: Data for variable i is MAR if the incidence of missing data points is unrelated to the existing data for variable i, given the other variables. Generally, this means that there is an external cause (not caused by the value of i) that causes values of i to be missing. [p 346]

missing completely at random: Data for variable i are MCAR if there is no correlation between the incidence of missing data and anything else in the data set. That is, the cause of missingness is entirely external and haphazard. [p 346]

missing not at random: Data for variable i is MNAR if there is a correlation between the incidence of missing data and the missing data's value. That is, the missingness is caused by the data's value. [p 346]

Monte Carlo method: Generating information about a distribution, such as parameter estimates, by repeatedly making random draws from the distribution. [p 356]

multicollinearity: Given a data set $\mathbf{X}$ consisting of columns $\mathbf{x}_1$, $\mathbf{x}_2$, $\ldots$, if two columns $\mathbf{x}_i$ and $\mathbf{x}_j$ are highly correlated, then the determinant of $\mathbf{X}'\mathbf{X}$ will be near zero and the value of the inverse $(\mathbf{X}'\mathbf{X})^{-1}$ unstable. As a result, OLS-family estimates will not be reliable. [p 275]

noncentral moment: Given a data vector $\mathbf{x}$ and mean $\overline{\mathbf{x}}$, the kth noncentral moment is $\frac{1}{n}\sum_{x\in\mathbf{x}} x^k$. In the continuous case, if x has distribution $p(x)$, then the kth noncentral moment of $f(x)$ is $\int_{-\infty}^{\infty} f(x)^k p(x)dx$. The only noncentral moment anybody cares about is the first—aka, the mean. [p 230]

non-ignorable missingness: See *missing not at random*. [p 346]

non-parametric: A test or model is non-parametric if it does not rely on a claim that the statistics/parameters in question have a textbook distribution (t, χ^2, Normal, Bernoulli, et cetera). However, all non-parametric models have parameters to tune, and all non-parametric tests are based on a statistic whose characteristics must be determined.

null pointer: A special pointer that is defined to not point to anything. [p 43]

object: A structure, typically implemented via a `struct`, plus any supporting functions that facilitate use of that structure, such as the `gsl_vector` plus the `gsl_vector_add`, `gsl_vector_ddot`, ..., functions.

object file: A computer-readable file listing the variables, functions, and types defined in a `.c` file. Object files are not executables until they go through *linking*. Bears no relation to *objects* or object-oriented programming. [p 51]

order statistic: The value at a given position in the sorted data, such as the largest number in a set, the second largest number, the median, the smallest number, et cetera.

Ordinary Least Squares: A model, fully specified on page 274, that contends that a dependent variable is the linear combination of a number of independent variables, plus an error term.

overflow error: When the value of a variable is too large for its type, unpredictable things may occur. For example, on some systems, `MAX_INT + 1 == -MAX_INT`. The IEEE standard specifies that if a `float` or `double` variable overflows, it be set to a special pattern indicating infinity. See also *underflow error*. [p 137]

path: A list of directories along which the computer will search for files. Most shells have a `PATH` environment variable along which they search for executable programs. Similarly, the preprocessor searches for header files (e.g., `#include <stdlib.h>`) along the directories in the `INCLUDEPATH` environment variable, which can be extended using the `-I` flag on the compiler command line. The linker searches for libraries to include using a libpath and its extensions specified via the `-L` compiler flag. [p 385]

pipe: A connection that directly redirects the output from one program to the input of another. In the *shell*, a pipe is formed by putting a `|` between two programs; in C, it is formed using the `popen` function. [p 395]

pivot table: See *crosstab*. [p 101]

plot: A graphic with two or three axes and function values marked relative to those axes. Not to be confused with a *graph*. [p 158]

pointer: A variable holding the location of a piece of data. [p 53]

POSIX: The Portable Operating System Interface standard. By the mid-1980s, a multitude of variants on the *UNIX* operating system appeared; the Institute of Electrical and Electronics Engineers convened a panel to write this standard so that programs written on one flavor of UNIX could be more easily ported to another flavor. Santa Cruz Operation's UNIX, International Business Machines' AIX, Hewlett-

Packard's HP-UX, Linux, Sun's Solaris, some members of Microsoft's Windows family, and others all more or less comply with this standard.

power: The likelihood of rejecting a false null. That is, if there is a significant effect, what are the odds that the test will detect it? This is one minus the likelihood of a *Type II error*. [p 335]

prime numbers: Prime numbers are what is left when you have taken all the patterns away. (Haddon, 2003, p 12) [p 61]

principal component analysis: A projection of data X onto a basis space consisting of n eigenvalues of $X'X$, which has a number of desirable properties. [p 265]

probability density function: The total area under a PDF for any given range is equal to the probability that a draw from the distribution will fall in that range. The PDF is always nonnegative. E.g., the familiar bell curve of a Normal distribution. Compare with *cumulative density function*. [p 236]

probability mass function: The distribution of probabilities that a given discrete value will be drawn. I.e., a PDF when the distribution is over discrete values. [p 236]

projection matrix: $X^P \equiv X(X'X)^{-1}X'$. $X^P v$ equals the projection of v onto the column space of X. [p 272]

profiler: A program that executes other programs, and determines how much time is spent in each of the program's various functions. It can thus be used to find the bottlenecks in a slow-running program. [p 215]

pseudorandom number generator: A function that produces a deterministic sequence of numbers that seem to have no pattern. Initializing the PRNG with a different *seed* produces a different streams of numbers. [p 357]

query: Any command to a database. Typically, the command uses the `select` keyword to request data from the database, but a query may also be a non-question command, such as a command to create a new table, drop an index, et cetera. [p 74]

random number generator: See *pseudorandom number generator*. [p 357]

regular expressions: A string used to describe patterns of text, such as 'two numbers followed by a letter'. [p 403]

scope: The section of code that is able to refer to a variable. For variables declared outside of a function, the scope is everything in a file after the declaration; for variables declared inside a block, the scope is everything after the declaration inside the block. [p 41]

score: Given a log *likelihood function* $\ln P(\boldsymbol{\theta})$, its score is the vector of its derivatives: $\mathbf{S} = (\partial \ln P/\partial \boldsymbol{\theta})$. [p 326]

seed: The value with which a *pseudorandom number generator* is initialized. [p 357]

segfault: An affectionate abbreviation for *segmentation fault*. [p 43]

segmentation fault: An error wherein the program attempts to access a part of the computer's memory that was not allocated to the program. If reading from unauthorized memory, this is a security hole; if writing to unauthorized memory, this could destroy data or create system instability. Therefore, the system catches segfaults and halts the program immediately when they occur. [p 43]

shell: A program whose primary purpose is to facilitate running other programs. When you log in to most text-driven systems, you are immediately put at the shell's input prompt. Most shells include facilities for setting variables and writing scripts. [p 393]

singular value decomposition: Given an $m \times n$ data matrix $\mathbf{X}$ (where typically $m >> n$), one can find the n eigenvectors associated with the n largest eigenvalues.[20] This may be done as the first step in a *principal component analysis*. SVD as currently practiced also includes a number of further techniques to transform the eigenvectors as appropriate. [p 265]

skew: The third *central moment*, used as an indication of whether a distribution leans to the left or right of the mean. [p 230]

source code: The human-readable version of a program. It will be converted into object code for the computer to execute.

stack: Each function runs in its own *frame*. When a program starts, it begins by establishing a `main` frame, and then if `main` calls another function, that function's frame is thought of as being laid on top of the `main` frame. Similarly for subsequent functions, so pending frames pile up to form a stack of frames. When the stack is empty, the program terminates. [p 38]

---

[20]This is assuming that $\mathbf{X}'\mathbf{X}$ has full rank.

standard deviation: The square root of the variance of a variable, often notated as σ. If the variable is Normally distributed, we usually compare a point's distance to the mean against $1\sigma, 2\sigma, \ldots$. For distributions that are not Normal (or at least bell-shaped), σ is of limited descriptive utility. See also *standard error* and *variance*. [p 222]

standard error: An abbreviation for the standard deviation of the error. [p 367]

statistic: A function that takes data as an input, such as the mean of **x**; the variance of the error term of a regression of **X** on **y**, or the OLS parameter $\hat{\beta} = (\mathbf{X}'\mathbf{X})^{-1}\mathbf{X}'\mathbf{y}$. [p 219]

string: An array of characters. Because the string is an array, it is handled using *pointer*-type operations, but there are also functions to print the string like the plain text it represents. [p 65]

structure: A set of variables that are intended to collectively represent one object, such as a person (comprising, e.g., a name, height, and weight) or a bird (comprising, e.g., a type and *pointers* to offspring). [p 31]

Structured Query Language: A standard language for writing database *queries*. [p 74]

switches: As with physical machinery, switches are options to affect a program's operation. They are usually set on the command line, and are usually marked by a dash, like -x. [p 208]

trimean: (first quartile + two times the median + third quartile)/4. (Tukey, 1977, p 46) [p 234]

threading: On many modern computers, the processor(s) can execute multiple chains of commands at once. For example, the data regarding two independent events could be simultaneously processed by two processors. In such a case, the single thread of program instructions can be split into multiple threads, which must be gathered together before the program completes. [p 119]

type: The class of data a variable is intended to represent, such as an integer, character, or *structure* (which is an amalgamation of subtypes). [p 27]

type casting: Converting a variable from one *type* to another. [p 33]

Type I error: Rejecting the null when it is true. [p 335]

Type II error: Accepting the null when it is false. See also *power*. [p 335]

unbiased statistic: The *expected value* of the statistic $\hat{\beta}$ equals the true population value: $E(\hat{\beta}) = \beta$. [p 220]

unbiased estimator: Let α be a test's *Type I* error, and let β be its *Type II* error. A test is unbiased if $(1 - \beta) \geq \alpha$ for all values of the parameter. I.e., you are less likely to accept the null when it is false than when it is true. [p 335]

underflow error: Occurs when the value of a variable is smaller than the smallest number the system can represent. For example, on any current system with finite-precision arithmetic, $2^{-10,000}$ is simply zero. See also *overflow error*. [p 137]

UNIX: An operating system developed at Bell Labs. Many call any UNIX-like operating system by this name (often by the plural, Unices), but UNIX properly refers only to the code written by Bell Labs, which has evolved into code owned by Santa Cruz Operation. Others are correctly called *POSIX*-compliant. The name does not stand for anything, but is a pun on a predecessor operating system, Multics.

variance: The second *central moment*, usually notated as σ^2. [p 222]

Weighted Least Squares: A type of *GLS* method wherein different observations are given different weights. The weights can be for any reason, such as producing a representative survey sample, but the method is often used for *heteroskedastic* data. [p 277]

BIBLIOGRAPHY

Abelson, Harold, Sussman, Gerald Jay, & Sussman, Julie. 1985. *Structure and Interpretation of Computer Programs*. MIT Press.

Albee, Edward. 1960. *The American Dream and Zoo Story*. Signet.

Allison, Paul D. 2002. *Missing Data*. Quantitative Applications in the Social Sciences. Sage Publications.

Amemiya, Takeshi. 1981. Qualitative Response Models: A Survey. *Journal of Economic Literature*, **19**(4), 1483–1536.

Amemiya, Takeshi. 1994. *Introduction to Statistics and Econometrics*. Harvard University Press.

Avriel, Mordecai. 2003. *Nonlinear Programming: Analysis and Methods*. Dover Press.

Axtell, Robert. 2006. Firm Sizes: Facts, Formulae, Fables and Fantasies. *Center on Social and Economic Dynamics Working Papers*, **44**(Feb.).

Barron, Andrew R, & Sheu, Chyong-Hwa. 1991. Approximation of Density Functions by Sequences of Exponential Families. *The Annals of Statistics*, **19**(3), 1347–1369.

Baum, AE, Akula, N, Cabanero, M, Cardona, I, Corona, W, Klemens, B, Schulze, TG, Cichon, S, Rietschel, M, Nathen, MM, Georgi, A, Schumacher, J, Schwarz, M, Jamra, R Abou, Hofels, S, Propping, P, Satagopan, J, Consortium, NIMH Genetics Initiative Bipolar Disorder, Detera-Wadleigh, SD, Hardy, J, & McMahon,

FJ. 2008. A genome-wide association study implicates diacylglycerol kinase eta (DGKH) and several other genes in the etiology of bipolar disorder. *Molecular Psychiatry*, **13**(2), 197–207.

Benford, Frank. 1938. The Law of Anomalous Numbers. *Proceedings of the American Philosophical Society*, **78**(4), 551–572.

Bowman, K O, & Shenton, L R. 1975. Omnibus Test Contours for Departures from Normality Based on $\sqrt{b_1}$ and b_2. *Biometrika*, **62**(2), 243–250.

Casella, George, & Berger, Roger L. 1990. *Statistical Inference*. Duxbury Press.

Chamberlin, Thomas Chrowder. 1890. The Method of Multiple Working Hypotheses. *Science*, **15**(366), 10–11.

Cheng, Simon, & Long, J Scott. 2007. Testing for IIA in the Multinomial Logit Model. *Sociological Methods Research*, **35**(4), 583–600.

Chung, J H, & Fraser, D A S. 1958. Randomization Tests for a Multivariate Two-Sample Problem. *Journal of the American Statistical Association*, **53**(283), 729–735.

Chwe, Michael Suk-Young. 2001. *Rational Ritual: Culture, Coordination, and Common Knowledge*. Princeton University Press.

Cleveland, William S, & McGill, Robert. 1985. Graphical Perception and Graphical Methods for Analyzing Scientific Data. *Science*, **229**(4716), 828–833.

Codd, Edgar F. 1970. A Relational Model of Data for Large Shared Data Banks. *Communications of the ACM*, **13**(6), 377–387.

Conover, W J. 1980. *Practical Nonparametric Statistics*. 2nd edn. Wiley.

Cook, R Dennis. 1977. Detection of Influential Observations in Linear Regression. *Technometrics*, **19**(1), 15–18.

Cox, D R. 1962. Further Results on Tests of Separate Families of Hypotheses. *Journal of the Royal Statistical Society. Series B (Methodological)*, **24**(2), 406–424.

Cropper, Maureen L, Deck, Leland, Kishor, Nalin, & McConnell, Kenneth E. 1993. Valuing Product Attributes Using Single Market Data: A Comparison of Hedonic and Discrete Choice Approaches. *Review of Economics and Statistics*, **75**(2), 225–232.

Dempster, A P, Laird, N M, & Rubin, D B. 1977. Maximum Likelihood from Incomplete Data via the EM Algorithm. *Journal of the Royal Statistical Society. Series B (Methodological)*, **39**(1), 1–38.

Efron, Bradley, & Hinkley, David V. 1978. Assessing the Accuracy of the Maximum Likelihood Estimator: Observed Versus Expected Fisher Information. *Biometrika*, **65**(3), 457–482.

Efron, Bradley, & Tibshirani, Robert J. 1993. *An Introduction to the Bootstrap*. Monographs on Statistics and Probability, no. 57. Chapman and Hall.

Eliason, Scott R. 1993. *Maximum Likelihood Estimation: Logic and Practice*. Quantitative Applications in the Social Sciences. Sage Publications.

Epstein, Joshua M, & Axtell, Robert. 1996. *Growing Artificial Societies: Social Science from the Bottom Up*. Brookings Institution Press and MIT Press.

Fein, Sidney, Paz, Victoria, Rao, Nyapati, & LaGrassa, Joseph. 1988. The Combination of Lithium Carbonate and an MAOI in Refractory Depressions. *American Journal of Psychiatry*, **145**(2), 249–250.

Feller, William. 1966. *An Introduction to Probability Theory and Its Applications*. Wiley.

Fisher, R A. 1934. Two New Properties of Mathematical Likelihood. *Proceedings of the Royal Society of London. Series A, Containing Papers of a Mathematical and Physical Character*, **144**(852), 285–307.

Fisher, Ronald Aylmer. 1922. On the Interpretation of χ^2 from Contingency Tables, and the Calculation of P. *Journal of the Royal Statistical Society*, **85**(1), 87–94.

Fisher, Ronald Aylmer. 1956. *Statistical Methods and Scientific Inference*. Oliver & Boyd.

Freedman, David A. 1983. A Note on Screening Regression Equations. *The American Statistician*, **37**(2), 152–155.

Friedl, Jeffrey E F. 2002. *Mastering Regular Expressions*. 2nd edn. O'Reilly Media.

Frisch, Æleen. 1995. *Essential System Administration*. O'Reilly & Associates.

Fry, Tim R L, & Harris, Mark N. 1996. A Monte Carlo Study of Tests for the Independence of Irrelevant Alternatives Property. *Transportation Research Part B: Methodological*, **30**(1), 19–30.

Gardner, Martin. 1983. *Wheels, Life, and Other Mathematical Amusements*. W H Freeman.

Gelman, Andrew, & Hill, Jennifer. 2007. *Data Analysis Using Regression and Multilevel/Hierarchical Models*. Cambridge University Press.

Gelman, Andrew, Carlin, John B, Stern, Hal S, & Rubin, Donald B. 1995. *Bayesian Data Analysis*. 2nd edn. Chapman & Hall Texts in Statistical Science. Chapman & Hall/CRC.

Gentle, James E. 2002. *Elements of Computational Statistics*. Statistics and Computing. Springer.

Gentleman, Robert, & Ihaka, Ross. 2000. Lexical Scope and Statistical Computing. *Journal of Computational and Graphical Statistics*, **9**(3), 491–508.

Gibbard, Ben. 2003. *Lightness*. Barsuk Records. In Death Cab for Cutie, *Transatlanticism*.

Gibrat, Robert. 1931. *Les Inégalités Économiques; Applications: Aux Inégalités des Richesses, a la Concentration des Entreprises, aux Populations des Villes, aux Statistiques des Familles, etc., d'une Loi Nouvelle, la Loi de L'effet Proportionnel*. Librarie du Recueil Sirey.

Gigerenzer, Gerd. 2004. Mindless Statistics. *The Journal of Socio-Economics*, **33**, 587–606.

Gill, Philip E, Murray, Waler, & Wright, Margaret H. 1981. *Practical Optimization*. Academic Press.

Givens, Geof H, & Hoeting, Jennifer A. 2005. *Computational Statistics*. Wiley Series in Probability and Statistics. Wiley.

Glaeser, Edward L, Sacerdote, Bruce, & Scheinkman, Jose A. 1996. Crime and Social Interactions. *The Quarterly Journal of Economics*, **111**(2), 507–48.

Goldberg, David. 1991. What Every Computer Scientist Should Know about Floating-point Arithmetic. *ACM Computing Surveys*, **23**(1), 5–48.

Gonick, Larry, & Smith, Woollcott. 1994. *Cartoon Guide to Statistics*. Collins.

Good, Irving John. 1972. Random Thoughts about Randomness. *PSA: Proceedings of the Biennial Meeting of the Philosophy of Science Association*, **1972**, 117–135.

Gough, Brian (ed). 2003. *GNU Scientific Library Reference Manual*. 2nd edn. Network Theory, Ltd.

Greene, William H. 1990. *Econometric Analysis*. 2nd edn. Prentice Hall.

Haddon, Mark. 2003. *The Curious Incident of the Dog in the Night-time*. Vintage.

Huber, Peter J. 2000. Languages for Statistics and Data Analysis. *Joural of Computational and Graphical Statistics*, **9**(3), 600–620.

Huff, Darrell, & Geis, Irving. 1954. *How to Lie With Statistics*. W. W. Norton & Company.

Hunter, John E, & Schmidt, Frank L. 2004. *Methods of Meta-Analysis: Correcting Error and Bias in Research Findings*. 2nd edn. Sage Publications.

Internal Revenue Service. 2007. *2007 Federal Tax Rate Schedules*. Department of the Treasury.

Kahneman, Daniel, Slovic, Paul, & Tversky, Amos. 1982. *Judgement Under Uncertainty: Heuristics and Biases*. Cambridge University Press.

Karlquist, A (ed). 1978. *Spatial Interaction Theory and Residential Location*. North Holland.

Kernighan, Brian W, & Pike, Rob. 1999. *The Practice of Programming*. Addison-Wesley Professional.

Kernighan, Brian W, & Ritchie, Dennis M. 1988. *The C Programming Language*. 2nd edn. Prentice Hall PTR.

Klemens, Ben. 2007. *Finding Optimal Agent-based Models*. Brookings Center on Social and Economic Dynamics Working Paper #49.

Kline, Morris. 1980. *Mathematics: The Loss of Certainty*. Oxford University Press.

Kmenta, Jan. 1986. *Elements of Econometrics*. 2nd edn. Macmillan Publishing Company.

Knuth, Donald Ervin. 1997. *The Art of Computer Programming*. 3rd edn. Vol. 1: Fundamental Algorithms. Addison-Wesley.

Kolmogorov, Andrey Nikolaevich. 1933. Sulla determinazione empirica di una legge di distributione. *Giornale dell' Istituto Italiano degli Attuari*, **4**, 83–91.

Laumann, Anne E, & Derick, Amy J. 2006. Tattoos and body piercings in the United States: A National Data Set. *Journal of the American Academy of Dermatologists*, **55**(3), 413–21.

Lehmann, E L, & Stein, C. 1949. On the Theory of Some Non-Parametric Hypotheses. *The Annals of Mathematical Statistics*, **20**(1), 28–45.

Maddala, G S. 1977. *Econometrics*. McGraw-Hill.

McFadden, Daniel. 1973. *Conditional Logit Analysis of Qualitative Choice Behavior*. In:Zarembka (1973). Chap. 4, pages 105–142.

McFadden, Daniel. 1978. *Modelling the Choice of Residential Location*. In:Karlquist (1978). Pages 75–96.

Nabokov, Vladimir. 1962. *Pale Fire*. G P Putnams's Sons.

National Election Studies. 2000. *The 2000 National Election Study [dataset].* University of Michigan, Center for Political Studies.

Newman, James R (ed). 1956. *The World of Mathematics.* Simon and Schuster.

Neyman, J, & Pearson, E S. 1928a. On the Use and Interpretation of Certain Test Criteria for Purposes of Statistical Inference: Part I. *Biometrika,* **20A**(1/2), 175–240.

Neyman, J, & Pearson, E S. 1928b. On the Use and Interpretation of Certain Test Criteria for Purposes of Statistical Inference: Part II. *Biometrika,* **20A**(3/4), 263–294.

Orwell, George. 1949. *1984.* Secker and Warburg.

Papadimitriou, Christos H, & Steiglitz, Kenneth. 1998. *Combinatorial Optimization: Algorithms and Complexity.* Dover.

Paulos, John Allen. 1988. *Innumeracy: Mathematical Illiteracy and its Consequences.* Hill and Wang.

Pawitan, Yudi. 2001. *In All Likelihood: Statistical Modeling and Inference Using Likelihood.* Oxford University Press.

Pearson, Karl. 1900. On the Criterion that a given System of Deviations from the Probable in the Case of a Correlated System of Variables is Such That it Can be Reasonably Supposed to Have Arisen from Random Sampling. *London, Edinburgh and Dublin Philosophical Magazine and Journal of Science,* July, 157–175. Reprinted in Pearson (1948, pp 339–357).

Pearson, Karl. 1948. *Karl Pearson's Early Statistical Papers.* Cambridge.

Peek, Jerry, Todino-Gonguet, Grace, & Strang, John. 2002. *Learning the UNIX Operating System.* 5th edn. O'Reilly & Associates.

Perl, Judea. 2000. *Causality.* Cambridge University Press.

Pierce, John R. 1980. *An Introduction to Information Theory: Symbols, Signals, and Noise.* Dover.

Poincaré, Henri. 1913. *Chance.* In Newman (1956), translated by George Bruce Halsted. Pages 1380–1394.

Polhill, J Gary, Izquierdo, Luis R, & Gotts, Nicholas M. 2005. The Ghost in the Model (and Other Effects of Floating Point Arithmetic). *Journal of Artificial Societies and Social Simulation,* **8**(1).

Poole, Keith T, & Rosenthal, Howard. 1985. A Spatial Model for Legislative Roll Call Analysis. *American Journal of Political Science,* **29**(2), 357–384.

Press, William H, Flannery, Brian P, Teukolsky, Saul A, & Vetterling, William T. 1988. *Numerical Recipes in C: The Art of Scientific Computing*. Cambridge University Press.

Price, Roger, & Stern, Leonard. 1988. *Mad Libs*. Price Stern Sloan.

Rumi, Jelaluddin. 2004. *The Essential Rumi*. Penguin. Translated by Coleman Barks.

Särndal, Carl-Erik, Swensson, Bengt, & Wretman, Jan. 1992. *Model Assisted Survey Sampling*. Springer Series in Statistics. Springer-Verlag.

Scheffé, Henry. 1959. *The Analysis of Variance*. Wiley.

Shepard, Roger N, & Cooper, Lynn A. 1992. Representation of Colors in the Blind, Color-Blind, and Normally Sighted. *Psychological Science*, **3**(2), 97–104.

Silverman, B W. 1985. Some Aspects of the Spline Smoothing Approach to Non-Parametric Regression Curve Fitting. *Journal of the Royal Statistical Society. Series B (Methodological)*, **47**(1), 1–52.

Silverman, Bernard W. 1981. Using Kernel Density Estimates to Investigate Multimodality. *Journal of the Royal Statistical Society, Series B (Methodological)*, **43**, 97–99.

Smith, Thomas M, & Reynolds, Richard W. 2005. A Global Merged Land Air and Sea Surface Temperature Reconstruction Based on Historical Observations (1880–1997). *Journal of Climate*, **18**, 2021–2036.

Snedecor, George W, & Cochran, Willian G. 1976. *Statistical Methods*. 6th edn. Iowa State University Press.

Stallman, Richard M, Pesch, Roland H, & Shebs, Stan. 2002. *Debugging with GDB: The GNU Source-level Debugger*. Free Software Foundation.

Stravinsky, Igor. 1942. *Poetics of Music in the Form of Six Lessons: The Charles Eliot Norton Lectures*. Harvard University Press.

Stroustrup, Bjarne. 1986. *The C++ Programming Language*. Addison-Wesley.

Student. 1927. Errors of Routine Analysis. *Biometrika*, **19**(1/2), 151–164.

Thomson, William. 2001. *A Guide for the Young Economist: Writing and Speaking Effectively about Economics*. MIT Press.

Train, Kenneth E. 2003. *Discrete Choice Methods with Simulation*. Cambridge University Press.

Tukey, John W. 1977. *Exploratory Data Analysis*. Addison-Wesley.

Vuong, Quang H. 1989. Likelihood Ratio Tests for Model Selection and Non-Nested Hypotheses. *Econometrica*, **57**(2), 307–333.

Wolfram, Stephen. 2003. *The Mathematica Book*. 5th edn. Wolfram Media.

Zarembka, P (ed). 1973. *Frontiers in Econometrics*. Academic Press.

.

gsl_cdf_chisq_P, 305
gsl_cdf_chisq_Pinv,
   306
χ^2 test, 309
   goodness-of-fit, 321
   scaling, 314
chmod (POSIX), 160
choose
   gsl_sf_choose, 238
Chung & Fraser (1958),
   375
Chwe (2001), 262
Cleveland & McGill
   (1985), 180
closed-form expression,
   422
CLT, 296, *419*
clustering, 289–291
CMF, 236, *419*, *see*
   cumulative mass
   function
Codd (1970), 95
coefficient of
   determination, 228
color, 180
column (POSIX), 399, 402
command-line utilities, 98
combinatorial optimization,
   338
Command line
   arguments on, 203
command-line programs,
   see POSIX commands
commenting out, 25
comments
   in C, 25–26
   in Gnuplot, 160
   in SQL, 78
commit (SQL), 84
comparative statics, 152
compilation, 48, 51
compiler, 18, *422*
conditional probability, 258
conditionals, 20
configuration files, 383
conjugate distributions,
   374, *422*
conjugate gradient, 341

Conover (1980), 323, 376
consistent estimator, 221,
   422
consistent test, 335, *422*
const (C), 65
constrained optimization,
   151
contour plot, 162
contrast, 309, *422*
Conway, John, 178
Cook's distance, 131
Cook (1977), 131
correlation coefficient, 229,
   422
counting words, 401
covariance, 228, *422*
Cox (1962), 354
cp (POSIX), 399
Cramér–Rao Inequality,
   335
Cramér–Rao lower bound,
   221, 229, 333, *423*
create (SQL), 84
Cropper *et al.* (1993), 283
crosstab, 101, *423*
csh
   redirecting stderr, 396
csh (POSIX), 382
cumulative density
   function, 236, *423*
cumulative mass function,
   236, *423*
current directory attack,
   385
cut (POSIX), 399, 401
CVS, *see* subversion

data
   conditioning, 139
   format, 75, 147
data mining, 316, *423*
data snooping, 316, *423*
data structures, 193
de Finetti, 330
debugger, 43, *423*
debugging, 43–47
decile, *see* quantile
declaration, *423*

of functions, 36
of gsl_matrix, 114
of gsl_vector, 114
of pointers, 57
of types, 31
of variables, 28–33
degrees of freedom, 222,
   423
delete (SQL), 86
Dempster *et al.* (1977), 347
dependency, *423*
dereferencing, 43
desc (SQL), 83
descriptive statistics, 1, *423*
designated initializers, 32,
   353
df, *419*
diff (POSIX), 399, 402
discards qualifiers from
   pointer target type, 201
discrete data, 123
distinct, 80, 81
do (C), 23
dos2unix (POSIX), 418
dot files, 383
dot product, 129
dot, graphing program, 182
double (C), 29, 135
doxygen (POSIX), 185
drop (SQL), 85, 86
dummy variable, 110–111,
   123, **281–283**, 316, *424*

e, 136
ed (POSIX), 403, 404
efficiency, 220, 334, *424*
Efron & Hinkley (1978),
   13, 349
Efron & Tibshirani (1993),
   231
egrep (POSIX), 403, 406
eigenvectors, 267
Einstein, Albert, 4
Electric Fence, 214
Eliason (1993), xi
else (C), 21
EMACS (POSIX), 387, 402,
   403